ENVIRONMENTAL SCIENCE, ENGINEERING AND TECHNOLOGY

GRASSLAND BIODIVERSITY AND CONSERVATION IN A CHANGING WORLD

ENVIRONMENTAL SCIENCE, ENGINEERING AND TECHNOLOGY

Additional books in this series can be found on Nova's website under the Series tab.

Additional e-books in this series can be found on Nova's website under the e-book tab.

ENVIRONMENTAL SCIENCE, ENGINEERING AND TECHNOLOGY

GRASSLAND BIODIVERSITY AND CONSERVATION IN A CHANGING WORLD

PIERRE MARIOTTE
AND
PAUL KARDOL
EDITORS

nova publishers
New York

For permission to use material from this book please contact us:
Telephone 631-231-7269; Fax 631-231-8175
Web Site: http://www.novapublishers.com

Library of Congress Cataloging-in-Publication Data

ISBN: 978-1-61761-619-8

Published by Nova Science Publishers, Inc. † New York

CONTENTS

PREFACE

Grasslands cover around 30% of the Earth's terrestrial surface and provide important ecosystem services including food production, water purification and carbon sequestration. On the other hand, grasslands are among the most endangered ecosystems, threatened by land-use change, mainly through agricultural intensification or abandonment, and ongoing climatic changes, such as warming and drought. Worldwide, ecological research aims at understanding the effects of land-use and climatic changes on grasslands in order to protect, conserve and restore these valuable ecosystems. The principal aim of this book is to give an overview of current knowledge and future challenges for ecological studies in grasslands. The book also provides important new perspectives for management and conservation of grasslands ecosystems. Both aboveground and belowground ecosystem components, as well as the linkages between these components, are taken into consideration.

The first chapter of this book presents current threats on grasslands by reviewing the influence of land-use and climate change on the integrity of grassland ecosystems, through changes in plant species composition and biodiversity (Chapter 1). While aboveground patterns and processes in grasslands are relatively well known, the importance of belowground components in shaping grassland ecosystem is less well understood. This is despite recognition that most of grassland biomass and biodiversity is belowground, and that soil communities provide key ecosystem functions. Chapter 2 redresses the importance of soil organisms by synthetizing their effects on grassland production and diversity, illustrated by a simplified soil food web with antagonist and mutualistic pathways. Soil microbial communities are particularly important for grassland functioning and need to be taken into account in management practices. As such, Chapter 3 highlights the importance of seasonal variability of soil microbial communities and gives perspectives to better manage disturbed grasslands. In addition to temporal variation, also spatial variability should be considered in grassland management. Chapter 4 presents a conceptual toolkit that accounts for spatial and temporal variability to facilitate restoration planning and management. Grazing is one of the most common practices in grassland management, with positive or negative effects on biodiversity and ecosystem functioning, depending on intensity and frequency. Chapter 5 investigates how grazing components (i.e. defoliation, trampling, animal excretions) affect spatial heterogeneity and as such affect the functioning of grassland ecosystems and the services they provide. Grazing can also be used as a tool to restore degraded ecosystems and Chapter 6 shows how grazing, as well as other management practices, affects plant diversity and biomass production in a unique long-term restoration experiment in China. In a historic perspective, over the past two centuries, grasslands have been transformed globally by conversion to intense agricultural use, while low-intensity grazed grasslands are disappearing. The ongoing effects of climate change are an additional threat to these endangered ecosystems as shown in Chapter 7. Climate change impacts are especially important in high elevation ecosystems, which are disproportionally affected, and Chapter 8 and 9 explore the influences of land-use and climate change on mountain and alpine grasslands. Both these chapters focus explicitly on changes in plant functional group composition, which are key drivers of ecosystem properties and functioning. In this light, Chapter 10 gives and overview of the role of the largely understudied functional group of bryophytes by showing how this group contributes to the biodiversity of grasslands and constitutes a key group in the functioning of grassland ecosystems.

We wish to thank all authors who contributed to this book. Thanks for your enthusiasm and commitment and for meeting the deadlines. We are also grateful to those who provided comments and feedback throughout the process of bringing this book together.

Pierre Mariotte[1] and Paul Kardol[2]
Editors
30 January 2014

[1]Department of Environmental Science, Policy and Management,
University of California, Berkeley, CA, US
email: *pierre.mariotte@hotmail.com*

[2]Department of Forest Ecology and Management,
Swedish University of Agricultural Sciences, 90183, Umeå, Sweden
email: *Paul.Kardol@slu.se*

In: Grassland Biodiversity and Conservation … ISBN: 978-1-61761-619-8
Editors: Pierre Mariotte and Paul Kardol

Chapter 1

TEMPERATE GRASSLANDS: OUT OF SIGHT, OUT OF MIND? CONSERVATION AND RESEARCH PRIORITIES FOR ONE OF THE WORLD'S MOST THREATENED ECOLOGICAL COMMUNITIES

Anett Richter*[*]*and Will Osborne
Institute for Applied Ecology, University of Canberra, Australia

ABSTRACT

This chapter reviews the influence of land-use change and predicted climate change on temperate grasslands – an overlooked ecological community at the global scale - and identifies future research strategies to halt the further loss of biodiversity in temperate grasslands. We discuss how previous research has contributed to our understanding of grassland ecology and describe how two threats to native temperate grasslands might affect the ecological integrity of these grasslands. Based on a consideration of research conducted in three native temperate grassland regions - North America, South Asia and South-eastern Australia - it is concluded that increasing legal protection, a wider public appreciation and more research is required to reverse the current trend of biodiversity loss and to improve the current lack of appreciation of temperate grasslands. In particular, there is a need for more detailed inventory surveys of the temperate grasslands biota and increased autecological research on the invertebrate fauna. The establishment of long term monitoring is required to provide a measurement of the success of conservation action and to relate the effects of historical and current threats to native grassland and component biodiversity.

Keywords: Native temperate grassland, conservation, ecosystem services

[*] Corresponding author: Email: Anett.Richter@canberra.edu.au.

INTRODUCTION

'Heaven knows, we've got plenty of grass' (former Victorian Conservation Minister Australia). Native (naturally occurring) temperate grasslands are ecological communities that have, at a global scale, undergone dramatic reduction and modification as a result of agricultural and, more recently, urban development (TGCI 2008). According to the Millennium Ecosystem Assessment (2005) nearly 80 percent of the world coverage of the temperate grassland biome has already been lost. Surviving intact remnants face an array of threats and are afforded poor conservation protection (Chape et al., 2008; TGCI 2008). At an international scale, these grasslands are afforded a very low level of protection when compared to other terrestrial biomes (Chape et al., 2008; TGCI 2008). Fortunately, in many countries, conservation efforts have commenced that provide some protection for specific grassland communities, particularly those that are endangered. Nonetheless, the level of protection applied to temperate grasslands at the global and regional level is not sufficient to stop further loss of biodiversity and ecosystem degradation. In this chapter we provide an international perspective on the conservation of native temperate grasslands. We provide an overview of their vulnerability to past and ongoing threats and identify priorities for research and conservation that will be better secure their future. In our review we draw attention to key studies from East Asia, North America and South-eastern Australia and focus on what we consider are major threats to their persistence: (1) the consequences of past habitat loss, fragmentation and degradation; (2) climate change; and (3) biotic invasion to the biodiversity in temperate grassland. We selected these geographically remote temperate regions to demonstrate the global pattern of large scale temperate grassland modification and the effects of new arising threats such as climatic change. These regions once supported extensive areas of temperate grassland and are representative of the extent of temperate grassland modification at the global scale. This brief review also allows us to consider key research topics that are currently underrepresented with respect to native grasslands and to identify future research and conservation priorities of international relevance. The review is not intended to be exhaustive; rather it reveals the knowledge gaps present in our understanding of native temperate grasslands as distinctive ecological communities, describes the threats posed to them from historic, ongoing and combined environmental changes, and considers the future research and conservation actions that are essential to maintaining the viability of these unique grasslands.

To do this we address the following questions:

1. What are the ecological characteristics of native temperate grasslands?
2. What are the historical and current distributions of temperate grasslands and what are the threats that they are exposed to?
3. How has research improved the understanding of the biodiversity and ecology within temperate grasslands?
4. What direction should future research take to improve the conservation of native temperate grasslands?

SETTING THE SCENE

'Knowing trees, I understand the meaning of patience. Knowing grass, I can appreciate persistence' (Unknown author).

In an international context, native temperate grasslands are recognised by many names. In North America they are known as short and tallgrass prairies, they are called steppes in parts of north-central Eurasia, veldt in South Africa, pampas in South America, puszta in Hungary and lowland native grassland (natural temperate grassland) in South-eastern Australia (ACT Government 2005; Gurevitch et al., 2002). The term 'temperate native grassland' is interchangeably used in the literature for 'natural temperate grassland' or 'native temperate grassland' or 'indigenous temperate grasslands' (Mark & McLennan 2005). We use the term 'native temperate grassland' throughout this overview as that is the term most broadly applied internationally. Most native temperate grasslands are located in the interior of large continents in the rain shadow of mountain ranges where continental climate brings harsh winter conditions along with hot dry summers (Chape et al., 2008). All native temperate grasslands share the characteristic of the absence of a mature tree cover and the overall dominance of a ground layer of perennial warm-season (C_4) and cold-season (C_3) grass species with a mixed forb layer (Duffey et al., 1974; Gurevitch et al., 2002). The dominance of grasses is explained by climatic conditions (mainly low rainfall and periodic droughts) and disturbance mechanisms such as fire (natural and anthropogenic origin) and grazing activities by native herbivores (Anderson 2006; Bredenkamp et al., 2002; Gurevitch et al., 2002; Walter 1968). All major groups of fauna are represented in temperate grasslands and many native grasslands have levels of diversity as high as those characteristics of tropical forests (Sala et al., 2001). The flora and fauna in temperate grasslands is often diverse, highly specialised and endemic (Groombridge 1992). For example, in North America, more than 250 native plant species are found in tallgrass prairies with the majority belonging to perennial grasses (Freeman 1998). Around 350-400 vascular plant species have been recorded in temperate grasslands in China (Peng et al., 2008). The most obvious and iconic animals in temperate grasslands are the large herbivores which have prominence in secondary production (Sala et al., 2001). Bison are the most well-known of the large ungulate grazers in North America that make a significant contribution to the functioning and production of the prairies through their impact on nutrient cycles, primary production and for the shaping of the composition of the flora (Frank et al., 1994; Frank & McNaughton 1992). Smaller-bodied vertebrates such as birds, rodents and reptiles also play an important role in shaping the biodiversity of temperate grasslands. Birds are one of the best documented groups in temperate grasslands (Brennan & Kuvlesky 2005; Chapman et al., 2004; Coppedge et al., 2001). The continental location of most temperate grasslands, such as the prairies in North America, collectively often constitute a transition zone for birds where eastern and western bird species are found (Costello 1969). Reptiles comprise another grassland group that takes advantage of the characteristics of the grassy ecosystem. Microhabitat structure (grass tussocks, rocks and soil cracks), sufficient dietary supply (micro and macroinvertebrates) and high level of solar radiation provide ideal settings for a rich reptile fauna in temperate grasslands (Collins 1993; Osborne et al., 1993b). They are however, a much less diverse group than birds and mammals (Kauffman et al., 1998). Invertebrates are one of the most diverse and abundant groups in grassland ecosystems. They are essential for the functioning of the ecosystems through their role as herbivores, pollinators, predators, parasitoids, and decomposers (Samways 2005). Despite the significance of the invertebrate fauna to the stability and functioning of grassland ecosystems (Mulder et al., 1999; Risser et al., 1981), this important and mega-diverse taxonomic group is often less appreciated in temperate grassland ecosystems as many species are inconspicuous or undertake lengthy stages of their life-histories below ground (Coupland 1979).

Historical Distribution of Temperate Grasslands

Native temperate grasslands were once a major global vegetation system that occurred widely within the temperate climate zone between 30–40° latitude on all continents except Antarctica (Figure 1) and covered approximately eight percent of the earth's terrestrial surface (Duffey et al., 1974; TGCI 2008). In the northern hemisphere, extensive areas of native temperate grassland occurred in North America, covering large areas from the Chihuahuan desert in the south to southern parts of Canada in the north and from the Rocky Mountains in the west to the adjunct deciduous forests of the eastern United States. Vast areas of temperate grasslands extended from the Ukraine to China on the Eurasian- Asian continent. In the southern hemisphere, substantial areas of temperate native grassland were present in the Campos or Pampa regions of South America (includes parts of southern Brazil, southern Paraguay, northeastern Argentina, and the whole of Uruguay). Large areas of temperate grassland also occurred in the temperate climate zone in south eastern Australia (Sala et al., 2001) (Figure 1). The conversion of temperate grassland into agriculture land (including intensive livestock grazing, pasture improvement and cropping), urban development with associated infrastructure, inappropriate fire regimes, and in some places excessive water extraction, have resulted in large scale losses and degradation of the vegetation (TGCI 2008). Today, much of this landscape of grassland plains, pampas and prairies has been lost or is highly degraded.

Land-Use Changes to Native Temperate Grassland — An International Perspective

Congruent with other ecological systems, habitat loss, fragmentation and degradation are listed as singular major threats that, in combination with the overarching threat of climate change, add pressure to the integrity of the biodiversity in temperate grasslands. Climate change acts as a singular major threat, but more importantly with its additive power, in combination with anthropogenic modification (Figure 2). The World Temperate Grassland Conservation Initiative (TGCI) is an established project developed by the Grasslands Protected Area Task Force of the World Commission on Protected Areas (WCPA) of the International Union for Conservation of Nature (IUCN). The TGCI considers that temperate grasslands are one of the worlds most threatened ecosystems because of the extensive losses of temperate grassland at a global scale. Based on the example of three international temperate regions, we highlight the severity of these large scale losses and degradations of temperate grasslands.

Figure 1. Extent of native temperate grasslands (black areas) in the temperate zone between 30 °–40° latitude in both the northern and southern hemispheres prior to large scale anthropogenic modifications. Modified after Coupland (1992) and Bailey (1998). Note that these areas are indicative of occurrence in a general region, which may support a variety of vegetation types.

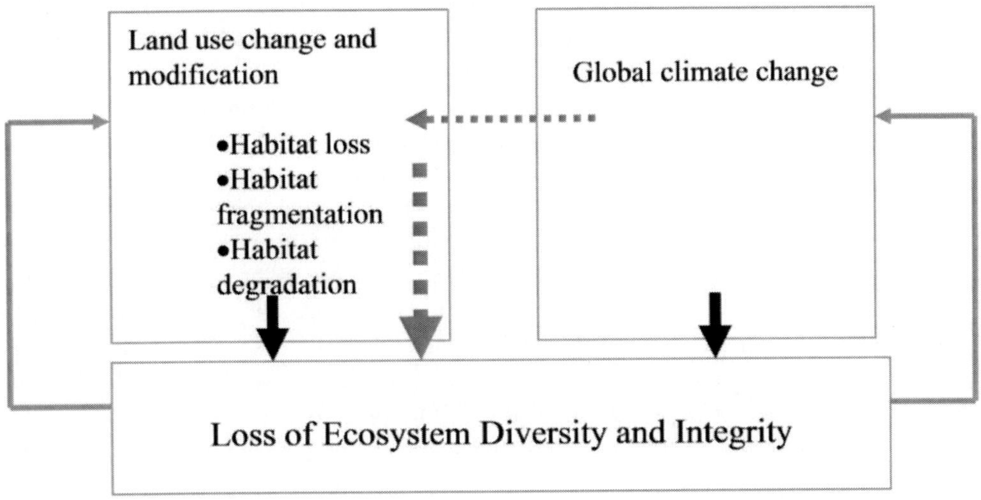

Figure 2. Schematic diagram modified from TGCI (2008) showing interactions between the major threats of land-use changes and global climate change to temperate grasslands ecosystem services. Global climate change acts as a singular factor and with its additive power.

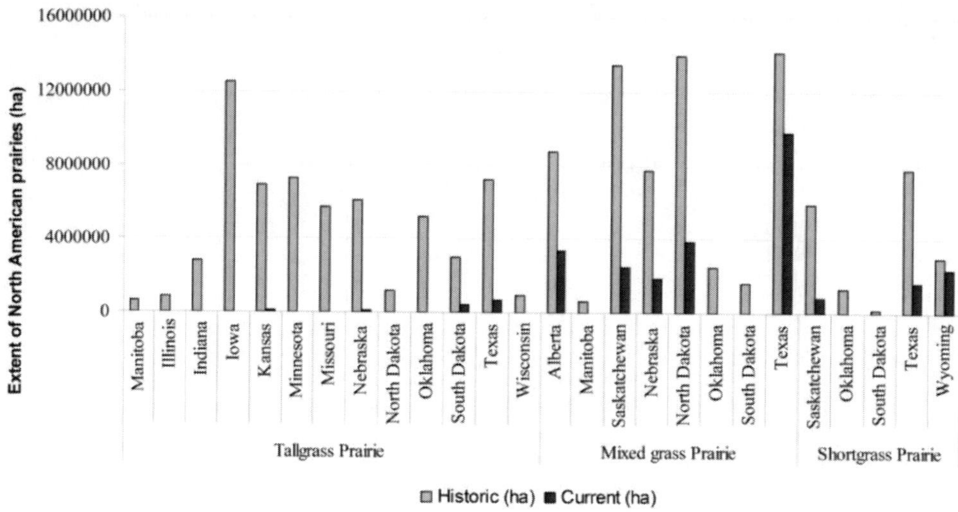

Figure 3. The extent of prairie declines in North America. Based on data provided by Samson and Knopf (1994). The figure summarises the estimated current areas and historic area of the tallgrass, mixed grass, and short grass prairies. For full details see Sampson and Knopf (1994).

East Asia has one of the largest areas of temperate grassland on Earth (Kehuiet al. 2007) (Figure 1). The East Asian temperate grasslands are recognised as being a fundamental natural resource that is important to economic growth and to the cultures in this region (Qiet al. 2007; Zhan et al., 2007). Most of the temperate grasslands in East Asia are densely populated by herders and farmers, along with their millions of livestock. For example, in China more than 196 million ha are classed as temperate grassland and meadow, but these areas also support 110 million people and their livestock (Peng et al., 2008). Overgrazing and the transformation of the natural temperate grasslands into farmland and urban land, combined with the effects of prolonged droughts, have led to large scale degradation of the original temperate grasslands. According to recent reports, most of the temperate grassland in China has been highly declined in their conditions. Estimates of the amount of remaining temperate grassland in reasonable condition range from 10 to 50 percent (see citation within Peng et al., 2008). Large scale modification of temperate grassland throughout temperate East Asia has caused adverse ecological effects at local and regional scales threatening the sustainability of natural temperate grasslands throughout East Asia (Parton et al., 1994; Reynolds 2005). Limited food supply and a lack of arable land elsewhere will lead to even greater pressure on these East Asian grasslands in the future (Reynolds 2005; Suttie 2005). Thus, the temperate grasslands in East Asia are under ongoing pressure, leading to further degradation and loss.

In North America, the first European settlers found the temperate grasslands (prairies) of the Central Grasslands and Great Plains to be in pristine condition with little anthropogenic modification (Pieper 2005). They referred to the grasslands as prairies – a word most likely derived from the French language for "meadows". Generally, depending on the quality of soil and rainfall, the prairies are further divided into three formations: the short grass prairies, the tallgrass prairies and mixed prairies. Following the introduction of European agricultural practices, most of the historical grasslands of North America were rapidly transformed by agriculture, particularly through cropping (Wiken et al., 2008). Today, most remaining native grasslands are comprised of short grass prairies (Samson & Knopf 1994) (Figure 3). It is estimated that only less than five percent of the pre-European settlement tall-grass prairies remain intact (Coppedge et al., 2001; Fletcher & Koford 2002; Vanderhoff et al., 1994). Not surprisingly, the extensive reduction and modification of North American temperate prairie grasslands into agricultural land has led to significant population declines and the extinction of many prairie dependant species (Hyde 1996). According to Sampson & Knopf (1994), in the United States 55 grassland species are listed as threatened or endangered with more than 700 additional species being candidates for becoming threatened or endangered. Approximately one third of the species considered as endangered by the Canadian Committee on the Endangered Wildlife are associated with grasslands (World Wildlife Fund Canada (WWFC) 1988). Today, the intrinsic values of the North American prairies for the provision of ecosystem services are recognised. However, a recent review of protected areas of temperate grassland in North America (federal, provincial and state parks and the central plains grasslands) indicated clearly that the level of protection is not adequate to conserve the full extent of the remaining intact prairie grasslands (TGCI 2008).

The third example is from Australia where native temperate grasslands were once widely distributed in the temperate climate zone, covering extensive areas in the south-east of the continent. Not surprisingly, these open (treeless) grasslands were first settled by Europeans as a consequence of receiving relatively consistent rainfall and the presence of suitable soils for agricultural development. Since European settlement in Australia enormous effort has been made by farmers to improving the productivity of natural grasslands, leading to the large scale transformation of native grassland into native pasture (grassland still dominated by native species of grass, but lacking many of the original native grasses and forbs) and exotic grasslands (grasslands comprised almost entirely of introduced species) (Kirkpatrick et al., 1995; Taylor 1998). Most native pastures have undergone modifications resulting from grazing by domestic stock and the addition of fertilizer and legumes (Mitchell 2003). Exotic grasslands include improved pastures and cultivated grasslands in agricultural and urban areas (Benson 1996). At present it is estimated that less than 100 000 ha of the former extensive temperate grassland in South-eastern Australia remain in a reasonably natural state, indicating that more than 95 percent has been destroyed or significantly altered (ACT Government 2005; Kirkpatrick et al., 1995; State of Victoria Department of Infrastructure 2000).

As has happened elsewhere, the biota in the native temperate grasslands in Australia is expected to have undergone significant declines and extinction as a result of such extensive large-scale loss of natural habitat and historical and ongoing degradation. Moreover, many of the remaining, often highly fragmented, native temperate grasslands have become increasingly threatened by urban development. This is in particular the case for the temperate grasslands in South Eastern Australia where rural expansion through the spread of small holding farms and the growth of cities have increased the pressure of remaining temperate grasslands to become developed into urban land (ACT Territory Plan 2008; State of Victoria Department of Infrastructure 2000). The cities of Melbourne (Victoria) and Canberra (Australian Capital Territory) provide examples of modern metropolitan areas situated in temperate grassland regions undergoing rapid urban expansion. Much of Melbourne is situated on the former grassy volcanic plains that surrounded Port Phillip Bay, whereas Canberra was built on the so-called Limestone Plains, an area of treeless grasslands. Both cities require a steady supply of land for residential and industrial development; generally supplied at the expense of surrounding grasslands (Sharp & Shorthouse 1996; Williams et al.,

2005). This constant high demand for urban development has led to ongoing losses of temperate grasslands in both cities and surroundings. Melbourne and Canberra have experienced losses of 36 percent of lowland native grassland and a 44 percent reduction in grassland area respectively over a 15-year period (1985–2000) and overall the destruction of 70 percent of temperate grassland over the 16 year period from 1980 to 1995 as a direct consequence of increased urban and rural development and weed invasion (Benson 1994; Morgan 1998; Stuwe and Parsons 1977) (Figure 4). Considerable effort by the Australian governments has been executed to legally protect and manage the remaining natural temperate grasslands in Melbourne and Canberra, which today contain the largest native grassland remnants in Australia (ACT Government 2005; Australian Government 1999). Despite these attempts, it is likely that further destruction and degradation of native grassland will occur in these remnants, especially those with high perimeter to area ratios, and in close proximity to major roads and residential and industrial areas (Williams et al., 2005).

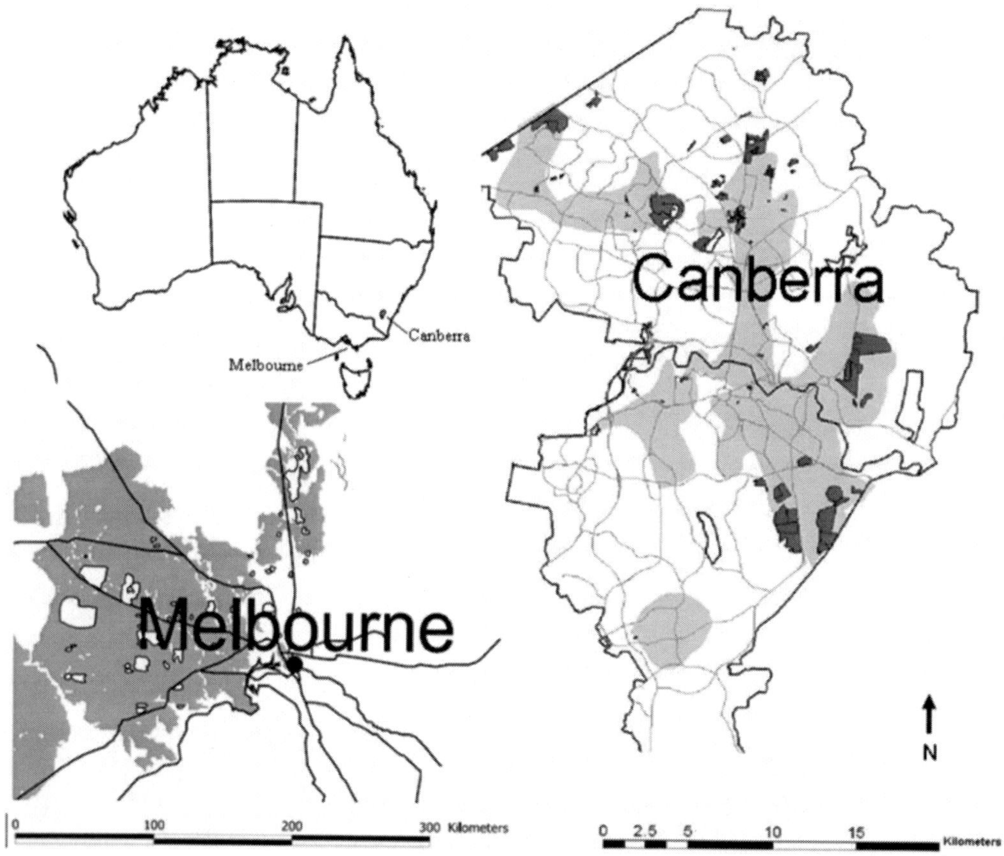

Figure 4. Extent of losses and fragmentation of native temperate grassland within the vicinity of the two Australian cities: Canberra and Melbourne. Light grey areas indicate extent of pre-European settlement native grassland distribution. Locations in Melbourne indicated as white (no fill) and in Canberra as dark grey fill show remaining native grassland fragments. Sources: ACT Government (2005) and modeled native vegetation data from the Department of Sustainability and Environment (DSE) Corporate Spatial Data Library (2005).

NATIVE TEMPERATE GRASSLAND

Ecosystem Services

Temperate grasslands, even in an unmodified condition, provide a range of ecological services for the well-being of humans. According to the Millennium Ecosystem Assessment (2005), the grassland biomes provide: a) provisioning services (e.g. food, fibre, biochemical and natural medicine, ornamental resources); b) regulatory services (e.g. pollination, seed dispersal, climate, air quality, water and erosion regulations); c) cultural services (e.g. spiritual, aesthetic and religious values, cultural heritage); and d) supportive services (e.g. primary production, nutrient and water cycling, provision of habitat, soil formation and retention). Temperate grasslands specifically contribute substantially to the storage of soil carbon and are recognised as significant contributors to the global carbon cycling processes (Bertin 2008; Fischlin et al., 2007). The ability of ecosystems to continuously provide these services is generally expected to be dependent on the maintenance of an intact and diverse biota as a result of the strong correlation between ecosystem process rates and biological diversity (Schläpfer 1999). Therefore, the maintenance of the genetic, biotic and landscape diversity of temperate grassland is important because of their role in securing future ecosystem services. Many of the services that grasslands provide have commercial value – these include for example the provision of meat, milk, wool, and leather derived from grazing animals. However, other services such as maintenance of the composition of the atmosphere, maintenance of the genetic library, amelioration of weather, and the conservation of soils are often perceived to be freely available, limitless and without any market value (Sala & Paruelo 1997). Heidenreich (2009) observes that, in terms of total economic value, temperate grassland remains one of the least understood biomes in terms of its value and its sustainable economic use, social-cultural services, as well as the many ecosystem goods and services that it contributes to human well-being (Heidenreich 2009). The perception that grassland services can be taken for granted and the underestimation of their economic value as providers of services have consequences for decision makers, researchers and the society as a whole. Sala & Paruelo (1997) concluded that: "Errors in valuation (in regard to native grasslands generally) may lead to inappropriate decisions on the fate and best use of natural resources for society. The ability of providing goods and services with market value is not necessarily related to the ability to provide other services, which currently may not have a market value. Ignoring the value of services with no market value may be seriously misleading."

Ecological Research

From an ecological perspective, temperate grasslands, in unmodified and modified condition, have provided an ideal setting for ecological research that has addressed fundamental principles in ecology and conservation biology (Golley 1993; Kang et al., 2007; Knapp et al., 1998). For example, two central questions in ecology: 1) the relationships between ecosystem productivity and stability, and 2) the relationship between ecosystem processes (e.g. productivity) and biodiversity, were examined in temperate grasslands (Bai et al., 2004; Chalcraft et al., 2009; Jennings et al., 2005; Tilman & Downing 1994; Tilman et al.,

2001). Bai et al., (2007) studied temperate grassland at field sites across the Mongolian temperate region and reported on the positive linear relationships between productivity and diversity at multiple organisational levels (e.g. vegetation type and biome) and across local, regional and landscape scales. Their results contrast with the more general view of there being no positive linear productivity-richness relationships, instead species richness typically peaks at an intermediate level of productivity (hump-shaped form of productivity-diversity relationship), as determined from observations in the terrestrial ecosystems of Europe, Africa and North America (Bai et al., 2007; Coupland 1992). Also, a range of experimental grassland studies — some of them conducted in the North American prairies — (see Table 1 in Diaz & Cabido 2001) revealed empirical evidence of relations between ecosystem processes and different components of plant diversity (species richness, functional richness and functional composition). Based on these findings, there is general consensus that ecosystem processes are governed by their functional traits, rather than by species richness (alpha diversity), and its abundance and distribution in a community (Diaz & Cabido 2001). By contrast, Pasari et al., 2013 analysed data from the world's longest running temperate grassland biodiversity-functioning field experiment (Cedar Creek Biodiversity II) to strengthen the hypothesis that local species richness (alpha diversity) has strong positive effects on most individual functions and multi-functionality. Their findings underpin earlier studies demonstrating the importance of species diversity for ecosystem functioning as more functional, temporal and environmental contexts are considered (Pasari et al., 2013). These studies are exemplars for the broader research undertaken in temperate grasslands, and highlight the importance of testing general ecological hypotheses in ecosystems that may share taxonomic similarities, e.g. the North American Prairies and the Eurasian steppe, but differ in climatic conditions, soil, plant and animal functional traits (Hector et al., 1999; Tilman 1997; Tilman et al., 2001; Tilman et al., 1996). A second field of ecological research that is being addressed in temperate grasslands is the exploration of response of the biota to environmental changes over time. The monitoring programs in temperate North America (International Biological Program IBP) and China (Chinese Academy of Science) were established in the 1960's and early 1970's (Hobbie et al., 2003) and provide ecologists with comprehensive and comparable data sets of enormous value. The analysis of long term data have improved enormously our understanding of the ecological requirements of many temperate grassland species and are now used to predict the ecological effects of changes in climate and land-use (Bai et al., 2004; Fan et al., 2008; Liu et al., 2006; Shi et al., 2002; TGCI 2008). The significant contributions provided by ecological research in temperate grassland to ecological theory have considerably influenced the ongoing debate about the relationship between biodiversity and the functionality of ecological systems in the light of the loss of biodiversity and have provided insight into the role of ecological processes involved in the maintenance of ecosystem stability.

Consequences of Climate Change

Climate is an important component of a set of global-scale environmental changes that drives evolutionary, ecological and demographic changes in species and populations (Hazler et al., 2006). For example, variation in rainfall determines physiological and phenological processes in organisms, controls the pattern of migration of highly mobile populations

(Newton 1998) and influences annual variation in population size in birds and mammals (Stenseth et al., 2002). Thus, changes in climatic variables, including temperature and rainfall, will ultimately directly and indirectly influence the distribution and the composition of species populations and communities as well as affect their inter- and intraspecific interactions (Hughes 2000; 2003; Parmesan 2007; Parmesan et al., 1999; Thomas et al., 2004). There is concern that temperate grasslands will be particularly responsive to climate change (IPCC 2001). However, the magnitude and direction of any responses are likely to vary considerably across locations depending on site conditions, biota and management treatment (White et al., 2012). Thus, the identification of site specific conditions or mechanisms that drives the predictable reposes remains unknown. The prediction of the effects of changing climate on ecological processes in temperate grasslands is of great conservation interest because temperate grasslands are considered to be particularly vulnerable to climate change as a result of a long evolutionary history between climate and grassland productivity, nutrient cycles and the evolution of species (Johnson et al., 2005; Niu et al., 2008; Sala et al., 1996; Soussana and Luescher 2007). Additive changes in spatial and temporal patterns of precipitation, temperature, and increasing atmospheric carbon dioxide as predicted under current global climate change scenarios, are expected to profoundly influence the diversity, productivity and stability in temperate grasslands (Borchert 1950; Sala et al., 1996; White et al., 2012). In particular, significantly more disturbances, especially increasing fires and more frequent periods of lower amounts of rainfall (droughts) are predicted to occur in the future in temperate grassland, placing greater stress on the survival of the biota and the integrity of the grassland ecosystem functions (Fischlin et al., 2007). Rising carbon dioxide levels and increased temperatures are likely to lead to an increase in plant productivity, resulting in changes in plant species composition and their demography (Williams et al., 2007). Generally, it is hypothesized that elevated atmospheric carbon dioxide levels will stimulate C_3 plant species to a greater extent than C_4 species and increasing CO_2 input may cause persistent increases in primary production, leading to greater storage of carbon in plant residues and soil organic matter (Hunt et al., 1991). However, research on C_3 and C_4 grasses in South-eastern Australian by Williams et al., (2007) found significantly reduced plant growth in a dominant C_3 grass exposed to elevated carbon dioxide and markedly warmer conditions. These results indicate that responses by plants species to increased atmospheric carbon dioxide can be variable and the subsequent feedback loops are unclear. Although native temperate grasslands experience droughts on a regular basis (IPCC 2001; Parton et al., 1994), native grasslands are highly sensitive to limited soil water availability associated with significant rainfall deficiencies (Kammer 2002; Sebastia et al., 2008). Increasing periods of drought may cause quantitative and qualitative changes in species composition and configuration by altering the density and frequency of the species involved (Frank 2007; Morecroft et al., 2002). This process can even persist for several years after the drought event (Frank 2007; Kammer 2002). Changes in precipitation and associated changes in the availability of water are likely to have effects on the composition and structure of temperate vegetation (Frank and McNaughton 1992), primary net production, consumption, nutrient flux and on the life history of species (e.g. changed phenology, poor survival, changes in body size; Bernays & Chapman 1973) associated with temperate grasslands (Georgeet al. 1992; Macias-Duarteet al. 2004; Manzano-Fischeret al. 1999). Drought conditions also affect the survival, and population dynamics of larger grassland species such as birds (Yarnell et al., 2007), reptiles (Sperry & Weatherhead 2008), small mammals (Cairns and Grigg 1993;

Caughley et al., 1985) and marsupials (Munn and Dawson 2010). Changes in climate may also reduce plant species survival indirectly as a consequence of changed herbivore-plant-insect interactions, changed predator-prey interactions and cascading effects (Goulson et al., 2005). One major economic and ecological concern related to the impacts of climate change is the increasing likelihood of pest population outbreaks (Goulson et al., 2005; Kreyling et al., 2008). Research on the factors that trigger pest outbreaks indicates that there is a close relationship between the timing of outbreaks and precipitation (Patterson et al., 1999). Under the predicted climate scenarios, pest outbreaks and the speed of invasion may increase with rising temperatures and changes in precipitation. Consequently, climatic anomalies such as extreme drought and floods have the ability to strongly affect the periodicity and severity of pest outbreaks and invasion. However, the prediction of disease outbreaks will be more difficult in periods of rapidly changing climate and unstable weather and the effectiveness of pesticides on targeted pests might change under changing climatic conditions (Harrington et al., 2001).

Invasions

Native temperate grasslands have been exposed to a long history of invasions of anthropogenic and natural origins. In less than 300 years the majority of temperate grasslands (except the Eurasian temperate grasslands) have been irreversibly modified by human kind through the introduction of non-native species. Some temperate grasslands such as those in Australia, Argentina, Uruguay and the Central Valley of California have been particularly affected by invasive species, whereas the grasslands in Eurasia, southern Africa and central America have changed comparatively little, except where ploughed (Costello 1944; Mack 1989). These differences have been attributed to the influences of two quite different evolutionary histories under which the temperate grasslands evolved: firstly the presence or absence of a substantial evolutionary history with or without large, heavy-hoofed grazing mammals. Sites with a long history of grazing by these mammals are thought to have developed ecological traits that impart a resistance to invasion. The other reason for the vulnerability for invasion is the dominance by caespitose grasses prior to European settlement. Caespitose grasses display little tolerance of grazing or trampling by large grazers and disappear from the floristic community when the pressure from disturbance increases (Mack 1989).

The Future of Native Temperate Grasslands — An Ongoing Concern

There are clear reasons for our overall concern about the future viability of native temperate grasslands. These grasslands have declined extensively at the global scale and remaining temperate grasslands, that are still intact, are threatened by ongoing modification (e.g. being highly fragmented and modified) and by the continuous pressure of land-use (e.g. agricultural, rural and urban development) on the remaining grasslands.

Table 1. Identification of potential research and conservation priorities for native temperate grasslands

Conservation/research action in native temperate grasslands		South East Asia	North America	South Africa	South Eastern Australia	South America
Legal protection and conservation		O	O	O	O	O
Long term monitoring		X	X	O	O	O
Species inventories		O	X	X	O	O
Research	Climate change	X	X	O	O	-
	Management	O	X	X	X	-
	Management/ Climate change	O	X	-	O	-
	Fragmentation/ Habitat Loss	-	X	X	X	-
	Grassland Restoration	-	X	-	X	-
	Invasion/ Pest outbreaks	X	X	-	O	-
Public awareness		O	X	O	O	O

Symbol X indicates existing information and attempts to address these issues; O indicates areas that are important to improve in future and are currently not sufficiently addressed; "minus" indicates that not sufficient information was found. All information presented in this table was obtained from contributions of invited authors at the World Temperate Grassland Conservation Initiative Workshop 2008 report (TGCI 2008).

Considering the single and additive effects of the past and ongoing threats and the status quo of remaining temperate grasslands, there is a justified concern that native temperate grasslands around the globe are at the risk of becoming one ecological community that is at the brink of extinction. Important ecosystem services that are provided by native temperate grasslands are expected to become reduced and limited as a result of these historic and ongoing alterations. The World Temperate Grassland Conservation Initiative (TGCI) expressed two key concerns regarding the future of temperate grasslands. First, the overall level of legal protection for the grasslands is insufficient. Only four percent of the temperate grassland biome is under protected status, resulting in the lowest level of protection of the world's 14 biomes. The second concern is the challenge of managing temperate grassland in the face of widespread livestock grazing, the introduction of exotic species, fire management and the loss of grasslands through urban and forest encroachment. In addition, grassland conservation is threatened by the overarching, yet not well understood, impact of global climatic changes (TGCI 2008).

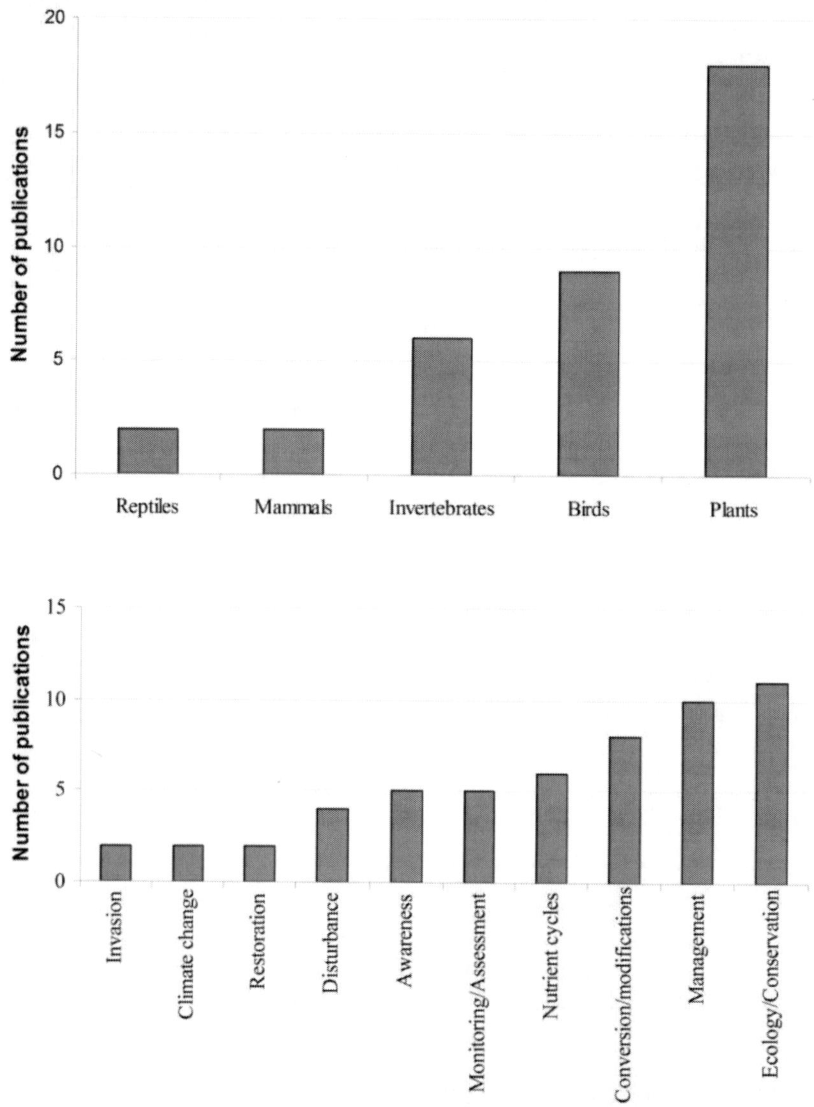

Figure 5. Representation of taxa (a) and research themes (b) in temperate grassland conservation. Results obtained from a search of 78 publications that were published in international journals. Publication list retrieved from the Web of Science and individually searched for the presence of each category. Studies that addressed organic matter and chemical cycles are excluded from the top graph.

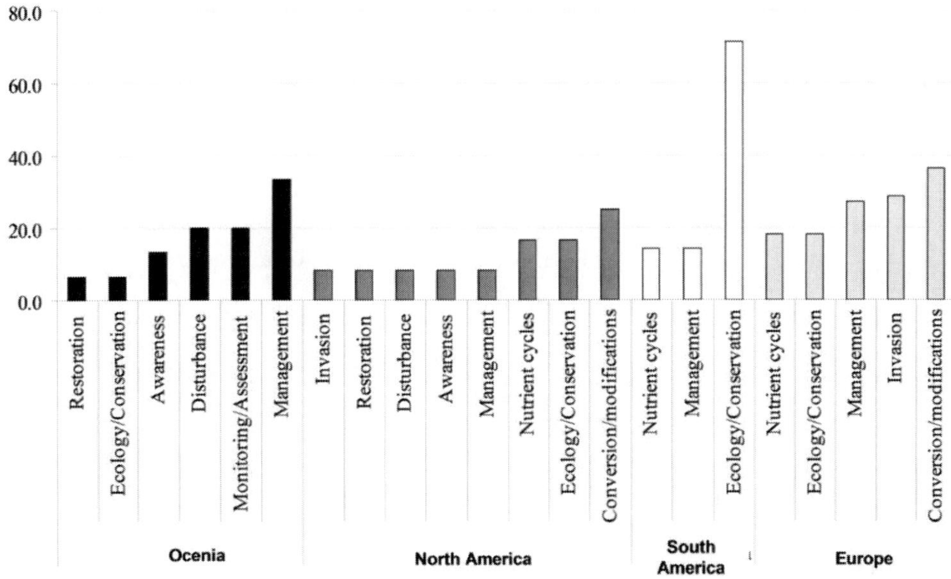

Figure 6. Proportion of literature (%) that addressed conservation research topics in native temperate grassland classified according to the continent where the study was conducted. The information is based on published literature found under the keyword search "temperate grassland conservation" at the Web of Science.

Identifying the Gaps in Native Temperate Grassland Conservation Research

Conservation research is one important tool that is essential for defining management prescriptions and for the identification of the relationship between the biota and the application of management under various conditions (climatic conditions, intensive management). To set future conservation research priorities it is important to identify key gaps in knowledge. To do this we examined 78 international publications that matched with the keyword search ("temperate grassland conservation") in the Web of Science and were published between 1990 and 2010. Based on its content, each publication was assigned to pre-selected research themes and the broad taxonomic group of the organism under study. Studies that address two or more topics were classified to two or more categories. The assignment of publication to research themes revealed a major focus of research on plants and birds (Figure 5). Despite their recognised contribution to biodiversity in grasslands, invertebrates were widely underrepresented in this literature. Less than 10 percent of all studies focused on invertebrates in temperate grassland conservation research. Most of the ecological studies that we examined were related to management, conversion and modification of grasslands (Figure 6). Surprisingly, topics such as climate change and invasion were scarcely addressed. We also compiled information from the TGCI report to further identify future conservation research in temperate grasslands at the global scale (TGCI 2008). Based on information provided by experts on temperate grassland conservation it was clear that considerable insight into temperate grassland ecology and the effects of climate change, invasion and management has been gained from North America and to a much smaller extent from Australia and South Africa (Figure 6, Table 1). In North America and Asia long-term ecological monitoring programs are established, however, in Australia, South America and South Africa there appears to be a lack of any coordinated monitoring program (TGCI 2008). All temperate grassland experts from the World Temperate Grassland Conservation Initiative agreed on the fact that all regions of native temperate grasslands experience a low public awareness (with the exceptions of the North American prairies) and increasing "public environmental education" is needed (Table 1). The development of mechanisms for improving the international collaboration related to temperate grassland conservation was identified as one major global strategy for future temperate grassland conservation (TGCI 2008). This can be achieved through multidisciplinary research networks that assist in gaining insight into ecological patterns and processes in temperate grasslands. Collaborative efforts will lead to a better understanding of the ecology of temperate grassland and such approaches should be developed and expanded to regions where such networks do not exist. The establishment of networks of monitoring and research sites in temperate grasslands as demonstrated by the LTER monitoring scheme (http://www.lternet.edu) provides great opportunity for developing an understanding of ecological processes in temperate grasslands over long temporal and broad spatial scales. Although aspects of the ecology of some temperate grassland ecosystems are reasonably well understood (e.g. in North America), temperate grasslands in other regions such as in South Eastern Australia, South Africa and South America still lack up to date species inventories and basic ecological information (Table 1). The lack of taxonomy for many species, in particular for invertebrates, is the greatest impediment for collecting information of the species composition in temperate grassland. Similar challenges involving the lack of species inventories also face tropical ecosystems in remote areas that are often considered as biodiversity hotspots. However, studies in tropical ecosystems that include the knowledge of indigenous people and experienced citizen scientists have assisted greatly with the mapping of flora and fauna and have provided insights into the feasibility of such an approach (Stringer et al., 2003). Similar attempts to encourage and educate the people that live in and around temperate grasslands should play a much stronger role in the future of temperate grassland conservation at a global scale. In Australia, well advised community groups such as Friends of Grassland have made major contributions in this way (Richter et al., 2009).

Conservation Research for Native Temperate Grassland — Where to From Here?

Our overview indicates that native temperate grasslands throughout the world have been exposed to an extensive array of threats and human activities. These impacts will continue to alter the extent and quality of temperate grasslands to such a level that there is concern about their long-term viability. Thus, keys to improving the long-term survival of the temperate grassland are: (a) increasing legal protection; (b) the establishment of reserves that are large enough to maintain the diversity of habitats and species that are characteristic for the region; and c) a significant increase in conservation-related research to better understand the functioning of native temperate grassland that in turn will assist with the management and conservation of remaining grasslands (Bomar 2001; Martin & Wilsey 2006; Piper et al., 2007; Ries et al., 2001; TGCI 2008). We found that there is an under representation of research that addresses the effects of the threats of climate change and biological invasion. Climate change and biological invasions are two major threats to the diversity and integrity of temperate grasslands. Research that contributes to our understanding of the ecological effects of climate change and biotic invasion, singularly and in combination, is essential to inform temperate grassland mangers how to best maintain and manage grassland remnants. Such research will play an important role in shaping future grassland conservation – and is made more imperative with the growing evidence that climate change is a reality, and that there is considerable uncertainty about its direct and indirect effects. We found little information about the application and potential for rehabilitation and restoration of temperate native grasslands. This limitation leads to the need for action that targets the expansion of the size of native temperate grassland reserved areas and for the better protection of existing temperate grassland remnants. The rehabilitation of degraded areas that are contiguous with intact grasslands and the establishment of corridors and adequate buffers will play an important role in future temperate grassland research related to adaptive management. Several studies undertaken in the North American prairies show that at least some trial restoration of native grasslands has been successful (Vogel et al., 2007). Other attempts have had limited success (Pywell et al., 2002). In particular, combined approaches involving several different treatments such as species introductions and the reduction of dominant competitive species (invasive species) will be required for successful restoration of grasslands (Hobbie et al., 2003). It is acknowledged that restoration and re-vegetation are costly in labour and time and dependant on a particular range of environmental conditions and a high level of cooperation among government agencies, non-profit organisations and private landowners. However, such approaches will be paramount to achieving conservation of temperate grasslands in the future. Until we have established successful ways to rehabilitate and restore temperate grasslands, land managers and conservationists urgently require scientific evidence about the tolerable levels of grassland losses, fragmentation and degradation among a range of taxa and various scales prior to the application of adaptive management tools to prevent further declines in species and communities.

SYNTHESIS

Our review revealed that, at the level of individual taxonomic groups, invertebrates have been underrepresented in temperate grasslands research during the past decades. This is in stark contrast to their proportional occurrence in these grasslands. Thus, research focused on temperate grassland invertebrates should be given much greater consideration. There is an opportunity to study long-term effects on all taxa by allying research programs with long term monitoring programs in temperate regions. This should assist in determining the level of impacts of environmental change on the grassland biota and on ecosystem functioning. Long-term observations over time and across landscape scales will provide a sound basis for detecting the generality in ecological questions, will assists in the assessment of biodiversity losses, allow early detection of changes, and aid in recognising the impacts of anthropogenic-induced climate change. Such findings are of enormous relevance to the broader community and are essential for future conservation successes in native temperate grasslands (Parton et al., 1994). Clearly, regions where long-term monitoring has not yet taken place would benefit from the establishment of monitoring sites to record changes and to guide subsequent appropriate management. We conclude that to achieve effective conservation of native temperate grasslands a multiple approach involving legal protection, practical restoration, and adaptive management combined with applied research will be required. Balancing these priorities will be a future challenge in native temperate grassland research and can be overcome by integrating multi-disciplinary research that assists with the development of protection and conservation policies (Lemaire et al., 2005). Through such integrative research, an understanding of the resilience of temperate grasslands to historical and ongoing threats can be gained. Future temperate grassland research and conservation will be needed to be set in the socio-political and economic context that will drive human decision-making processes and lead to better on-ground conservation action. National and international networks and collaborations (such as the LTER network or the TGCI initiative) build suitable platforms to achieve such an inter-disciplinary approach and should be considered in future.

Acknowledgments

We would like to express our thank you to all who have contributed to this review. First, we thank all members of the Institute for Applied Ecology at the University of Canberra and the community group Friends of Grassland (Australian Capital Territory, Australia) who discussed issues associated with the conservation of native temperate grasslands during tea time, field visits and at meetings. Second, thank you to the Australian government who provided geographical data on the distribution of remaining native grasslands and funded research in native grassland in Canberra. Third, we thank Professor Charles J. Krebs who encouraged us to write about the native temperate grasslands and always motivated us also to publish in books. Fourth, we thank the two editors of this book, Pierre Mariotte and Paul Kardol, who reviewed our chapter and provided valuable suggestions to improve the chapter. Finally, thanks to our families who ensured that we had the time and the settings to write this review.

References

ACT Government. (2005). A Vision Splendid of the Grassy Plains Extended: ACT Lowland Native Conservation Strategy. Action Plan No. 28. Canberra.

Australian Government. (1999). Environment Protection and Biodiversity Act 1999. Natural temperate grassland of the Victorian Volcanic Plain. A nationally threatened ecological community. Eds. Department of the Environment, Heritage and the Arts.

ACT Territory Plan. (2008). ACT Territory Plan. Eds. ACT Land and Planning Authority. Canberra. Australia.

Anderson, R.C. (2006). Evolution and origin of the Central Grassland of North America: climate, fire, and mammalian grazers. *Journal of the Torrey Botanical Society*, *133*, 626-647.

Bai, Y., Han, X., Wu, J., Chen, Z. & Li, L. (2004). Ecosystem stability and compensatory effects in the Inner Mongolia grassland. *Nature*, *431*, 181-184.

Bai, Y.F., Wu, J.G., Pan, Q.M., Huang, J.H., Wang, Q.B., Li, F.S., Buyantuyev, A. & Han, X.G. (2007). Positive linear relationship between productivity and diversity: evidence from the Eurasian Steppe. *Journal of Applied Ecology*, *44*, 1023-1034.

Benson, J. (1994). The native grasslands of the Monaro region: Southern Tablelands of NSW. *Cunninghamia*, *3*, 609-650.

Benson, J. (1997). Conservation of native lowland grassland remnants in south-eastern Australia. In *Conservation Outside Nature Reserves*. (Eds. P. Hale, D. Lamb) 424-428. Centre for Conservation Biology, University of Queensland. Brisbane, Australia.

Bernays, E.A. & Chapman, R.F. (1973). The role of food plants in the survival and development of *Chortoicetes terminifera* (Walker) under drought conditions. *Australian Journal of Zoology*, *21*, 575-592.

Bertin, R.I. (2008). Plant phenology and distribution in relation to recent climate change. *The Journal of the Torrey Botanical Society*, *135*, 126-146.

Bomar, C.R. (2001). Comparison of grasshopper (Orthoptera: Acrididae) communities on remnant and reconstructed prairies in western Wisconsin. *Journal of Orthopteran Research*, *10*, 105-112.

Borchert, J.R. (1950). The climate of the central North American grassland. *Ann Assoc Am Geogr*, *40*, 1-39.

Bredenkamp, G.J., Spada, F. & Kazmierczak, E. (2002). On the origin of northern and southern hemisphere grasslands. *Plant Ecology*, *163*, 209-229.

Brennan, L.A. &Kuvlesky, W.P. (2005). North American grassland birds: An unfolding conservation crisis? *Journal of Wildlife Management*, *69*, 1-13.

Brye, K.R., West, C.P. & Gbur, E.E. (2004). Soil quality differences under native tallgrass prairie across a climosequence in Arkansas. *American Midland Naturalist*, *152*, 214-230.

Cairns, S.C. & Grigg, G.C. (1993). Population dynamics of Red Kangaroos (*Macropus rufus*) in relation to rainfall in the South Australian pastoral zone. *Journal of Applied Ecology*, *30*, 444-458.

Caughley, G., Grigg, G.C. & Smith, L. (1985). The effect of drought on kangaroo populations. *Journal of Wildlife Management*, *49*, 679-685.

Chalcraft, D.R., Wilsey, B.J., Bowles, C. & Willig, M.R. (2009). The relationship between productivity and multiple aspects of biodiversity in six grassland communities. *Biodiversity and Conservation*, *18*, 91-104.

Chape, S., Spalding, M. & Jenkins, M.D. (2008). *The world's protected areas*. Prepared by the UNEP World Conservation Monitoring Centre. University of California Press. Berkeley. USA.

Chapman, R.N., Engle, D.M., Masters, R.E. & Leslie, D.M. (2004). Grassland vegetation and bird communities in the southern Great Plains of North America. *Agriculture, Ecosystems & Environment*, *104*, 577-585.

Collins, J.T. (1993). *Amphibians and reptiles in Kansas*. University of Kansas Museum of Natural History. Lawrence. USA.

Conner, J.R., Hamilton, W.T., Sheehy D.P., Stuth, J.W. & Kreuter U.P. (1998). Grassland-based Livestock Production in Temperate Zones. *World Animal Review, 90*, 6-12.

Coppedge, B.R., Engle, D.M., Masters, R.E. & Gregory, M.S. (2001). Avian response to landscape change in fragmented southern Great Plains grasslands. *Ecological Applications*, *11*, 47-59.

Costello, D.F. (1944). Natural revegetation of abandoned plowed land in the mixed prairie association of northeastern Colorado. *Ecology*, *25*, 312–326.

Costello, D.F. (1969). *The prairie world.* Thomas Y. Crowell Company. New York.

Coupland, R.T. (1979). The nature of grasslands. In *Grassland Ecosystems of the World: Analysis of Grasslands and their Uses*. Ed. R.T. Coupland. Cambridge University Press. Cambridge.

Coupland, R.T. (1992). Approach and generalization. In *Ecosystems of the World 8A*, *Natural Grasslands, Introduction and Western Hemisphere*. Ed. R.T. Coupland. Elsevier. New York.

Diaz, S. & Cabido, M. (2001). Viva la difference: plant functional diversity matters to ecosystem processes. *Trends in Ecology and Evolution*, *16* (11), 646-655.

Duffey E., Morris, M.G., Sheail, J., Ward, L.K., Wells, D.A. & Wells, T.C.E. (1974). *Grassland ecology and wildlife management.* Chapman and Hall. London.

Dukes, J.S. & Mooney, H.A. (1999). Does global change increase the success of biological invaders? *Trends in Ecology and Evolution*, *14*, 135-139.

Fan, J.W., Zhong, H.P., Harris, W., Yu, G.R., Wang, S.Q., Hu, Z.M. &Yue, Y.Z. (2008). Carbon storage in the grasslands of China based on field measurements of above- and below-ground biomass. *Climatic Change*, *86*, 375-396.

Fischlin, A., Midgley, G.F., et al. (2007). Ecosystems, their properties, goods, and services. Climate change 2007: impacts, adaptation and vulnerability. In *Contribution of Working Group II to the Fourth Assessment Report of the Intergovernmental Panel on Climate Change*. Eds M.L. Parry, O.F. Canziani, J.P. Palutikof, P.J. van der Linden & C.E. Hansen. Cambridge University Press. Cambridge.

Fletcher, R.J. & Koford, R.R. (2002). Habitat and landscape associations of breeding birds in native and restored grasslands. *Journal of Wildlife Management*, *66*, 1011-1022.

Frank, D.A. (2007). Drought effects on above- and belowground production of a grazed temperate grassland ecosystem. *Oecologia*, *152*, 131-139.

Frank, D.A., Inouye, R.S., Huntly, N., Minshall, G.W. & Anderson, J.E. (1994). The biogeochemistry of a north-temperate grassland with native ungulates - nitrogen dynamics in Yellowstone- National- Park. *Biogeochemistry*, *26*, 163-188.

Frank, D.A. & McNaughton, S.J. (1992). The ecology of plants, large mammalian herbivores, and drought in Yellowstone-National-Park. *Ecology*, *73*, 2043-2058.

Freeman, C.C. (1998). The flora of the Konza Prairie: a historical review and contemporary pattern. In *Grassland dynamics: Long-term Ecological Research in Tallgrass Prairie.* Eds. A. Knapp, J. Briggs, D. Hartnett & S. Collins. Oxford Press. New York

George, T.L., Fowler, A.C., Knight, R.L. & McEwen, L.C. (1992). Impacts of a severe drought on grassland birds in western North-Dakota. *Ecological Applications*, *2*, 275-284.

Golley, F.B. (1993). *A History of the Ecosystem Concept in Ecology: More than the Sum of the Parts.* Yale University Press. New Haven, Connecticut.

Goulson, D., Derwent, L.C., Hanley, M.E., Dunn, D.W. & Abolins, S.R. (2005). Predicting calyptrate fly populations from the weather, and probable consequences of climate change. *Journal of Applied Ecology*, *42*, 795-804.

Grant, T.A., Madden, E. & Berkey, G.B. (2004). Tree and shrub invasion in northern mixed-grass prairie: implications for breeding grassland birds. *Wildlife Society Bulletin*, *32*, 807-818.

Groombridge, B. (1992). *Global Biodiversity*. Chapman & Hall. London.

Gurevitch, J., Scheiner, S.M. & Fox, G.A. (2002). *The Ecology of Plants*. Sinauer Associate Incorporated. Sunderland.

Harrington, R., Fleming, R.A. & Woiwod, I.P. (2001). Climate change impacts on insect management and conservation in temperate regions: can they be predicted? *Agricultural and Forest Entomology*, *3*, 233-240.

Hazler, K.R., Amacher, A.J., Lancia, R.A. &Gerwin, J.A. (2006). Factors influencing acadian flycatcher nesting success in an intensively managed forest landscape. *Journal of Wildlife Management*, *70*, 532-538.

Hector, A., Schmid, B., et al. (1999). Plant diversity and productivity experiments in European grasslands. *Science*, *286*, 1123-1127.

Heidenreich, B. (2009). What are global temperate grasslands worth? A case for their protection. An analysis of current research on the total economic value of indigenous temperate grasslands. Report prepared for the Temperate Grassland Conservation Initiative. Canada: download: http://cmsdata.iucn.org/downloads/grasslandsreview.pdf October 2013.

Hobbie, J.K. Carpenter, S.R., Grimm, N.B., Gosz, J.R. & Seastedt, T.R. (2003). The US long term ecological research program. *Bioscience*, *53*, 21-32.

Hughes, L. (2000) Biological consequences of global warming: is the signal already apparent? *Trends in Ecology and Evolution, 15*, 56-61.

Hughes, L. (2003) Climate change and Australia: Trends, projections and impacts. *Austral Ecology*, *28*, 423-443.

Hunt, H.W., Trlica, M.J., et al. (1991). Simulation model for the effects of climate change on temperate grassland ecosystems. *Ecological Modeling*, *53*, 205-246.

Hyde, M.K. (1996). *Eucalyptus oderata* woodland in South Australia. In *Report to Australian Heritage Commission*. Unpublished report (Canberra, Australia).

IPCC. (2001). *Climate change 2001: synthesis report. A contribution of working groups I, II, and III to the Third Assessment Report of the Intergovernmental Panel on Climate Change*. Cambridge, United Kingdom, and New York. USA.

Jennings, W.D. Williams, J.W. & Stromberg, M.R. (2005). Diversity and productivity of plant communities across the Inland Northwest, USA. *Oecologia*, *143*, 607-618.

Johnson, C., Millett, B., Gilmanov, T., Voldspeth, R., Guntenspergen, G. & Naugle, D. (2005). Vulnerability of northern prairie wetlands to climate change. *Bioscience*, *55*, 863-872.

Kammer, P.M. (2002). Developmental responses of subdominant grassland species to current weather conditions and their relevance for annual vegetation changes. *Folia Geobotanica*, *37*, 185-204.

Kang, L., Han, X.G., Zhang, Z.B. & Sun, O.J. (2007). Grassland ecosystems in China: review of current knowledge and research advancement. *Philosophical Transactions of the Royal Society B-Biological Sciences*, *362*, 997-1008.

Kauffman, D.W., Fay, P.A, Kauffman, G. & Zimmermann, J.L. (1998). Diversity of terrestrial macrofauna. In *Grassland Dynamics: Long-term Ecological Research in*

Tallgrass Prairie. Eds A. Knapp, J. Briggs, D. Hartnett, D. Collins. Oxford University Press. New York.

Kehui, O., Kun, W., Derong, L., Zhenming, L. & Bing, L. (2007). Establishment of sown pastures in the hilly red soil region of the subtropics in southern China. *Tropical Grasslands*, *41*, 92-99.

Kirkpatrick, J., McDougall, K. & Hyde, M. (1995). *Australian's Most Threatened Ecosystem-The Southeastern Lowland Native Grasslands*. Surry Beatty & Sons. Chipping Norton. Australia.

Knapp, A.K., Briggs, J.M., Harnett, D.C. & Collins, S.L. (1998) *Grassland Dynamics: Long-term Ecological Research in Tallgrass Prairie*. Oxford University Press. New York. USA.

Kreyling, J., Beierkuhnlein, C., Ellis, L. & Jentsch, A. (2008). Invasibility of grassland and heath communities exposed to extreme weather events- additive effects of diversity resistance and fluctuating physical environment. *Oikos*, *117*, 1542-1554.

Lemaire, G., Wilkins, R. & Hodgson, J. (2005). Challenges for grassland science: managing research priorities. *Agriculture Ecosystems & Environment*, *108*, 99-108.

Liu, P., Huang, J., Han, X., Sun, O.J. & Zhou, Z. (2006). Differential responses of litter decomposition to increased soil nutrients and water between two contrasting grassland plant species of Inner Mongolia, China. *Applied Soil Ecology*, *34*, 266-275.

Lunt, I. (1997). Germinable soil seed banks of anthropogenic native grasslands and grassy forest remnants in temperate south-eastern Australia. *Plant Ecology*, *130*, 21-34.

Lunt, I., Barlow, T. & Ross, J. (1998). *Plains Wandering: Exploring the Grassy Plains of South-eastern Australia*. Victorian National Park Association Inc. and Trust for Nature Victoria. Melbourne. Australia.

Macias-Duarte, A., Montoya, A.B., Hunt, W.G., Lafon-Terrazas, A. & Tafanelli, R. (2004). Reproduction, prey, and habitat of the Aplomado Falcon (*Falco femoralis*) in desert grasslands of Chihuahua, Mexico. *Auk*, *121*, 1081-1093.

Mack, N.R. (1989). Temperate grasslands vulnerable to plant invasion: characteristics and consequences. In *Biological Invasions: a Global Perspective*. Ed: J.A Drake. John Wiley & Sons. New York.

Manzano-Fischer, P., List, R. & Ceballos, G. (1999). Grassland birds in prairie-dog towns in northwestern Chihuahua, Mexico. In *Ecology and Conservation of Grassland Birds of the Western Hemisphere*. Eds. P.D. Vickery and J.R. Herkert. Cooper Ornithological Society. Camarillo. CA.

Mark, A.F. & McLennan, B. (2005). The conservation status of New Zealand's indigenous grasslands. *New Zealand Journal of Botany*, *43*, 245-270.

Martin, L.M. & Wilsey, B.J. (2006). Assessing grassland restoration success: relative roles of seed additions and native ungulate activities. *Journal of Applied Ecology*, *43*, 1098-1109.

Mitchell, M. (2003). Native grasses: an example of using on-farm biodiversity. In *Land-use change – YES – But will biodiversity be OK?*. Eds. J. Crosthwaite, Q. Farmar-Bowers and C. Hollier. Department of Sustainability and Environment. Victoria. Australia.

Millennium Ecosystem Assessment (2005). *Ecosystem and human well-being: current state and trends.* World Resources Institute. Washington DC.

Morecroft, M.D., Bealey, C.E., Howells, E., Rennie, S. & Woiwod, I.P. (2002). Effects of drought on contrasting insect and plant species in the UK in the mid-1990s. *Global Ecology and Biogeography*, *11*, 7-22.

Morgan, J.W. (1998). Patterns of invasion of an urban remnant of a species-rich grassland in southeastern Australia by non-native plant species. *Journal of Vegetation Science*, *9*, 181-190.

Mulder, C.P.H., Koricheva, J., Huss-Danell K., Hogberg, P. & Joshi, J. (1999). Insects affect relationships between plant species richness and ecosystem processes. *Ecology Letters*, *2*, 237-246.

Munn, A.J. & Dawson, T.J. (2010). Mechanistic explanations for the drought-related mortality of juvenile red kangaroos (*Macropus rufus*): implications for population dynamics and modeling. In *Biology of Kangaroos, Wallabies and Rat-kangaroos*. Eds G. Coulson, D. Eldridge CSIRO Publishing. Victoria. Australia.

Newton, I. (1998). *Population Limitations in Birds*. Academic Press. London.

Niu, S.L., Wu, M.Y., Han, Y., Xia, J.Y., Li, L.H. & Wan, S.Q. (2008). Water-mediated responses of ecosystem carbon fluxes to climatic change in a temperate steppe. *New Phytologist*, *177*, 209-219.

Osborne, W.S., Kukolic, K.D. & Williams, K.D. (1993b). *Conservation of Reptiles in Lowland Native Grasslands in the Southern Tablelands of New South Wales and the Australian Capital Territory*. Transactions of the Royal Society of NSW. Sydney. Australia.

Parmesan, C. (2007). Influences of species, latitudes and methodologies on estimates of phenological response to global warming. *Global Change Biology*, *13*, 1860-1872.

Parmesan, C., Ryrholm, N., et al. (1999). Poleward shifts in geographical ranges of butterfly species associated with regional warming. *Nature*, *399*, 579-583.

Parton, W.J., Ojima, D.S. & Schimel, D.S. (1994). Environmental change in grasslands-assessment using models. *Climatic Change*, *28*, 111-141.

Pasari, J.R., Levi, T., Zavaleta, E.S. & Tilman, D. (2013). Several scales of biodiversity affect ecosystem multifunctionality. *PNAS, 25*, 10219–10222.

Patterson, D.T., Westbrook, J.K., Joyce, R.J.V., Lingren, P.D. & Rogasik, J. (1999). Weeds, insects, and diseases. *Climatic Change*, *43*, 711-727.

Peng, L., Ning, W. & Tao, L. (2008). Temperate grassland region: China. In *The world temperate grassland conservation initiative workshop. Life in a working landscape: towards a conservation strategy for the world's temperate grasslands*. Compendium of regional templates on the status of temperate grassland conservation and protection. Hohot. China.

Pieper, R.D. (2005). Grasslands of Central North America. In *Grasslands of the World*. Eds. J.M. Sattie, S.G. Reynolds, C. Batello. Food and agriculture organisation of the United Nations. Rome.

Piper, J.K., Schmidt, E.S. & Janzen, A.J. (2007). Effects of species richness on resident and target species components in a prairie restoration. *Restoration Ecology*, *15*, 189-198.

Pywell, R.F., Bullock, J.M., Hopkins, A., Walker, K.J. Sparks, T.H., Burke, M.J.W. & Peel, S. (2002). Restoration of species-rich grassland on arable land: assessing the limiting processes using a multi-site experiment. *Journal of Applied Ecology*, *39*, 294-309.

Qi, Y.C., Dong, Y.S., Liu, J.Y., Domroes, M., Geng, Y.B, Liu, L.X., Liu, X.R. & Yang, X.H. (2007). Effect of the conversion of grassland to spring wheat field on the CO2 emission characteristics in Inner Mongolia. *China. Soil & Tillage Research*, *94*, 310-320.

Reynolds, S.G. (2005). Introduction and grassland perspectives. In *Grasslands of the World*. Eds. J.M. Suttie, S.G. Reynolds, C. Batello. Food and agricultural organisation of the United Nations. Rome.

Richter, A., Osborne, W.S., Robertson, G. & Hnatiuk, S. (2009). Community monitoring of golden sun moths in the Australian Capital Region, 2008-2009. Report prepared for WWF Australia. University of Canberra. Institute for Applied Ecology. Canberra. Australia.

Ries, L., Debinski, D.M. & Wieland, M.L. (2001). Conservation value of roadside prairie restoration to butterfly communities. *Conservation Biology*, *15*, 401-411.

Risser, P.W., Birney, E.C., Blocker, H.D., May, S.W., Parton, W.J. & Wiens, J.A. (1981). The true prairie ecosytem. Hutchinson Ross: Stroudsburg. USA.

Sala, O.E., Austin, A. T. &Vivanco, L. (2001). Temperate grassland and shrubland ecosystems. *Encyclopedia of Biodiversity*, *5*, 627-635.

Sala, O.E. & Paruelo J.M. (1997). *Ecosystem Services in Grasslands*. Ecosystem services. Washington DC. Island Press.

Sala, O.E., Lauenroth, W.K. & Burke, I.C. (1996). Carbon budgets of temperate grasslands and the effects of global change. In *Global Change: effects on coniferous forests and grasslands*. Eds. A.I. Breymeyer, D.O. Hall, J.M. Melillo, G.I. Agren. John Wiley & Sons Ltd. Chichester.

Samson, F.B. & Knopf, F.L. (1994). Prairie conservation in North America. *Bioscience*, *44*, 418-421.

Samways, M.J. (2005). *Insect diversity conservation*. Cambridge University Press. Cambridge.

Schläpfer, F. (1999). Expert estimates about effects of biodiversity on ecosystem processes and services. *Oikos*, *84*, 346-352.

Sebastia, M.T., Kirwan, L. & Connolly, J. (2008). Strong shifts in plant diversity and vegetation composition in grassland shortly after climatic change. *Journal of Vegetation Science*, *19*, 299-227.

Sharp, S. & Shorthouse, D.J. (1996). Native grasslands and threatened species in the Australian Capital Territory. Steps towards protection and recovery. In *Back from the brink: Refining the threatened species recovery process*. Eds. S. Stephens, S. Maxwll. Surrey Beatty & Sons. Chipping Norton.

Shi, D., Wan, X., Davis, S.A., Pech, R.P. & Zhang, Z. (2002). Simulation of lethal control and fertility control in a demographic model for Brandt's vole *Microtus brandti*. *Journal of Applied Ecology*, *39*, 337-348.

Soussana, J.F. & Luescher, A. (2007). Temperate grasslands and global atmospheric change: a review. *Grass and Forage Science*, *62*, 127-134.

Sperry, J.H. & Weatherhead, P.J. (2008). Prey-mediated effects of drought on conditions and survival of a terrestrial snake. *Ecology*, *89*, 2770-2776.

State of Victoria Department of Infrastructure. (2000). Housing Melbourne, Residential Forecast 2000. Ed. Department of Infrastructure. Melbourne. Victoria.

Stenseth, N.C., Mysterud, A., Ottersen, G., Hurell, J.W., Chan, K.S. & Lima, M. (2002). Ecological effects of large-scale climate fluctuations. *Science*, *297*, 1292-1296.

Stringer, I.A.N., Bassett, S.M., McLean, M.J., McCartney, J. & Rarrish, G.R. (2003). Biology and conservation of the rare New Zealand land snail *Paryphanta busbyi watti* (Mollusca, Pullmonata). *Invertebrate Biology*, *122*, 241-251.

Stuwe, J. & Parsons, R.F. (1977). *Themeda australis* grasslands on the Basalt Plains, Victoria: floristics and management effects. *Austral Ecology*, *2*, 467-476.

Suttie, J.M. (2005). Grazing management in Mongolia. In *Grasslands of the World*. Eds. J.M. Suttie, S.G. Reynolds, C. Batello. Food and agricultural organisation of the United Nations. Rome.

Taylor, S.C. (1998). South-eastern Australian temperate lowland native grasslands: protection levels and conservation. *Parks*, *8*, 21-26.

TGCI. (2008). Life in a working landscape: towards a conservation strategy for the world's temperate grasslands Grassland Protected Area Task Force of the World Commission on Protected Areas (WCPA). Vancouver. Canada.

Thomas, C.D., Cameron, A., et al. (2004). Extinction risk from climate change. *Nature*, *427*, 145-148.

Tilman, D. (1997). Community invasibility, recruitment limitation, and grassland biodiversity. *Ecology*, *78*, 81-92.

Tilman, D. & Downing, J.A. (1994). Biodiversity and stability in grasslands. *Nature*, *367*, 363-365.

Tilman, D., Reich, P.B., Knops, J., Wedin, D., Mielke, T. & Lehman, C. (2001). Diversity and productivity in a long-term grassland experiment. *Science*, *294*, 843-845.

Tilman, D., Wedin, D. & Knops, J. (1996). Productivity and sustainability influenced by biodiversity in grassland ecosystems. *Nature*, *379*, 718-720.

Van Auken, O.W. (2000). Shrub invasions of North American semiarid grasslands. *Annual Review of Ecology and Systematics*, *31*, 197-215.

Vanderhoff, J.L., Robel, R.J. & Kemp, K.E. (1994). Numbers and extent of black tailed prairie dogs in Kansas. *Transactions of the Kansas Academy of Science*, *97*, 36-73.

Vogel, J.A., Debinski, D.M., Koford, R.R. & Miller, J.R. (2007). Butterfly, responses to prairie restoration through fire and grazing. *Biological Conservation*, *140*, 78-90.

Walter, H. (1968). *Vegetation of the Earth from an Ecopysiological point of you*, *Vol. II. Temperate and arctic zones*. VEB Gustav Fisher Verlag (in German). Jena. Germany.

White, S.R., Carlyle, C.N., Fraser, L.H. & Cahill, J.F. (2012). Climate change experiments in temperate grasslands: synthesis and future directions. *Biological Letters*, *8*, 484-487.

Wiken, E.B., Gauthier, D.A., Schrag, A.M., Hoth, J., Agguirre, C. &Lafon, A (2008). Temperate grassland region: North America. IUCN and World Commission on Protected Areas (WCPA). Vancouver. Canada.

Williams, A.L., Wills, K.E., Janes, J.K., Schoor, J.K.V., Newton, P.C.D. & Hovenden, M.J. (2007). Warming and free-air CO2 enrichment alter demographics in four co-occurring grassland species. *New Phytologist*, *176*, 365-374.

Williams, N.S.G., McDonnell, M.J. & Seager, E.J. (2005). Factors influencing the loss of an endangered ecosystem in an urbanising landscape: a case study of native grasslands from Melbourne, Australia. *Landscape and Urban Planning*, *71*, 35-49.

World Wildlife Fund Canada (WWFC). (1988). *Prairies Conservation Action Plan*. Ed. WWF Canada. Toronto, Ontario. Canada.

Yarnell, R.W., Scott, D.M., Chimimba, C.T. & Metcalfe, D.J. (2007). Untangling the roles of fire, grazing and rainfall on small mammal communities in grassland ecosystems. *Oecologia*, *154*, 387-402.

Zhan, W.L., Chen, S.P., Chen, J., Wei, L., Han, X.G. & Lin, G.H. (2007). Biophysical regulations of carbon fluxes of a steppe and a cultivated cropland in semiarid Inner Mongolia. *Agricultural and Forest Meteorology*, *146*, 216-229.

In: Grassland Biodiversity and Conservation …
Editors: Pierre Mariotte and Paul Kardol
ISBN: 978-1-61761-619-8

Chapter 2

SOIL ORGANISM EFFECTS ON GRASSLAND PRODUCTION AND DIVERSITY

Andrew Kulmatiski[1*], Jes Hines[2] and Nico Eisenhauer[2]
[1]Department of Wildland Resources and the Ecology Center,
Utah State University, Logan, UT, US
[2]Institute of Ecology, Friedrich Schiller University of Jena, Germany

ABSTRACT

Roughly half of grassland biomass and most of grassland biodiversity is belowground, yet because soils are dark, dense and diverse, relatively little is known about how soil organisms affect grassland production and diversity. Here we summarize the various effects that soil organisms have on grassland plant growth and diversity. Soil organisms are often considered to have a net negative effect on plant growth. This perspective has been supported by the success of rotation cropping, increased plant growth in sterilized soils and more recently by a preponderance of negative effects measured in plant-soil feedback experiments. At the same time, however, symbionts and saprobes are known to have strong positive effects on plant growth. To provide a more objective test of the effects of different soil organisms on grassland plant growth, we use a literature review of studies where soil organisms were added or removed from the soil and grassland plant growth responses were measured. Soil organism additions increased plant growth by 260%. This reflected the fact that saprobes and symbionts increased plant growth by 38 and 890%, respectively. Predators of saprobes and symbionts also increased plant growth (by 39%), presumably by increasing nutrient cycling rates. Predators of plant pathogens and herbivores also increased plant growth by 15%, presumably by suppressing antagonist effects on plant growth. Of all the trophic categories examined, only herbivores and pathogens decreased plant growth and this effect was relatively small (12%). Results produced a surprisingly positive perspective of soil organism effects on grassland production and diversity. We suggest a simplified soil food web perspective with two pathways that reflect potential positive and negative effects on plant performance: an antagonist pathway and a mutualist pathway. This simplified food web shows that soil organisms have generally positive effects on grassland productivity, but

* Corresponding author: Email: andrew.kulmatiski@usu.edu.

soil organism effects on grassland diversity remain largely untested. Factors that reduce soil trophic diversity are likely to result in decreased aboveground primary productivity in grasslands.

Keywords: Belowground processes, diversity, pathogen niche, plant, plant soil feedback, soil organism, trophic cascade

INTRODUCTION

In this chapter, we explore how soil organisms affect grassland production and diversity. We use a trophic guild framework because soil biodiversity is extremely high and the 'species concept' is not appropriate for many soil organisms. Many bacteria, for example, transmit plasmid DNA to other soil organisms (Heinemann & Sprague 1989). Thus, during their short lifetimes of perhaps hours, an individual bacterium can become genotypically and phenotypically unique from its original 'self'. Beyond the archaea and bacteria, grassland soils contain a large diversity of other organisms including annelids, arthropods, fossorial mammals, fungi, molluscs, nematodes and protozoa. Not only is phylogenetic diversity high in soils, but functional diversity is often high within phylogenetic groups. Nematodes, for example, can be root pathogens, bacterivores, fungivores or arthropod pathogens: species in this taxonomic group thus can occupy almost any feeding guild in the soil. Because phylogenetic and functional diversity is so high in soils, throughout this review we focus on the trophic guild of different soil organisms. More specifically, we discuss the role of saprobes, symbionts, predators of saprobes or symbionts, herbivores, plant pathogens, predators of herbivores and plant pathogens, predators of predators and ecosystem engineers. To further simplify this system, we place these trophic guilds into one of two 'pathways': the antagonist pathway that has grassland plant herbivores and pathogens at its foundation and the mutualist pathway that has saprobes and symbionts at its foundation. We use a literature review to evaluate how soil organisms in these different trophic positions and pathways affect plant growth in grasslands. Very little research has examined the effects of soil organisms on plant diversity (Wurst et al., 2008), so we rely on conceptual models to help understand how the soil organisms are likely to affect grassland diversity.

ANTAGONIST PATHWAYS

Antagonist Effects on Grassland Plant Growth

Soil organisms are often assumed to have net negative effects on plant growth in many ecosystems (Troelstra et al., 2001; Kulmatiski et al., 2008), and these effects have often translated into ecosystem management practices. In agricultural systems, the negative impacts of soil organisms on plant growth led to the widespread use of rotation cropping. In one of the most dramatic examples of soil pathogen effects, in less than 50 years, the introduced fungus *Cryphonectria parasitica* essentially eliminated what was probably the dominant life form in the eastern United States, the American Chestnut, *Castanea dentata* (Anagnostakis 1982). Though such dramatic examples are not as common in diverse grassland communities, extensive and growing evidence suggests that soil pathogens are strong drivers of productivity in grasslands (Kulmatiski et al., 2008; Petermann et al., 2008; Van Der Heijden et al., 2008; Maron et al., 2011; Schnitzer et al., 2011; Wehner et al., 2011; Latz et al., 2012). For example, a review of PSF experiments found that 75% of plant species realize negative PSFs, or grow worse on soils that have previously grown conspecifics (Figure 1; Kulmatiski et al., 2008). In other words, most plants appear to accumulate belowground pathogens or herbivores that decrease their own growth over time. These negative PSF effects were associated with a ~60% decrease in plant growth; an effect similar in magnitude to the effect of aboveground herbivory (Kulmatiski et al., 2008). These values reflected the growth of all types of plants, but most data were derived from grassland plant species and grasses and forbs realized the most negative of any life forms (i.e., relative to shrubs and trees). We expect that effective management of soil organisms will lead to increased productivity in natural grasslands.

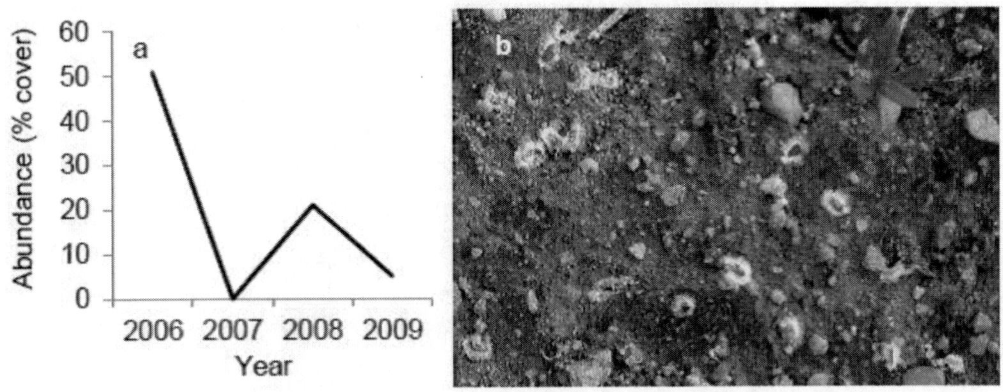

Figure 1. Example of a negative plant-soil feedback. The native annual plant, *Collomia grandiflora*, was grown in a common garden experiment. This plant is common at low densities across the landscape, but when grown in the common garden attained high densities and abundance in the first year of the experiment (a). In the second year of the experiment, no *Collomia* seeds survived to germinate and nearly all seeds observed in the plots had been attacked by a fungal pathogen (b) (Kulmatiski, unpubl. data). In panel b, the white areas are the hyphae from a fungal pathogen feeding on the red *Collomia* seeds. This example shows how plants can accumulate pathogens with time. Interestingly, this data suggests that after eliminating all *Collomia* growth in the second year of the experiment, abundance of the fungal pathogen in the soil declined to a level that allowed some *Collomia* growth in the third year of the experiment (a; photograph by A. Kulmatiski).

Although their effects are not always as widely recognized as aboveground antagonists, many different groups of soil organisms have negative effects on grassland plants. Fungal, bacterial, oomycete and viral pathogens (Raaijmakers et al., 2009), as well as soil herbivores can all limit grassland plant growth, though it should be noted that herbivory does not always decrease plant growth (Frank et al., 2002). Aphids (Hutchinson & Campbell 1994), earthworms (Curry & Schmidt 2007), nematodes (Van der Stoel et al., 2006), Collembola (Endlweber et al., 2009), beetle larvae (Schallhart et al., 2012), fly larvae (Treonis et al., 2005) and fossorial mammals (Figure 2; Whitford & Kay 1999) can all feed on grassland plant roots and decrease plant growth. One reason why the effects of soil organisms may receive less recognition is their small size. Soil pathogens tend to be orders of magnitude smaller than soil herbivores, which in turn, also tend to be orders of magnitude smaller than aboveground herbivores. As a result, belowground herbivores can have pervasive effects on plant growth that are not as easily seen as large-scale disturbances caused, for example, by cattle or elephants (Cumming et al., 1997). It is also important to remember that while specialist soil pathogens and herbivores are likely to decrease the growth of one plant species they are unlikely to have large negative effects on the growth of the whole plant community.

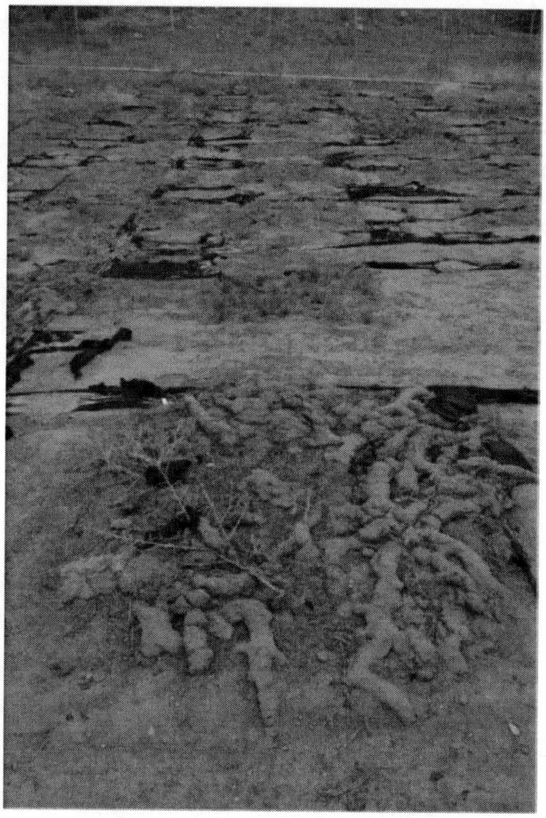

Figure 2. Research on soil organisms often focuses on soil archeae, bacteria and fungi, but larger soil organisms can also affect plant growth and community development. The maze of 'tubes' in the foreground of this photo are created by pocket gophers, *Thomomys talpoides*, burrowing through soil in a common garden experiment. We can see strong species-specific preference of the pocket gopher for the plot containing the taprooted plant *Centaurea diffusa* (foreground) relative to plots growing other plant species (background; photograph by A. Kulmatiski). This activity has been correlated with invasive plant growth (Kyle et al., 2007, Kyle et al., 2008). Across the landscape at these study sites, pocket gophers disturb 1 – 6% of soil each year (Kulmatiski 2006).

Antagonists May Increase Productivity in Grassland Communities Relative to Monocultures

While soil antagonists can generally be expected to decrease the growth of individual plants or plant species, counter-intuitively, this same effect may increase overyielding (i.e., plants can grow more in communities than would be predicted from plant growth in monoculture) (Maron et al., 2011; Schnitzer et al., 2011; Kulmatiski et al., 2012). This can occur when species-specific pathogens accumulate and suppress growth of particular plant species more in monocultures, where the plant is more likely to contact its own pathogens compared to when the species is grown in a more diverse plant community, where it is more likely to be exposed to pathogen-free neighboring plants (Figure 3a; Kulmatiski et al., 2012). As a result, the biomass of plant species that have strong negative PSF when grown in monoculture is expected to be an underestimation of the biomass produced by the same species in more diverse communities. Most plant species are affected negatively by PSFs, so this process is likely to provide a common explanation for enhanced production or overyielding in diverse plant communities relative to plant monocultures (Figure 3b).

The idea that soil pathogens may explain diversity-productivity relationships has been suggested several times (Maron et al., 2011; Schnitzer et al., 2011; Eisenhauer et al., 2012b; Kulmatiski et al., 2012). Soil pathogen effects, however, cannot explain why so many plant species underyield in plant communities (Figure 3b). Competitive suppression has long been used as an explanation for underyielding, however, a recent model of PSF effects on plant community productivity suggested that positive PSFs can also explain underyielding (i.e. the production of less biomass in diverse communities compared to monocultures; Figure 3a). More specifically, plants with positive PSFs are less likely to encounter the soil organisms that increase their growth in diverse plant communities than in monocultures. Interestingly, roughly 25% of plants appear to realize positive PSFs in PSF experiments (Kulmatiski et al., 2008), and roughly 25% of plants appear to underyield in diversity-productivity experiments (Figure 3b; Hector et al., 1999).

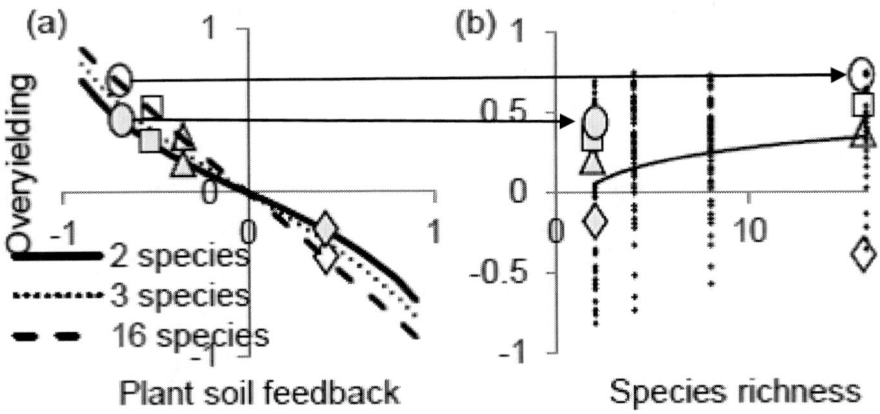

Figure 3. PSFs and overyielding. (a) Theoretical models predict that plants with negative PSFs will overyield in communities and plants with positive PSFs will underyield in communities. This effect will increase with species richness because plants in 2-species communities will encounter half 'self' soil while species in 16-species communities will encounter 1/16th 'self' soil. (b) Biomass data from Hector et al., (1999) converted to overyielding values. If one were to draw from a random pool of typical PSF values (i.e., 3/4ths negative with a mean value of -0.6; Kulmatiski et al., 2008) as indicated by the open and filled symbols (circle, square, triangle and diamond) in panel (a), the theoretical model of Kulmatiski et al., (2012) would predict both the general pattern of overyielding in diversity-productivity experiments and also much of the variability in those experiments.

This counter-intuitive relationship between plant pathogens and productivity in plant communities highlights a problem with many experiments using soil organisms. Plant monocultures are expected to be highly susceptible to soil pathogens. As a result, it is not clear how experiments on plant monocultures or low-diversity plant communities relate to plant communities. This is particularly a problem, for example, for the interpretation of soil sterilization experiments using plant monocultures. Soil sterilization experiments often demonstrate that sterilization increases plant growth, but plant growth is often measured using plant monocultures. Sterilization experiments using plant monocultures are, therefore, likely to over-estimate pathogen effects in plant communities.

Again, to be clear, PSFs are predicted to explain why many plants would produce more biomass in communities than they would in monoculture (i.e., overyield). Because of this relationship between PSF and overyielding, it would be easy to assume that negative PSFs are tightly correlated with community biomass. This may or may not be true. There are two reasons for this. First, plant community biomass will depend just as much on the intrinsic growth capacity of different plants as on their PSFs (Kulmatiski et al., 2011). For example, a plant that grows to 1 g in a month will have little effect on plant community biomass if it grows with other plant species that can grow 10 g in a month unless PSF effects are comparably large (i.e., PSF values of 10). Second, and perhaps most importantly, much of the PSF effect on plant biomass is assumed to occur as species richness increases from one to roughly five plant species. In this range of species richnesses, 'self' soils can be assumed to account for 100, 50, 33, 25 and 20% of the soils. In communities with more than five plant species (i.e., most natural plant communities), 'self' soils become less and less important in determining community biomass. This is because most soils encountered by a plant root will have been cultivated by 'other' plant species. As a result, negative 'self' effects are minimized in diverse communities. In other words, negative PSFs are likely to be very important in determining plant biomass in one, two or three-species plant communities, but negative PSFs are likely to not be very important to plants in more diverse communities because 'self' soils will always represent a small proportion of the soil encountered by a plant's roots.

These dynamics are perhaps best illustrated with an example. Using the biomass-explicit, multi-species model reported by Kulmatiski et al., (2012), here we show the community-level biomass and overyielding predicted for 16-species plant communities with different community-level PSF values (i.e., -0.5 or -0.6) and the different intrinsic growth rates used to create those PSF values (i.e., 0.43, 0.5, 0.51, 0.6, 1.0, 1.07, 1.2 and 1.29 g m^{-2} $month^{-1}$). To be clear a PSF value of -0.5 could result because a plant grows to 0.5 grams on 'self' soils and 1.0 grams on 'other' soils or because a plant grows to 0.6 grams on 'self' soils and 1.2 grams on 'other' soils. PSF values are calculated as [growth on 'self' soils – growth on 'other' soils) / maximum (growth on 'self' or growth on 'other)]. In this example, total community biomass after 40 days simulated growth can be greater in communities with more negative PSFs (i.e., the first vs. third and second vs. fourth bars), but this effect can be overcomed by an increase in intrinsic growth rates of the plant species in the community (i.e., the second vs. first bar and the fourth vs. third bar; Figure 4). In other words, if one plant grows 20% faster than another, then a PSF that slows plant growth by 5% will not have a significant effect on competition between the two species.

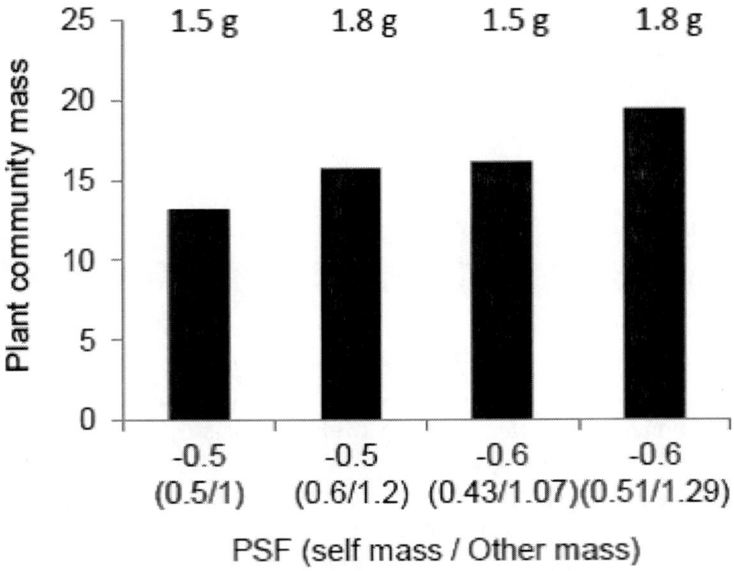

Figure 4. Relationship between plant-soil feedbacks (PSF) and plant community biomass for plants with different intrinsic growth rates (shown in parentheses on the X axis). The sum of plant growth on 'self' and 'other' soils is shown above each bar. Model results show that negative PSFs increase biomass within any 'level' of intrinsic plant growth (e.g., 1.5 g or 1.8 g) and that greater intrinsic growth rates are similarly important in determining community-level biomass as PSF values. Note that the increase in PSF values from -0.5 to -0.6 represented a 20% increase in PSF value. Similarly, the increase in plant biomass in the community from 1.5 to 1.8 represented a 20% increase. Data derived from model simulations from the model described by Kulmatiski et al., (2012).

Plant Defenses

While soil herbivores and pathogens certainly can affect plant growth, particularly in low-diversity communities (Eisenhauer 2012), plants are not defenseless against soil enemies (Walker et al., 2003). How different plants respond to soil enemies can determine plant community composition. Plants have long been known to release volatile organics, like ethylene, that can induce the production of secondary defense compounds by plants over large areas (100s of meters). Recently many new and complex communication pathways have been identified in soils. For example, some plants have been found to release anti-quorom sensing compounds like acyl-homoserine lactone that can inhibit bacterial communication and population growth (Teplitski et al., 2000; Adonizio et al., 2006; Hughes & Sperandio 2008). Indeed, the more soil systems are studied, the more we learn about the role of inter-kingdom communication among plants and soil organisms (vanLoon 2007; Hughes & Sperandio 2008). Communication works both ways in soils and many soil organisms have developed ways of recognizing roots (Prusty et al., 2004) and inducing plant responses (Arkhipova et al., 2005). Many bacteria, for example, have been found to release plant growth hormones, such as indol acetic acid, to stimulate root growth and exudation (Mao et al., 2007). Similarly, some bacterial endophytes use 1-aminocyclopropane-1-carboxylate deaminase to suppress ethylene concentrations, decrease plant stress and increase root growth (Hardoim et al., 2008). We are only beginning to describe the complex array of communication that occurs in soils and understanding how this communication affects plant community dynamics (Raaijmakers et al., 2009). Further study of these communication pathways has the potential to reveal new mechanisms that influence plant community development – mechanisms that could potentially be used as subtle, targeted manipulations of grassland productivity or diversity.

Antagonists May Increase Diversity

The natural enemies of plants are often thought of from a negative or antagonistic perspective because these organisms attack and decrease plant biomass. This same process, however, can increase plant diversity by suppressing the growth of dominant plants through negative frequency dependence (Connell 1970; Janzen 1970; Bever 1994; Bever et al., 1997). Soil pathogens and herbivores often co-evolve with their hosts, and as a result, the effects of soil antagonists are often species-specific (Figures 1 & 2). As a plant species becomes more abundant, its predators and/or pathogens can also become more abundant. This results in negative frequency dependence which slows the growth of abundant species and leads to species replacements (Connell 1970; Janzen 1970). It is easy to envision, for example, that the fungal pathogen observed in Figure 1 creates a low abundance of *Collomia grandiflora* that allows the growth of other plant species in a soil that could otherwise be dominated by *Collomia grandiflora* (the ~50% ground cover observed in 2006 is near the maximum for the semi-arid study site; Kulmatiski 2006).

Despite strong conceptual support for the influence of PSFs on local community diversity (Bever 1999, 2003; Bonanomi et al., 2005; Eppstein & Molofsky 2007; Fukami & Nakajima 2013; Suding et al., 2013), the influence of PSFs on communities at larger spatial scales remains unclear. Until recently, it was thought that negative PSFs 1) increased local (alpha) diversity by suppressing dominants and encouraging species replacements, but 2) lead to convergence of communities, or decreased regional (gamma) diversity. Reciprocally, positive PSFs were thought to decrease alpha diversity by encouraging priority effects and local dominance (Hartnett & Wilson 1999), but increase gamma diversity by increasing community divergence and development of alternative stable state communities. This thinking, however, was based on models that did not account for variability in PSF effects among multiple species. For example, most PSF experiments measure plant growth on self-cultivated soil vs. plant growth on a homogenized soil cultivated by many other plant species to produce a 'self' vs. 'other' PSF measurement. This approach greatly simplifies experiments by decreasing the number of 'other' soil replicates needed, however, plants do not respond the same way to all 'other' soils. For example, plant A may grow well on plant B's soils, very well on plant C's soils and poorly on plant D's soils (Jiang et al., 2010; Kulmatiski et al., 2011). When this variation is taken into account, negative feedbacks are predicted to slow community convergence and allow for transient, long-lasting alpha, beta and gamma diversity (Fukami & Nakajima 2013). Thus, at both the local (Kulmatiski et al., 2011) and regional scales (Fukami & Nakajima 2013) there have been calls to measure PSFs for each species in a community. We can hope that 'self' vs. 'other' approaches will suffice in describing PSF effects on grassland productivity and diversity (as in Klironomos 2002), but until factorial species-level PSF measurements are made and their effects are tested it will be difficult to know how important differences in growth among 'other' soil types are.

MUTUALIST PATHWAYS

Many soil organisms have beneficial effects on grassland plants. Even in low-diversity, agricultural systems which are are prone to pest outbreaks, there is growing recognition of the potential benefits of some soil organisms on plant growth. Recent research, for example, is exploring ways that soil organisms can increase nutrient uptake, increase root growth, induce plant defenses that protect plants from further attack (Walker et al., 2011; Zamioudis & Pieterse 2012), protect from disease and reduce the need for fertilizers and irrigation (Compant et al., 2005; Jiang et al., 2012, Latz et al., 2012). Here we summarize some of the positive roles that soil organisms have on plant growth in grasslands.

Decomposers

Decomposition processes driven by soil microbes and invertebrate detritivores can have strong effects on grassland diversity by influencing nitrogen partitioning (Kahmen et al., 2006; Harrison et al., 2007; Ashton et al., 2010). Decomposers have been estimated to be responsible for 0-50% of primary productivity (Van Der Heijden et al., 2008), though it is certainly easy to argue that plant growth would cease altogether without the nutrient recycling performed by soil decomposers. The breakdown of organic matter is a gradual process with many intermediate steps, making a large number of chemical compounds available (Bardgett 2005), and consequently relaxing competition between plants that are able to take advantage of different forms of nitrogen. McKane et al., (2002) showed in a ^{15}N tracer field experiment that plant species varied in timing, depth and chemical form of N uptake. Furthermore, Eisenhauer et al., (2010) demonstrated that functionally dissimilar decomposers unlock different soil N pools for plants, which had strong extended effects on plant and herbivore performance. A similar idea has been proposed for the mineralization and partitioning of phosphorus forms (Bardgett & Wardle 2010). Accordingly, patterns of detritivore diversity are likely to govern the spatial availability of varying N forms (Bardgett 2005; Bardgett & Wardle 2010; Eisenhauer 2012). We can also expect that detritivore taxa with different activity patterns induce temporal variability in process rates. Indeed, Eisenhauer et al., (2008a) found that anecic earthworms were active in temperate grassland during summer at times when other earthworm species showed limited activity due to dry soil conditions. Information is limited regarding which functional traits are responsible for spatial and temperal differences in nutrient processing among detritivores. Therefore, current attempts to establish trait databases for invertebrates may be a promising avenue to understanding the impacts of detritivore communities on grassland diversity.

Symbionts

In addition to detritivores, arbuscular mycorrhizal fungi (AMF) have been shown to facilitate plant coexistence (Maherali & Klironomos 2007; Wagg et al., 2011), and to maintain plant diversity (Van der Heijden et al., 1998) by influencing variation in resources (as described above) as well as by increasing the availability of limiting resources

(Klironomos et al., 2000). For instance empirical studies demonstrated functional complementarity among AMF species with differing strategies for acquiring phosphorus (Smith et al., 2000), an important resource limiting plant and microbial growth (Koide 2000). AMF diversity, therefore, has been shown to increase the relative performance of subordinate species, which can increase overall plant diversity (Van der Heijden et al., 1998) and enhance plant community productivity (Klironomos et al., 2000, Maherali & Klironomos 2007).

In discussing the interactions between plants and soil organisms we have now noted that negative PSFs can encourage species replacements and plant diversity (Bever et al., 1997). Yet, here we add that symbionts can also serve as a force that encourages diversity (Van der Heijden et al., 1998). This is possible because symbionts can encourage species replacements as long as a symbiont provides a greater benefit to an alternate plant than to the host plant (Figure 5; Bever 1999). For example, two plants could be predicted to coexist even if both plants are grown in soils with symbionts and no pathogens. This can occur if each plant accumulates symbionts that improve the growth of another plant species more than they increase their own growth. As a more specific example, nitrogen-fixing plants that increase their own growth through the accumulation of rhizobacteria have innately positive plant-soil feedbacks that also create a soil which benefit other plants (Eisenhauer & Scheu 2008). In this case each plant can grow in a nitrogen-enriched soil, yet each plant can dedicate less of its own photosynthate to rhizobacteria. While this may seem like an exceptional example, recent research suggest that such positive interactions may be dominant drivers of plant-soil organism interactions (Eisenhauer 2012). Further support was found in studies that have recorded that increasing mycorrhizal diversity can enhance plant diversity and productivity (Van der Heijden et al., 1998; Jonsson et al., 2001, Maherali & Klironomos 2007; Van Der Heijden et al., 2008).

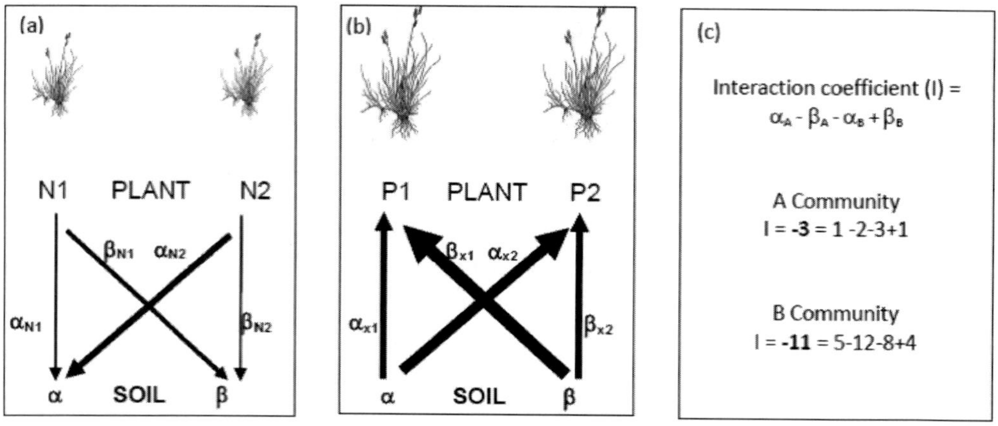

Figure 5. Hypothetical relationship between plant-soil feedbacks (PSF), plant diversity and plant productivity. Plants can accumulate pathogens creating negative individual PSFs and small plant productivity (a) or plants can accumulate symbionts creating positive PSFs and large productivity (b). In either case, coexistence and plant diversity are encouraged by negative community-level PSFs (c). The interaction coefficient (c) summarizes this community-level PSF value (Bever 1997). The key to coexistence in the positive PSF system (b) is that each plant (i.e., P1 and P2) must grow better on 'other' soils than on 'self' soils. Thus, P1 and P2 can realize negative PSF even though they may both accumulate symbionts that increase their own growth.

Many other symbioses are critical to plant growth and diversity (vanLoon 2007). Plants shunt large portions (10-40%) of their photosynthate belowground and much of this becomes root exudates. These exudates create a nutrient-rich rhizosphere environment, which is colonized by soil symbionts. These exudates can add moisture to shallow, nutrient-rich soils that encourages nitrogen mineralization (Cardon et al., 2013). Root exudates also act as chemical signals. Many soil symbionts are motile and rely on chemotaxis to locate growing roots (Compant et al., 2005; Hardoim et al., 2008), which then protect them from infection by pathogenic organisms. Alternatively, symbionts can produce inhibitory allelochemicals or induce systemic resistance in host plants (Nowak & Shulaev 2003). Endophyte fungi, for example, can produce alkaloids that decrease herbivory (Rúa et al., 2013). Such PSF effects were shown to increase plant performance due to reduced effects of plant antagonists in soils with diverse plant communities (Latz et al., 2012). Therefore, Latz et al., (2012) suggest that instead of using pesticides to control pathogen effects on community productivity, soil resistance to soil-borne pathogens can be increased by increasing grassland diversity. In fact, in agricultural settings, there is a growing interest in describing and developing 'weed-suppressive' soils (Kremer & Li 2003). More broadly, there is growing recognition of the potential use of soil organisms to control weeds and increase target plant growth (Compant et al., 2005).

Structural Heterogeneity

The structures built by soil invertebrate ecosystem engineers increase the heterogeneity of habitats, possibly allowing different plant species to co-exist (Eisenhauer et al., 2008b). Ecosystem engineers structure their environment by changing the distribution of materials and energy via non-trophic interactions with abiotic and biotic components of the respective ecosystem (Jones et al., 1994; Jones et al., 1997). These processes affect the distribution, establishment, abundance and diversity of species via the modification, maintenance, creation or destruction of habitats (Jones et al., 1997; Wright et al., 2004). Through bioturbation soil invertebrates, such as ants, earthworms and termites (Folgarait 1998; Lavelle et al., 1998), influence physical properties of soil such as aggregate stability, soil structure, infiltration of water, and aeration of deeper soil layers, as well as biotic properties, such as microbial biomass and activity, nutrient cycling, density of other soil invertebrates, plant community composition, and gas fluxes (Folgarait 1998; Eisenhauer et al., 2007; Wu et al., 2013).

Soil invertebrates (e.g., collembola and earthworms) have been reported to strongly influence the biomass and architecture of roots (Endlweber & Scheu 2007; Eisenhauer et al., 2012a; Eisenhauer et al., 2012b). In a review of earthworm effects on plant growth, Scheu (2003) found that earthworms significantly affected root biomass in 88% of the studies, with an increase of root biomass of 50%. Earthworm burrows are known to represent 'hotspots' of microbial activity and nutrient availability (Maraun et al., 1999; Tiunov and Scheu 1999) and may allow deeper rooting depth due to preferential root growth in earthworm burrows (Eisenhauer et al., 2012b). Indeed, a recent study investigating the effects of decomposer species richness on plant diversity showed that grassland communities were able to build more root biomass in deeper soil layers in the presence of the earthworm species *Lumbricus terrestris* (Eisenhauer et al., 2012a). Although plant communities showed a relative increase in root biomass with increasing plant diversity and higher plant interspecific competition in the upper soil layer, the opposite pattern was found for microcosms with earthworms. The elevated proportion of deep roots in the presence of earthworms explained a significant fraction of the plant complementarity effect on plant shoot biomass production. The presence of anecic earthworms allowed plants to root deeper, which may have increased biotope space (Dimitrakopoulos & Schmid 2004).

Vertical earthworm burrows, such as those formed by *L. terrestris*, represent channels allowing deeper root penetration, particularly in hard, unbroken ground with high nutrient content, microbial enzyme activity and fast carbon turnover rates (Tiunov & Scheu 1999). Middens and earthworm burrows can increase the small-scale spatial heterogeneity of resource availability (Figure 6), allowing more plant species to coexist (Eisenhauer et al., 2008a) and increase plant complementarity (Eisenhauer 2012). Similar mechanisms may apply to other soil invertebrate ecosystem engineers; however, further studies are needed to relate the structures formed by soil invertebrates to grassland performance and diversity. One promising approach is the recently developed method of a DNA-based separation of plant roots (Mommer et al., 2010), which may allow investigating which plant species are able to use which kinds of burrows/structures in order to circumvent competition with other plant species. Such changes in root depth distribution may be important for plant complementarity and thus the positive relationship between plant diversity and ecosystem functioning (Mueller et al., 2012). Moreover, spatial statistical analyses may help connecting structures built by soil invertebrates to the location of elevated plant germination and growth.

Figure 6. Earthworms as ecosystem engineers. Earthworms can transform their environment through bioturbation. Some earthworm species, such as *Lumbricus terrestris*, build vertical, permanent burrows with hill-like structures on the soil surface (called middens). Middens and earthworm burrows increase the small-scale spatial heterogeneity. The left panel shows an earthworm midden with accumulated litter material (photograph by N. Eisenhauer). Germination of some plant species benefits from such nutrient-rich patches. The right panel shows plants germinating out of a vertical earthworm burrow built by *L. terrestris* in a transparent planar cosm (photograph by S. Radsick).

Predators

Predators in both the antagonist and the mutualist pathways can be expected to increase grassland productivity by suppressing herbivores and pathogens, and by increasing nutrient cycling rates (Clarholm 1985; Hedlund & Ohrn 2000; Moore et al., 2003a; Knight et al., 2006; Lenoir et al., 2007; Nieminen 2008a; Sin et al., 2008). For example, in the herbivore-predator food chain, predators are typically thought to improve plant growth by suppressing herbivory. However, it is also possible for predators to help plants in the symbiont-predator food chain. For example, foraging on fungal symbionts can help select for mycorrhizae that induce greater benefits (Klironomos et al., 1996; Klironomos 2003), or less costs to host plants. That is, although symbionts and saprobes often benefit plants by exchanging nitrogen and phosphorous for sugars (Allen 1991), this exchange is not altruistic: there is selective pressure for plants to keep their sugars and for their symbionts to keep their nutrients (Klironomos 2003; Huddleston 2011). As a result, symbionts may be better viewed as 'frenemies' or in detente. Predators that mineralize the cell contents of saprobes and symbionts thereby increase nutrient availability to plants and they do this without requiring additional sugars from plants (Hedlund & Ohrn 2000; Moore et al., 2003b; Lenoir et al., 2007; Nieminen 2008b). This relationship could be summarized as 'the enemy of my frenemy is my friend'. This 'predator-mediated nutrient enrichment' has also been observed in aboveground systems (Sin et al., 2008) and may be a more broadly important mechanism in grasslands than has been recognized.

LITERATURE SYNTHESIS

In the above sections, we presented simplified categories of antagonistic and mutualistic plant-soil interactions in grasslands. These interactions are, of course, not as 'clear cut' as this. As noted, herbivores can actually stimulate plant growth (Frank et al., 2002). Further, soil organisms that are typically thought of as 'symbionts' can become 'antagonistic'. For example, mycorrhizae have been found to decrease grassland plant growth when nutrient availability is high, plant-fungal combinations are not optimal from the plant perspective, or when common mycorrhizal networks shunt resources from vigorous to less vigorous plants (Johnson et al., 1997; Booth 2004). Similarly, bacterial stimulation of plant root growth and exudation can increase or decrease plant growth (Barea et al., 1976). In some cases, increased root exudation can increase bacterial growth, nutrient cycling and plant growth (Glick et al., 1999). In other cases, stimulated root growth can shunt photosynthate away from above ground plant parts, decreasing carbon fixation and plant growth (Lugtenberg & Kamilova 2009). Therefore, to generalize the direction and magnitude of effects that soil organisms have on plant growth in grasslands, we performed a literature search for experiments that either decreased or increased soil organism abundance (i.e., sterilize or inoculate). Experiments testing the effects of soil organisms on plant diversity are extremely rare so we focused on examining soil organism effects on grassland plant biomass.

Publications were located by searching keywords in Web of Science for the following terms: "plant and soil and grassland" and each of the following individually "aphid, collembolan, decomposer, entomopathogenic, fungicide, earthworm, enchytraeid, herbivor*, nematode, inoculation, mesofauna, microarthropods, mite, predatory nematode, oligochaeta, predaceous, predator, protozoa, rhizosphere, root feeder, root herbivore, springtail, steriliz* and trophic cascade". Results were sorted by topic to include: Environmental Sciences, Plant Sciences, Biodiversity Conservation and Forestry. To focus on grassland ecosystems where management goals were more likely to be sustained production of natural resources and maintenance of ecosystem processes, we included studies from natural and semi-natural grasslands and we excluded studies from intensively managed agricultural ecosystems used to provide food and shelter for human use. Using similar rationale, we excluded studies focusing on bioremediation. After reviewing over 4,000 manuscript titles, we found 81 studies with 631 lines of data that met our search criteria.

Consistent with a trophic conceptual framework, we combined the effects of symbionts, saprobes and earthworms and we separated predators of saprobes and symbionts as a distinct compartment. In the antagonist pathway we examined the combined effects of plant pathogens and plant herbivores and separately examined the effects of the predators of these organisms. Low-abundance treatments were assigned as 'controls' and high-abundance treatments were assigned as 'treatments'. For each experiment, only the last reported plant growth measurement was considered. Aboveground plant mass was the most commonly reported value though total biomass and seed production were occasionally measured. To standardize responses among studies we recorded the proportional treatment effect as 100*[(treatment - control) / control]. Negative values indicate that adding soil organisms decreased plant growth and positive values indicate that adding soil organisms increased plant growth.

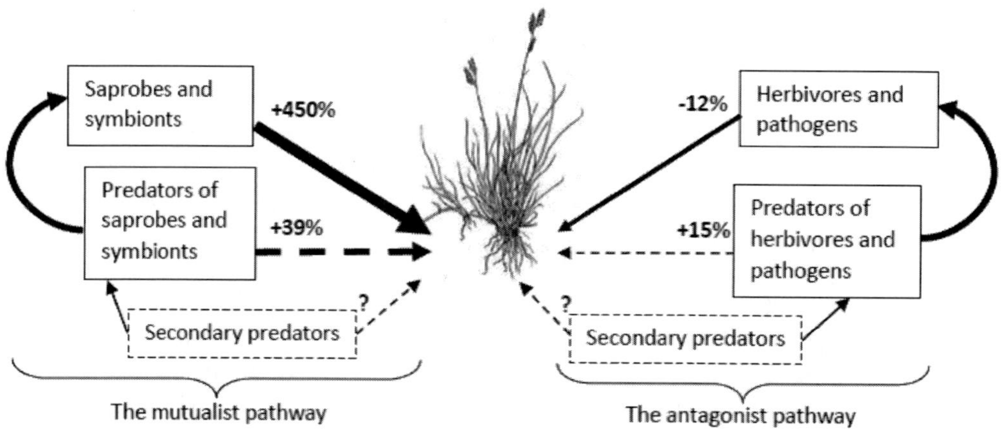

Figure 7. A simplified belowground foodweb showing effect sizes of consumers on plant growth from the literature review. Lines with numbers indicate the percentage change in plant growth that was caused by a particular group of soil organisms. Lines without numbers indicate hypothetical pathways through which observed experimental effects were likely to occur. Secondary predators, or large predators of predators are shown in a hatched box because they may play important roles in soil food webs, but very few studies have examined the effects of these organisms.

Despite the complexity of interactions among diverse soil organisms, our review suggests that most soil organisms are beneficial to plant growth (Figure 7). Across the whole dataset, plant growth was 260 % greater in soils with greater soil organism abundance than in soils with lower soil organism abundance. When grouped into trophic positions, saprophytes and mycorrhizae were found to increase plant growth by 450 % (162 observations from 21 studies). Only herbivores and plant pathogens were found to have a negative effect on plant growth, (decreased growth by 12 %; 134 observations from 24 studies). Predators of saprophytes and mycorrhizae increased plant growth by 39% (40 observations from 12 studies) and predators of herbivores and plant pathogens increased plant growth by 15% (60 observations from 12 studies). Studies adding large, top-level predators were not found. In contrast to previous reviews (Mikola & Setala 1998; Sackett et al., 2010), this review suggests that belowground trophic cascades do affect plant growth in grasslands.

Categorizing effects by phylogeny is of dubious use with soil organisms because most phylogenetic groups can be found in many different trophic levels. Nematodes, for example, span a gradient from microbivores (bacterial feeders and fungal feeders) to plant pathogens to predators and omnivores depending on the species. The net effect of all nematode additions was to decrease plant growth by 6 % (75 experiments from 15 studies). This suggested that most of the nematodes studied were plant pathogens or herbivores. Fungi similarly play various roles as saprobes, pathogens and predators (mycorrhizae were placed in a separate category). Fungicide treatments, however, indicated that the net effect of fungi is to decrease plant growth by 11 % (51 experiments from five studies). This suggested that most of the fungi studied were plant pathogens or potentially diseases of predators. Collembolan, on the other hand, also play various roles, but had a generally positive effect, increasing plant growth by 9 % (45 experiments from 10 studies). This suggested that the collembola studied were largely saprotrophic or fungal grazers. Enchytraeids had a fairly large negative effect on plant growth, though because all of the 24 experiments reviewed were from one study (Ladygina et al., 2010) these results are unreliable from the perspective of a review.

Soil sterilization treatments increased plant growth by 8 % (38 experiments in four studies). This suggested that the net effect of all soil organisms on plant growth is negative. However, soil sterilization treatments have been criticized because they result in a nutrient pulse that increases plant growth and represent treatments that are not relevant to natural systems (Troelstra et al., 2001; Kulmatiski et al., 2008). Further, short-term increases in plant growth on sterile soils are unlikely to be maintained without fertilization because decomposer organisms are needed to mineralize organically-bound nutrients. In support of this contention, experiments in which at least three groups of soil organisms from at least two trophic levels were added increased plant growth by 61 %. These multi-trophic level additions suggested that soil organism diversity has an inherently positive effect on plant growth.

This review summarized the effect of adding soil organisms, but their survival was rarely reported. It is quite reasonable to expect that many organisms that are added to soils may die because the soil is already occupied by soil organisms that are well-suited to the environment (Kardol et al., 2009). That soil organism additions produced the observed effects is impressive and likely underestimates the true effect of soil organisms on plant growth. Alternatively, there was likely to be a bias by researchers to use soil organisms and experimental designs that were expected to produce significant results. For example most studies examined the growth response of a single plant species. Multi-species plant communities are less likely to show treatment responses (Bradford et al., 2002). This is a common problem in literature reviews. As a consequence it is reasonable to compare the effects observed here to those observed in other similar reviews, but it should be recognized that all reviews are likely to exaggerate effects.

Applications and Conclusion

Our review suggests significant effects of soil organism trophic group and diversity on plant biomass and plant diversity in grasslands. This is consistent with previous reviews that suggest that decreasing soil diversity may decrease plant productivity (Brussaard et al., 2007). We find that soil organisms' effects on plant biomass that are reported in the literature are mostly positive and predominately realized through mutualist as opposed to antagonistic pathways. This is an important finding because it suggests that factors such as natural disturbances and anthropogenic activities which reduce soil diversity in grasslands are likely to decrease plant productivity and diversity. Due to the difficulties of working belowground, species-specific effects on grassland biodiversity remain poorly resolved overall, but this suggests that great advances in understanding of basic drivers can be expected in the coming years. We recommend using screening approaches that use new genetics techniques and PSF experiments to develop prioritized lists of potential microbial biocontrol agents. Much work remains to isolate, culture and test the effects of these organisms as well as how they interact in influencing plant communities.

Soil organisms should be regarded as crucial determinants of the productivity and biodiversity of grasslands. This notion may also inspire sustainable management strategies since recent research has been exploring ways that soil organisms can increase nutrient uptake, increase root growth, induce plant defenses that protect plants from further attack (Walker et al., 2011; Zamioudis and Pieterse 2012), protect from disease, and reduce the need for fertilizers and irrigation (Compant et al., 2005; Jiang et al., 2012). Recent advances in the understanding of the driving factors of soil communities as well as resulting feedback effects on plant growth may facilitate future directed management strategies of grasslands, e.g., by favoring mutualist pathways in soil. For instance, maintaining high-diversity grasslands may favor soil surppressiveness against soil-borne pathogens, limiting the necessity to apply pesticides (Latz et al., 2012). Moreover, in agricultural settings, there is growing recognition of the potential use of soil organisms to control weeds and increase target plant growth (Kremer & Li 2003; Compant et al., 2005). By continuing to unlock understanding of soil communities we are likely to identify new biocontrol agents and soil manipulations that can improve natural resource management (Kulmatiski 2011).

References

Adonizio, A. L., Downum, K., Bennett, B. C. & Mathee, K. (2006). Anti-quorum sensing activity of medicinal plants in southern Florida. *Journal of Ethnopharmacology*, *105*, 427-435.

Allen, M. F. (1991). *The Ecology of Mycorrhizae*. Cambridge University Press, Cambridge, UK.

Anagnostakis, S. L. (1982). Biological control of chestnut blight. *Science*, *215*, 466-471.

Arkhipova, T. N., Veselov, S. U., Melentiev, A. I., Martynenko, E. V. & Kudoyarova, G. R. (2005). Ability of bacterium Bacillus subtilis to produce cytokinins and to influence the growth and endogenous hormone content of lettuce plants. *Plant and Soil*, *272*, 201-209.

Ashton, I. W., Miller, A. E., Bowman, W. D. & Suding, K. N. (2010). Niche complementarity due to plasticity in resource use: plant partitioning of chemical N forms. *Ecology*, *91*, 3252-3260.

Bardgett, R. (2005). *The biology of soil: a community and ecosystem approach.* Oxford University Press, Oxford.

Bardgett, R. & Wardle, D. A. (2010). *Aboveground-belowground linkages, biotic interactions, ecosystem processes and global change.* Oxford University Press, Oxford.

Barea, J. M., Navarro, E. & Montoya, E. (1976). Production of Plant Growth Regulators by Rhizosphere Phosphate-solubilizing Bacteria. *Journal of Applied Bacteriology*, *40*, 129-134.

Bever, J. D. (1994). Feedback between plants and their soil communities in an old field community. *Ecology*, *75*, 1965-1977.

Bever, J. D. (1999). Dynamics within mutualism and the maintenance of diversity: inference from a model of interguild frequency dependence. *Ecology Letters*, *2*, 52-62.

Bever, J. D. (2003). Soil community feedback and the coexistence of competitors: conceptual frameworks and empirical tests. *New Phytologist*, *157*, 465-473.

Bever, J. D., Westover, K. M. & Antonovics, J. (1997). Incorporating the soil community into plant population dynamics: the utility of the feedback approach. *Journal of Ecology*, *85*, 561-573.

Bonanomi, G., Giannino, F. & Mazzoleni, S. (2005). Negative plant-soil feedback and species coexistence. *Oikos*, *111*, 311-321.

Booth, M. G. (2004). Mycorrhizal networks mediate overstorey-understorey competition in a temperate forest. *Ecology Letters*, *7*, 538-546.

Bradford, M. A., Jones, T. H., Bardgett, R. D., Black, H. I. J., Boag, B., Bonkowski, M., Cook, R., Eggers, T., Gange, A. C., Grayston, S. J., Kandeler, E., McCaig, A. E., Newington, J. E., Prosser, J. I., Setala, H., Staddon, P. L., Tordoff, G. M., Tscherko, D. & Lawton, J. H. (2002). Impacts of soil faunal community composition on model grassland ecosystems. *Science*, *298*, 615-618.

Brussaard, L., de Ruiter, P. C. & Brown, G. G. (2007). Soil biodiversity for agricultural sustainability. *Agriculture, Ecosystems & Environment*, *121*, 233-244.

Cardon, Z. G., Stark, J. M., Herron, P. M. & Rasmussen, J. A. (2013). Sagebrush carrying out hydraulic lift enhances surface soil nitrogen cycling and nitrogen uptake into inflorescences. *Proceedings of the National Academy of Sciences.*

Clarholm, M. (1985). Interactions of bacteria, protozoa and plants leading to mineralization of soil nitrogen. *Soil Biology and Biochemistry*, *17*, 181-187.

Compant, S., Duffy, B., Nowak, J., Clément, C. & Barka, E. A. (2005). Use of plant growth-promoting bacteria for biocontrol of plant diseases: Principles, mechanisms of action, and future prospects. *Applied and Environmental Microbiology*, *71*, 4951-4959.

Connell, J. H. (1970). On the role of natural enemies in preventing competitive exclusion in some marine animals and in rain forest trees. *Dynamics of populations: Proceedings of the advanced study institute on dynamics of numbers of populations* (eds P. J. Den Boer & G. R. Gradwell), 298-310. Center for Agricultural Publishing and Documentation, Wageningen.

Cumming, D., Fenton, M., Rautenbach, I., Taylor, R., Cumming, G., Cumming, M., Dunlop, J., Ford, A., Hovorka, M. & Johnston, D. (1997). Elephants, woodlands and biodiversity in southern Africa. *South African Journal of Science*, *93*, 231-236.

Curry, J. P. & Schmidt, O. (2007). The feeding ecology of earthworms – A review. *Pedobiologia, 50*, 463-477.

Dimitrakopoulos, P. G. & Schmid, B. (2004). Biodiversity effects increase linearly with biotope space. *Ecology Letters, 7*, 574-583.

Eisenhauer, N. (2012). Aboveground-belowground interactions as a source of complementarity effects in biodiversity experiments. *Plant and Soil, 351*, 1-22.

Eisenhauer, N., Hörsch, V., Moeser, J. & Scheu, S. (2010). Synergistic effects of microbial and animal decomposers on plant and herbivore performance. *Basic and Applied Ecology, 11*, 23-34.

Eisenhauer, N., Marhan, S. & Scheu, S. (2008). Assessment of anecic behavior in selected earthworm species: Effects on wheat seed burial, seedling establishment, wheat growth and litter incorporation. *Applied Soil Ecology, 38*, 79-82.

Eisenhauer, N., Milcu, A., Sabais, A. C. & Scheu, S. (2008). Animal ecosystem engineers modulate the diversity-invasibility relationship. *PLoS One, 3*, e3489.

Eisenhauer, N., Partsch, S., Parkinson, D. & Scheu, S. (2007). Invasion of a deciduous forest by earthworms: Changes in soil chemistry, microflora, microarthropods and vegetation. *Soil Biology & Biochemistry, 39*, 1099-1110.

Eisenhauer, N., Reich, P. B. & Isbell, F. (2012). Decomposer diversity and identity influence plant diversity effects on ecosystem functioning. *Ecology, 93*, 2227-2240.

Eisenhauer, N., Reich, P. B. & Scheu, S. (2012). Increasing plant diversity effects on productivity with time due to delayed soil biota effects on plants. *Basic and Applied Ecology.*

Eisenhauer, N. & Scheu, S. (2008). Earthworms as drivers of the competition between grasses and legumes. *Soil Biology and Biochemistry, 40*, 2650-2659.

Endlweber, K., Ruess, L. & Scheu, S. (2009). Collembola switch diet in presence of plant roots thereby functioning as herbivores. *Soil Biology and Biochemistry, 41*, 1151-1154.

Endlweber, K. & Scheu, S. (2007). Interactions between mycorrhizal fungi and Collembola: effects on root structure of competing plant species. *Biology and Fertility of Soils, 43*, 741-749.

Eppstein, M. J. & Molofsky, J. (2007). Invasiveness in plant communities with feedbacks. *Ecology Letters, 10*, 253-263.

Folgarait, P. J. (1998). Ant biodiversity and its relationship to ecosystem functioning: a review. *Biodiversity & Conservation, 7*, 1221-1244.

Frank, D. A., Kuns, M. M. & Guido, D. R. (2002). Consumer control of grassland plant production. *Ecology, 83*, 602-606.

Fukami, T. & Nakajima, M. (2013). Complex plant–soil interactions enhance plant species diversity by delaying community convergence. *Journal of Ecology, 101*, 316-324.

Glick, B., Patten, C., Holguin, G. & Penrose, D. (1999). Overview of plant growth-promoting bacteria. *Biochemical and Genetic Mechanisms Used by Plant Growth Promoting Bacteria", Glick, BR, Patten, CL, Holguin, G. and Penrose, DM*, 1-13.

Hardoim, P. R., van Overbeek, L. S. & Elsas, J. D. v. (2008). Properties of bacterial endophytes and their proposed role in plant growth. *Trends in Microbiology, 16*, 463-471.

Harrison, K. A., Bol, R. & Bardgett, R. D. (2007). Preferences for different nitrogen forms by coexisting plant species and soil microbes. *Ecology, 88*, 989-999.

Hartnett, D. C. & Wilson, G. W. T. (1999). Mycorrhizae influence plant community structure and diversity in tallgrass prairie. *Ecology, 80*, 1187-1195.

Hector, A., Schmid, B., Beierkuhnlein, C., Caldeira, M. C., Diemer, M., Dimitrakopoulos, P.G., et al. (1999). Plant diversity and productivity experiments in European grasslands. *Science*, *286*, 1123-1127.

Hedlund, K. & Ohrn, M. S. (2000). Tritrophic interactions in a soil community enhance decomposition rates. *Oikos*, *88*, 585-591.

Heinemann, J. A. & Sprague Jr, G. F. (1989). Bacterial conjugative plasmids mobilize DNA transfer between bacteria and yeast. *Nature*, *340*, 205-209.

Huddleston, J. E. (2011). Symbiosis: Market economics in plant–fungus relationships. *Nat Rev Micro*, *9*, 698-699.

Hughes, D. T. & Sperandio, V. (2008). Inter-kingdom signalling: communication between bacteria and their hosts. *Nat Rev Micro*, *6*, 111-120.

Hutchinson, W. D. & Campbell, C. D. (1994). Economic impact of the sugarbeet root aphid (Homoptera: Aphididae) on sugarbeet yield and quality in southern Minn. *Journal of Economic Entomology*, *87*, 465-475.

Janzen, D. H. (1970). Herbivores and the number of tree species in tropical forests. *The American Naturalist*, *104*, 501-528.

Jiang, F., Chen, L., Belimov, A. A., Shaposhnikov, A. I., Gong, F., Meng, X., et al. (2012). Multiple impacts of the plant growth-promoting rhizobacterium Variovorax paradoxus 5C-2 on nutrient and ABA relations of Pisum sativum. *Journal of Experimental Botany*, *63*, 6421-6430.

Jiang, L., Han, X., Zhang, G. & Kardol, P. (2010). The role of plant–soil feedbacks and land-use legacies in restoration of a temperate steppe in northern China. *Ecological Research*, *25*, 1101-1111.

Johnson, N., GRAHAM, J. H. & Smith, F. (1997). Functioning of mycorrhizal associations along the mutualism–parasitism continuum. *New Phytologist*, *135*, 575-585.

Jones, C. G., Lawton, J. H. & Shachak, M. (1994). Organisms as ecosystem engineers. *Oikos*, 373-386.

Jones, C. G., Lawton, J. H. & Shachak, M. (1997). Positive and negative effects of organisms as physical ecosystem engineers. *Ecology*, *78*, 1946-1957.

Jonsson, L. M., Nilsson, M. C., Wardle, D. A. & Zackrisson, O. (2001). Context dependent effects of ectomycorrhizal species richness on tree seedling productivity. *Oikos*, *93*, 353-364.

Kahmen, A., Renker, C., Unsicker, S. B. & Buchmann, N. (2006). Niche complementarity for nitrogen: an explanation for the biodiversity and ecosystem functioning relationship? *Ecology*, *87*, 1244-1255.

Kardol, P., Bezemer, T. M. & Van Der Putten, W. H. (2009). Soil Organism and Plant Introductions in Restoration of Species-Rich Grassland Communities. *Restoration Ecology*, *17*, 258-269.

Klironomos, J. N. (2003). Variation in plant response to native and exotic arbuscular mycorrhizal fungi. *Ecology*, *84*, 2292-2301.

Klironomos, J. N., McCune, J., Hart, M. & Neville, J. (2000). The influence of arbuscular mycorrhizae on the relationship between plant diversity and productivity. *Ecology Letters*, *3*, 137-141.

Klironomos, J. N., Rillig, M. C. & Allen, M. F. (1996). Below-ground microbial and microfaunal responses to Artemisia tridentata grown under elevated atmospheric CO_2. *Functional Ecology*, *10*, 527-534.

Knight, T. M., Chase, J. M., Hillebrand, H. & Holt, R. D. (2006). Predation on mutualists can reduce the strength of trophic cascades. *Ecology Letters*, *9*, 1173-1178.

Koide, R. T. (2000). Functional complementarity in the arbuscular mycorrhizal symbiosis. *New Phytologist*, *147*, 233-235.

Kremer, R. J. & Li, J. (2003). Developing weed-suppressive soils through improved soil quality management. *Soil and Tillage Research*, *72*, 193-202.

Kulmatiski, A. (2006). Exotic plants establish persistent communities. *Plant Ecology*, *187*, 261-275.

Kulmatiski, A., Beard, K. H. & Heavilin, J. (2012). Plant-soil feedbacks provide an additional explanation for diversity-productivity relationships. *Proceedings of the Royal Society B-Biological Sciences*, *279*, 3020-3026.

Kulmatiski, A., Beard, K. H., Stevens, J. & Cobbold, S. (2008). Plant-soil feedbacks: A meta-analytical review. *Ecology Letters*, *11*, 980-992.

Kulmatiski, A., Heavilin, J. & Beard, K. H. (2011). Testing predictions of a three-species plant-soil feedback model. *Journal of Ecology*, *99*, 542-550.

Kyle, G. P., Beard, K. H. & Kulmatiski, A. (2007). Reduced soil compaction enhances the establishment of non-native plant species. *Plant Ecology*, *193*, 223-232.

Kyle, G. P., Kulmatiski, A. & Beard, K. H. (2008). Influence of pocket gopher mounts on nonnative plant establishment in a shrubsteppe ecosystem. *Western North American Naturalist*, *68*, 374-381.

Ladygina, N., Henry, F., Kant, M. R., Koller, R., Reidinger, S., Rodriguez, A., et al. (2010). Additive and interactive effects of functionally dissimilar soil organisms on a grassland plant community. *Soil Biology & Biochemistry*, *42*, 2266-2275.

Latz, E., Eisenhauer, N., Rall, B. C., Allan, E., Roscher, C., Scheu, S. & Jousset, A. (2012). Plant diversity improves protection against soil-borne pathogens by fostering antagonistic bacterial communities. *Journal of Ecology*, *100*, 597-604.

Lavelle, P., Barois, I., Blanchart, E., Brown, G., Brussaard, L., Decaens, T., et al. (1998). Earthworms as a resource in tropical agroecosystems. *Nature & Resources*, *34*, 26-41.

Lenoir, L., Persson, T., Bengtsson, J., Wallander, H. & Wiren, A. (2007). Bottom-up or top-down control in forest soil microcosms? Effects of soil fauna on fungal biomass and C/N mineralisation. *Biology and Fertility of Soils*, *43*, 281-294.

Lugtenberg, B. & Kamilova, F. (2009). Plant-Growth-Promoting Rhizobacteria. *Annual Review of Microbiology*, *63*, 541-556.

Maherali, H. & Klironomos, J. N. (2007). Influence of phylogeny on fungal community assembly and ecosystem functioning. *Science*, *316*, 1746-1748.

Mao, X., Hu, F., Griffiths, B., Chen, X., Liu, M. & Li, H. (2007). Do bacterial-feeding nematodes stimulate root proliferation through hormonal effects? *Soil Biology & Biochemistry*, *39*, 1816-1819.

Maraun, M., Alphei, J., Bonkowski, M., Buryn, R., Migge, S., Peter, M., Schaefer, M. & Scheu, S. (1999). Middens of the earthworm Lumbricus terrestris (Lumbricidae): microhabitats for micro- and mesofauna in forest soil. *Pedobiologia*, *43*, 276-287.

Maron, J. L., Marler, M., Klironomos, J. N. & Cleveland, C. C. (2011). Soil fungal pathogens and the relationship between plant diversity and productivity. *Ecology Letters*, *14*, 36-41.

McKane, R. B., Johnson, L. C., Shaver, G. R., Nadelhoffer, K. J., Rastetter, E. B., Fry, B., et al. (2002). Resource-based niches provide a basis for plant species diversity and dominance in arctic tundra. *Nature*, *415*, 68-71.

Mikola, J. & Setala, H. (1998). No evidence of trophic cascades in an experimental microbial-based soil food web. *Ecology*, *79*, 153-164.

Mommer, L., van Ruijven, J., de Caluwe, H., Smit-Tiekstra, A. E., Wagemaker, C. A. M., Ouborg, N. J., Bogemann, G. M., van der Weerden, G. M., Berendse, F. & de Kroon, H. (2010). Unveiling below-ground species abundance in a biodiversity experiment: a test of vertical niche differentiation among grassland species. *Journal of Ecology*, *98*, 1117-1127.

Moore, J. C., McCann, K., Setala, H. & De Ruiter, P. C. (2003). Top-down is bottom-up: Does predation in the rhizosphere regulate aboveground dynamics? *Ecology*, *84*, 846-857.

Moore, J. C., McCann, K., Setälä, H. & De Ruiter, P. C. (2003). Top-down is bottom-up: Does predation in the rhizosphere regulate aboveground dynamics? *Ecology*, *84*, 846-857.

Mueller, K. E., Tilman, D., Fornara, D. A. & Hobbie, S. E. (2012). Root depth distribution and the diversity–productivity relationship in a long-term grassland experiment. *Ecology*, *94*, 787-793.

Nieminen, J. K. (2008a). Linking food webs to ecosystem processes: Piecewise linear models of soil microcosms. *Ecological Modelling*, *217*, 87-94.

Nieminen, J. K. (2008b). Soil animals and ecosystem processes: How much does nutrient cycling explain? *Pedobiologia*, *51*, 367-373.

Nowak, J. & Shulaev, V. (2003). Priming for transplant stress resistance in in vitro propagation. *In Vitro Cellular & Developmental Biology-Plant*, *39*, 107-124.

Petermann, J. S., Fergus, A. J., Turnbull, L. A. & Schmid, B. (2008). Janzen-Connell effects are widespread and strong enough to maintain diversity in grasslands. *Ecology*, *89*, 2399-2406.

Prusty, R., Grisafi, P. & Fink, G. R. (2004). The plant hormone indoleacetic acid induces invasive growth in Saccharomyces cerevisiae. *Proceedings of the National Academy of Sciences*, *101*, 4153-4157.

Raaijmakers, J. M., Paulitz, T. C., Steinberg, C., Alabouvette, C. & Moenne-Loccoz, Y. (2009). The rhizosphere: a playground and battlefield for soilborne pathogens and beneficial microorganisms. *Plant and Soil*, *321*, 341-361.

Rúa, M. A., McCulley, R. L. & Mitchell, C. E. (2013). Fungal endophyte infection and host genetic background jointly modulate host response to an aphid-transmitted viral pathogen. *Journal of Ecology*, *101*, 1007-1018.

Sackett, T. E., Classen, A. T. & Sanders, N. J. (2010). Linking soil food web structure to above- and belowground ecosystem processes: a meta-analysis. *Oikos*, *119*, 1984-1992.

Schallhart, N., Tusch, M. J., Wallinger, C., Staudacher, K. & Traugott, M. (2012). Effects of plant identity and diversity on the dietary choice of a soil-living insect herbivore. *Ecology*, *93*, 2650-2657.

Scheu, S. (2003). Effects of earthworms on plant growth: patterns and perspectives: The 7th international symposium on earthworm ecology· Cardiff· Wales· 2002. *Pedobiologia*, *47*, 846-856.

Schnitzer, S. A., Klironomos, J. N., HilleRisLambers, J., Kinkel, L. L., Reich, P. B., Xiao, K., et al. (2011). Soil microbes drive the classic plant diversity-productivity pattern. *Ecology*, *92*, 296-303.

Sin, H., Beard, K. H. & Pitt, W. C. (2008). An invasive frog, *Eleutherodactylus coqui*, increases new leaf production and leaf litter decomposition rates through nutrient cycling in Hawaii. *Biological Invasions*, *10*, 335-345.

Smith, F. A., Jakobsen, I. & Smith, S. E. (2000). Spatial differences in acquisition of soil phosphate between two arbuscular mycorrhizal fungi in symbiosis with Medicago truncatula. *New Phytologist*, *147*, 357-366.

Suding, K. N., Harpole, W. S., Fukami, T., Kulmatiski, A., MacDougall, A. S., Stein, C. & van Der Putten, W. H. (2013). Consequences of plant-soil feedbacks in invasion. *Journal of Ecology*.

Teplitski, M., Robinson, J. B. & Bauer, W. D. (2000). Plants secrete substances that mimic bacterial N-acyl homoserine lactone signal activities and affect population density-dependent behaviors in associated bacteria. *Molecular Plant-Microbe Interactions*, *13*, 637-648.

Tiunov, A. V. & Scheu, S. (1999). Microbial respiration, biomass, biovolume and nutrient status in burrow walls of *Lumbricus terrestris* L.(Lumbricidae). *Soil Biology and Biochemistry*, *31*, 2039-2048.

Treonis, A. M., Grayston, S. J., Murray, P. J. & Dawson, L. A. (2005). Effects of root feeding, cranefly larvae on soil microorganisms and the composition of rhizosphere solutions collected from grassland plants. *Applied Soil Ecology*, *28*, 203-215.

Troelstra, S. R., Wagenaar, R., Smant, W. & Peters, B. A. M. (2001). Interpretation of bioassays in the study of interactions between soil organisms and plants: involvement of nutrient factors. *New Phytologist*, *150*, 697-706.

Van Der Heijden, M. G. A., Bardgett, R. D. & Van Straalen, N. M. (2008). The unseen majority: soil microbes as drivers of plant diversity and productivity in terrestrial ecosystems. *Ecology Letters*, *11*, 296-310.

Van der Heijden, M. G. A., Klironomos, J. N., Ursic, M., Moutoglis, P., Streitwolf-Engel, R., Boller, T., Wiemken, A. & Sanders, I. R. (1998). Mycorrhizal fungal diversity determines plant biodiversity, ecosystem variability and productivity. *Nature*, *396*, 69-72.

Van der Stoel, C. D., Duyts, H. & Van der Putten, W. H. (2006). Population dynamics of a host-specific root-feeding cyst nematode and resource quantity in the root zone of a clonal grass. *Oikos*, *112*, 651-659.

vanLoon, L. C. (2007). Plant responses to plant growth-promoting rhizobacteria. *European Journal of Plant Pathology*, *119*, 243-254.

Wagg, C., Jansa, J., Stadler, M., Schmid, B. & Van Der Heijden, M. G. (2011). Mycorrhizal fungal identity and diversity relaxes plant-plant competition. *Ecology*, *92*, 1303-1313.

Walker, T. S., Bais, H. P., Grotewold, E. & Vivanco, J. M. (2003). Root exudation and rhizosphere biology. *Plant Physiology*, *132*, 44-51.

Walker, V., Bertrand, C., Bellvert, F., Moenne-Loccoz, Y., Bally, R. & Comte, G. (2011). Host plant secondary metabolite profiling shows a complex, strain-dependent response of maize to plant growth-promoting rhizobacteria of the genus Azospirillum. *New Phytologist*, *189*, 494-506.

Wehner, J., Antunes, P. M., Powell, J. R., Caruso, T. & Rillig, M. C. (2011). Indigenous arbuscular mycorrhizal fungal assemblages protect grassland host plants from pathogens. *PLoS One*, *6*, e27381.

Whitford, W. G. & Kay, F. R. (1999). Biopedturbation by mammals in deserts: a review. *Journal of Arid Environments*, *41*, 203-230.

Wright, J. P., Gurney, W. S. & Jones, C. G. (2004). Patch dynamics in a landscape modified by ecosystem engineers. *Oikos*, *105*, 336-348.

Wu, H., Lu, X., Wu, D., Song, L., Yan, X. & Liu, J. (2013). Ant mounds alter spatial and temporal patterns of CO_2, CH_4 and N_2O emissions from a marsh soil. *Soil Biology and Biochemistry*, *57*, 884-891.

Wurst, S., Allema, B., Duyts, H. & van der Putten, W. H. (2008). Earthworms counterbalance the negative effect of microorganisms on plant diversity and enhance the tolerance of grasses to nematodes. *Oikos*, *117*, 711-718.

Zamioudis, C. & Pieterse, C. M. J. (2012). Modulation of Host Immunity by Beneficial Microbes. *Molecular Plant-Microbe Interactions*, *25*, 139-150.

In: Grassland Biodiversity and Conservation ... ISBN: 978-1-61761-619-8
Editors: Pierre Mariotte and Paul Kardol

Chapter 3

SEASONAL VARIATION OF SOIL MICROBIAL COMMUNITIES IN SEMI-NATURAL AND IMPROVED GRASSLANDS IN NORTHERN JAPAN

Zabed Hossain*[*1] and Shu-ichi Sugiyama[2]
[1]Department of Botany, University of Dhaka, Dhaka, Bangladesh
[2]Faculty of Agriculture and Life Science, Hirosaki University, Hirosaki, Japan

ABSTRACT

Although soil microbes are often used as an important indicator of soil quality, our knowledge about the seasonal variability of soil microbial communities is limited. This chapter examined seasonal variation of the soil microbial communities in semi-natural and improved grasslands by analyzing phospholipid fatty acid (PLFA) profiles. Data were collected from semi-natural and improved grasslands of Shiriyazaki reserved area located in the northern Honshu island of Japan at five different seasonal periods during 2006. PLFAs of total microbes, bacteria, fungi and mycorrhiza were affected by seasons. Data showed significant differences in total microbial, total bacterial, Gram-positive bacterial and mycorrhizal PLFA biomass among the seasons in the semi-natural short-type grasslands. Bacterial biomass in the short-type grassland was significantly negatively correlated with soil moisture content. In the improved grasslands, total bacterial biomass was affected by seasons but not correlated with physico-chemical soil properties. Fungal biomass, on the other hand, showed a large seasonal effect only in the semi-natural tall-type grasslands and was correlated with soil C/N ratio. Semi-natural tall-type grasslands showed significantly higher fungal biomass and fungal to bacterial biomass ratio than the improved grassland. Because bacterial-dominated grasslands are linked with higher nitrogen leaching, improved grasslands would be associated with higher N losses. While a large part of Shiriyazaki grasslands has been converted into improved grasslands for economical reasons, our results showed that this type of management is not sustainable and suggest that grazing should be restored to improve ecosystem functioning and stability of these grasslands.

---

[*] Corresponding author: Zabed Hossain, email: zabed@du.ac.bd.

Keywords: Grasslands, microbial communities, phospholipid fatty acids (PLFA), seasonal variation

INTRODUCTION

Soil microbial communities play a critical role in breaking down organic matter and releasing mineral nutrients in terrestrial ecosystems (Wardle 2002, Bardgett 2005, Defrieri et al., 2011). Soil microbial biomass influences nutrient mineralization and is a source of major plant nutrients including, nitrogen (N) and phosphorus (P) (Bardgett et al., 2003, Kaye and Hart 1997, Lipson et al., 1999, Balota et al., 2003, Dick 1992, Miller et al., 2005). For example, Lipson et al., (1999) reported that microbial population dynamics were related to plant N availability in alpine dry meadows. Soil microbial community composition also influences soil ecosystem processes, such as decomposition and N mineralization (Orwin et al., 2006, Hossain et al., 2010). Further, soil microbes have been regarded as an indicator of soil quality (Anderson 2003), soil management (Balota et al., 2003, Powlson et al., 1987) and soil degradation (Balota et al., 2003, Brookes 1995). However, although soil microbial communities show seasonal variability in their structure and function (Bardgett 2005, Bardgett et al., 1999, Gray et al., 2011, Castro et al., 2010) the results of previous studies are contrasting. Indeed, some studies reported higher amount of soil microbial biomass in early spring (Bardgett et al., 1999, Grayston et al., 2001), while other studies showed a peak of microbial biomass in autumn and winter (Spedding et al., 2004, Lipson et al., 1999) or in summer (Hamel *et al.,* 2006). These results indicate the involvement of several factors in driving seasonal variability of soil microbial communities; such as changes in resource availability through litter input, rhizodeposition and root senescence, as well as seasonal changes in temperature and precipitation (Bardgett et al., 2005, Gray et al., 2011, Lipson et al., 2002). Gray et al., (2011) reported that microbial biomass was greater under low-precipitation conditions. They also found that temperature was the most significant predictor of shifts in soil microbial community composition, with fungal biomass increasing and bacterial biomass decreasing in warmed plots. These variations in soil microbial communities should have important consequences for the soil functioning such as organic matter decomposition and nutrient cycling.

Grasslands play an important role by providing food, forage, and livestock, by serving aesthetic values like recreation and by maintaining biodiversity and atmospheric carbon balance through sequestering organic carbon from the atmosphere (Peeters 2009). Flanagan et al., (2002) reported a large seasonal variation in the components of the annual carbon budget, such as carbon uptake and loss, which could be related with soil respiration in the northern temperate grassland ecosystem. Grasslands constitute one-third of all terrestrial ecosystems on Earth (Schlesinger 1997) and have a high soil organic carbon content (Adams et al., 1990), and as such are an important component of the global carbon cycle. Grasslands support uniquely large and active soil biological communities (Bardgett and Cook 1998). Management activities such as sowing, fertilization and grazing are common practices in grasslands in order to maximize the productivity (Peeters 2009). However, depending on its intensity, management practices are likely to alter belowground microbial communities, which in turn may affect plant community composition and ecosystem functioning. Japan grasslands include semi-natural grasslands (Yazaki et al., 2004, Mutoh et al., 1985) and improved grasslands, both mainly composed of temperate species such as orchard grass (*Dactylis glomerata*), tall fescue (*Festuca arundinacea*), timothy (*Phleum pratense*) and white clover (*Trifolium repens*). Improved grasslands have been mainly developed in order to increase livestock production. In Japan, Shiriyazaki reserved grassland area has been maintained under grazing by indigenous horses for over 300 years without chemical fertilizers, (Sugiyama et al., 2008). The improved grassland, which lies next to the reserved area, was established 30 years ago by sowing selected pasture species (*Lolium perenne*, *Dactylis glomerata*, and *Trifolium repense*) and has received chemical fertilizers every year. Differences in management activities have exerted differences in the composition of soil microbial communities of Shiriyazaki grasslands: the less disturbed semi-natural grasslands dominated by the grass *Miscanthus sinensis* supports fungal-dominated communities, while the improved grasslands favor bacterial-dominated communities (Hossain et al., 2007). However, there has been limited data on the seasonal variability of soil microbial communities in these grasslands although such information is relevant for the management and conservation of these grasslands.

The objectives of this chapter were (1) to examine the seasonal variation of microbial communities by analyzing PLFA profiles in semi-natural and improved grasslands of Shiriyazaki in the northern Honshu island of Japan, (2) to test the factors that are responsible for the seasonal variation of microbial communities, and also (3) to examine whether similar patterns of seasonal variation occur among the different grassland types, i.e., semi-natural tall- and short-type grasslands and improved grasslands.

MATERIALS AND METHODS

Study Site

Shiriyazaki reserved grassland area, consisting of ca 5km^2 (41°25'N and 141°26'E), is located on northern Honshu Island, Japan. A historical record describes that this grassland was developed by 1700 after the introduction of horses for agricultural use. Since then, no management, such as application of chemical or organic fertilizers, or sowing of pasture species, has been applied. Although horses were threatened to the extinction in 1960' with the prevalence of tractors, this grassland is now protected as a reserved area by a local government, and about forty horses are grazing naturally in this area. This semi-natural grassland is located in a cape faced with Pacific Ocean. The sea-side area is heavily grazed compared with the inside. Consequently, two floristically distinct communities developed depending on grazing intensities: short-grass and tall-grass (Figure 1). Short-grass community exposed to heavy grazing is characterized by the dominance of the C_4 grass *Zoysia japonica* and low vegetation height, while the tall-grass community subjected to less intensive grazing, is characterized by the dominance of the C_4 tallgrass, *Miscanthus sinensis*, the C_3 grass *Brachypodium sylvaticum*, and numerous forb species. Improved pasture, located 1 km distant from the semi-natural grassland, was established for beef production about 30 years ago by sowing of herbage species such as *Dactylis glomerata* and *Trifolium repens* and has been maintained by cow grazing with annual chemical fertilization (N/P/K; 100/120/100 Kg ha^{-1}).

Figure 1. Semi-natural Tall-type (a), Short-type (b) and Improved (c) grasslands of Shiriyazaki area in the northern Honshu island of Japan.

Soil Sampling

In each of the tall-type, short-type, and improved grassland, soil samples (5 cm diameter, 5 cm depth; HS-25, Fujiwara Ltd, Tokyo, Japan) were taken in 3 plots (50 × 50 cm) spaced out by 50 m to 100 m. Samplings were done in the same plots at five different periods in 2006: 18 April, 4 June, 21 July, 26 September, and 23 November. The first sampling was done before the onset of regrowth of constituent plants after winter, whereas the final sampling was done when most plant production had stopped because of low temperatures. Monthly total precipitation and monthly mean temperature varied greatly across the study period (Figure 2). After transfer to the laboratory, soils were sieved through a 2-mm mesh sieve to remove large rocks, roots, and macrofauna. Soils were analyzed for microbial communities in the laboratory within 24 h after collection.

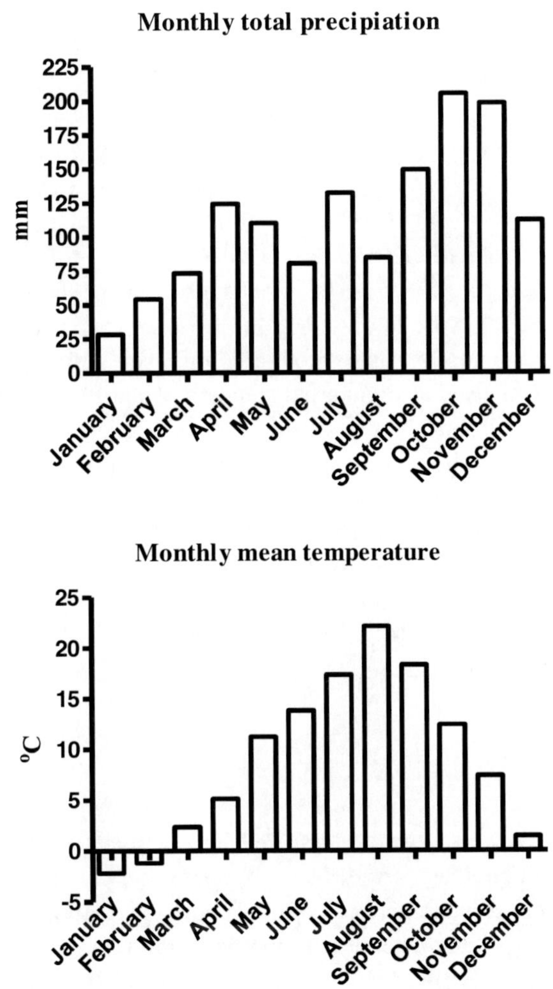

Figure 2. Monthly total precipitation (mm) and monthly mean temperature (°C) during the study period (2006) in the Shiriyazaki grassland area in northern Honshu Island, Japan. (Source: Japan Meteorological Agency).

Soil Physico-Chemical Properties

Dissolved inorganic N concentrations (NO_3^- and NH_4^+) were determined using the Kjeldahl method (Black 1965). Nitrogen was extracted from 5 g fresh soil in 50 ml 2 mol KCl. Soil extractable P concentrations were determined using a colorimetric method after extraction from 1 g dry soil in 0.002 mol H_2SO_4 solution and reaction with ammonium paramolybdate. Soil pH was measured using distilled water (using a 1:2 soil:water suspension). Soil moisture content was determined according to the weight loss after drying 10 g soil at 60 °C for 24 h. Soil total N (%) and total C (%) content was analyzed using an automatic high-sensitivity N-C analyzer (Sumigraph NC-90A; Sumika Chemical Analysis, Co., Japan).

PLFA Analysis

PLFAs were analyzed following the method described by Hossain et al. (2007), which is based on a modification of the procedure described by Frostegård et al. (1993). Briefly, lipids were extracted from fresh soils equivalent to 10 g dry weight according to procedure as described by Bligh and Dyer (1959). PLFAs were analyzed by using a gas chromatograph (GC-14B; Shimadzu Corp., Kyoto, Japan). In addition, PLFA mixtures extracted from soils were analyzed using GC-MS (PQ2010; Shimadzu Corp., Kyoto, Japan) to identify each PLFA peak. We also used a commercial bacterial fatty acid methyl esters mixture for peak identification (BAME; Supelco, Bellefonte, USA). PLFAs of each soil sample were identified from the chromatographic retention time through comparison with nonadecanoic acid (19:0; Supelco, Bellefonte, USA), used as an internal standard. Total PLFA (μg PLFA g^{-1} soil) was determined as the total microbial biomass using both identified and unidentified PLFAs in each sample; the identified individual PLFA was expressed in units of nanomoles of PLFA per gram dry soil. We used the fatty acid nomenclature described by Frostegård et al. (1993) with PLFAs i14:0, a14:0, i15:0, a15:0, 15:0, 2OH-14:0, 3OH-14:0, i16:0, i17:0, a17:0, 17:0, cy17:0, 2OH-16:0, 18:1ω7c, cy19:0 as total bacterial; branched PLFAs i14:0, a14:0, i15:0, a15:0, i16:0, i17:0 and a17:0 as Gram-positive bacterial; cyclopropyl PLFAs cy17:0 and cy19:0 as Gram-negative bacterial; 16:1ω5c as mycorrhizal; and 18:2ω6 as fungal biomarkers (Bossio and Scow 1998, Waldrop *et al.,* 2000, Frostegård *et al.,* 1993).

Statistical Analysis

Repeated measures ANOVA were conducted to examine the effects of grassland and season on microbial communities as well as on physical and chemical properties of soil. In addition, correlation analyses were performed to examine the relationships between microbial communities and the soil physico-chemical properties within and across the grassland types. Data analyses were conducted using JMP 4.0 (SAS Institute, Cary, NC, USA).

Table 1. Effects (F ratio) of grassland type and season on soil physico-chemical properties of the semi-natural tall-type and short-type grasslands, and improved grasslands of Shiriyazaki in the northern Honshu Island, Japan. Data shown for grassland types are mean value across season and data shown for season are mean values across grassland types

	Grassland type			Season					*F* ratio	
	Semi-natural		Improved							
	Tall-type	Short-type		April	June	July	September	November	Grassland type	Season
pH	6.9^{b}	7.3^{a}	7.3^{ab}	6.7^{b}	7.5^{b}	7.6^{a}	6.8^{ab}	7.3^{a}	3.1	13.5***
Moisture	41.8^{a}	37.5^{ab}	36.7^{b}	40.3^{a}	40.1^{a}	41.9^{a}	31.0^{b}	39.2^{a}	0.9	2.50***
DIN†	2.4^{b}	2.3^{b}	3.7^{a}	2.1^{a}	2.9^{a}	3.7^{a}	2.1^{a}	3.2^{a}	3.3	2.7*
P†	6.4^{b}	7.4^{b}	14.5^{a}	8.0^{a}	9.3^{a}	11.2^{a}	7.6^{a}	11.0^{a}	26.8***	1.0
%N	1.1^{a}	0.9^{a}	0.9^{a}	0.8^{a}	1.0^{a}	1.0^{a}	0.9a	1.0^{a}	0.3	2.6
%C	12.6^{a}	10.3^{a}	9.5^{a}	9.5^{a}	11.5^{a}	11.2^{a}	10.6^{a}	11.2^{a}	0.6	2.3
C/N	11.9^{a}	11.4^{a}	10.8^{b}	11.5^{a}	11.2^{a}	11.4^{a}	11.5^{a}	11.4^{a}	1.4	1.0

†Soil dissolved inorganic nitrogen (DIN) in mg N/100 g soil and available phosphorus (P) in mg P_2O_5/100 g soil. Different letters within grassland type and season represent significant differences (multiple comparison using Tukey–Kramer HSD). *, **, and *** indicate significant difference at $P < 0.05$, 0.01 and 0.001 levels, respectively.

RESULTS

Soil Physico-Chemical Properties

Soil pH and moisture content were significantly affected by season (F = 13.5, P = 0.0001) and not by grassland (Table 1). In both of the semi-natural tall- and short-type grasslands, the highest pH values 7.47 and 7.93 respectively, were recorded in July and the lowest values 6.42 and 6.40 were recorded in April. On the other hand, in the improved grassland, the highest pH value (7.48) was recorded in November and the lowest (6.93) was in September. Available N content was significantly affected by seasons (F = 2.7, P = 0.05) while P content was affected by grassland type (F = 26.8, P = 0.0001). Nitrogen contents ranged from 2.3 $mg.100g^{-1}$ soil in short-type to 3.7 $mg.100g^{-1}$ soil in improved grassland. The highest content of available N was recorded in July (3.7 $mg.100g^{-1}$ soil) and the lowest was in April and November (2.1 $mg.100g^{-1}$ soil). The highest P content was recorded in the improved grassland (14.5 $mg.100g^{-1}$ soil) and the lowest was in tall-type grassland (6.4 $mg.100g^{-1}$ soil). Soil total C (%), and C/N ration did not differ significantly among grasslands and seasons. The C/N ratio was higher in the semi-natural tall-type (11.9) and short-type (11.4) than in the improved grassland (10.8).

PLFA Profiles

As shown in Table 2, total PLFA, total bacterial and Gram-positive bacterial PLFA were significantly affected by season (F = 4.5, *P* = 0.01) and not by the grassland type. Gram-negative bacterial PLFA was not significantly affected by either grassland or season. Fungal, mycorrhizal and fungal/bacterial PLFA concentrations were significantly affected by both grassland and season (Figure 3). Short-type grasslands showed significantly higher concentrations of total PLFA, total bacterial, Gram-positive bacterial and mycorrhizal PLFA in September compared to the other months of the year. Total bacterial PLFA concentration also showed significant seasonal variation in the improved grasslands with the highest in July and the lowest in November. The tall-type grassland, on the other hand, showed significantly higher fungal PLFA concentrations and fungal to bacterial ratio in July than the other months while the other two grassland types did not show any significant seasonal variation (Figure 3).

Correlation analyses showed that total PLFA (r = -0.909, *P* = 0.03), bacterial (r = -0.924, *P* = 0.02), Gram-positive (r = -0,914, *P* = 0.03) and Gram-negative (r = -0.953, *P* = 0.01) PLFA were significantly and negatively correlated with soil moisture content in the short-type grassland (Figure 4). Fungal PLFA (r = -0.88, *P* = 0.05) and fungal to bacterial PLFA ratio (r = -0.87, *P* = 0.06) were negatively correlated with soil C/N ratio in the tall-type grasslands (Figure 5). When correlation analyses were done across grasslands, total PLFA (r = -0.539, *P* < 0.05), bacterial PLFA (r = 0.591, *P* < 0.05) and Gram-positive bacterial PLFA (r = -0.590, *P* < 0.05) were correlated with soil moisture content. Fungal PLFA showed significant positive correlations with total N (r = 0.597, *P* > 0.05) and total C (r = 0.549, *P* > 0.05). Soil total N and total C were significantly and positively correlated (r = 0.992, *P* > 0.0001).

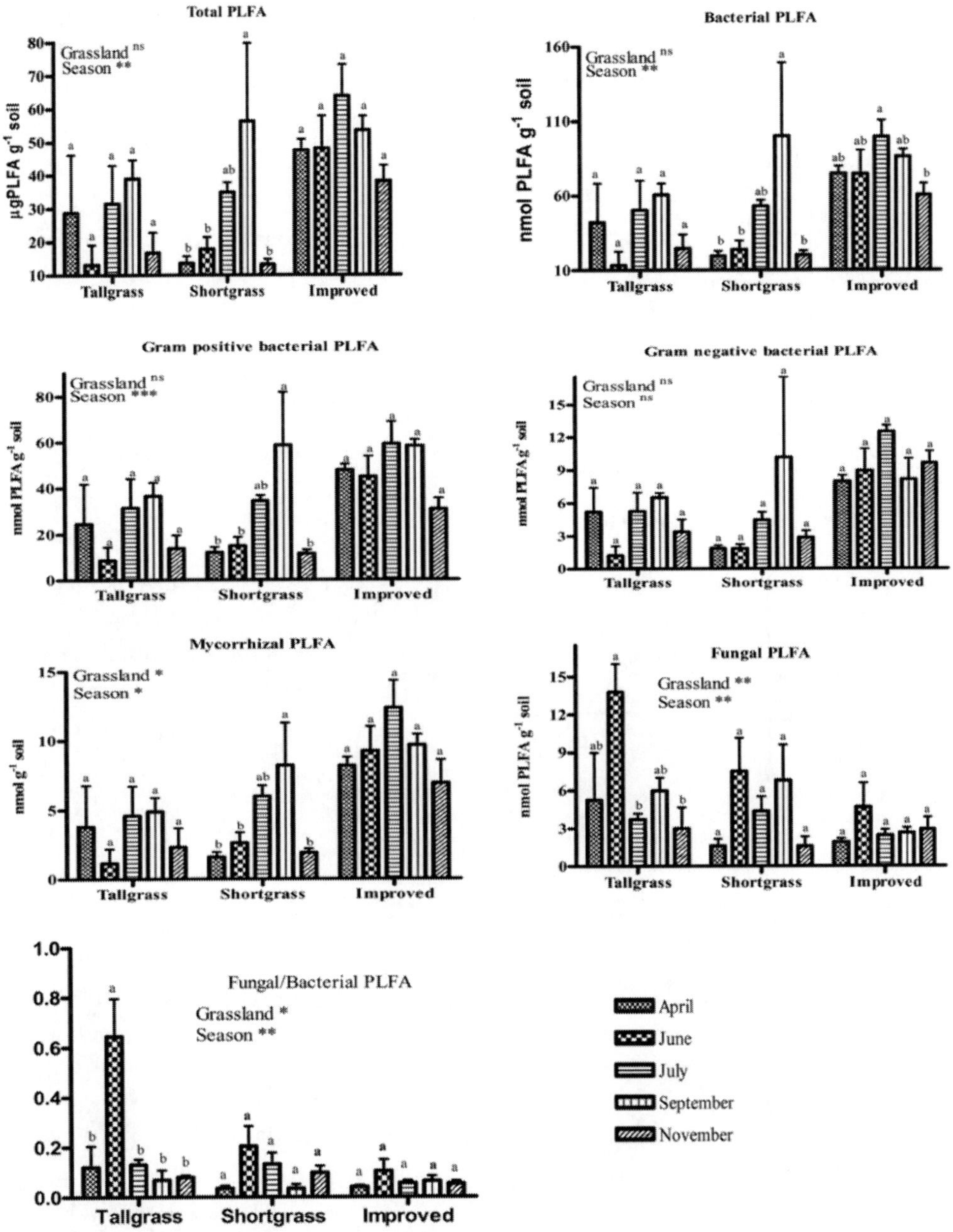

Figure 3. Seasonal variation of the soil microbial community structure determined by phospholipid fatty acid (PLFA) profiling of the semi-natural tall-type and short-type grasslands, and improved grasslands of Shiriyazaki in northern Honshu Island, Japan. Different letters within each grassland type represent significant differences (multiple comparison using Tukey–Kramer HSD). *, **, and *** indicate significant difference at $P < 0.05$, 0.01 and 0.001 levels, respectively. ns = not-significant.

Table 2. Effects (F ratio) of grassland and season on the soil microbial community structure as determined by phospholipid fatty acid (PLFA) profiling in soils of semi-natural tall-type and short-type, and improved grasslands of Shiriyazaki in the northern Honshu Island, Japan

	Grassland	Season
Total PLFA	3.69	4.7**
Total bacteria	3.4	5.4**
Gram-positive bacteria	3.3	5.0***
Gram-negative bacteria	1.4	1.5
Fungi	10.1**	6.5**
Mycorhizae	5.2*	3.1*
Fungi/Bacteria	7.3*	5.3**

*, **, and *** indicate significant difference at $P < 0.05$. 0.01 and 0.001 levels, respectively.

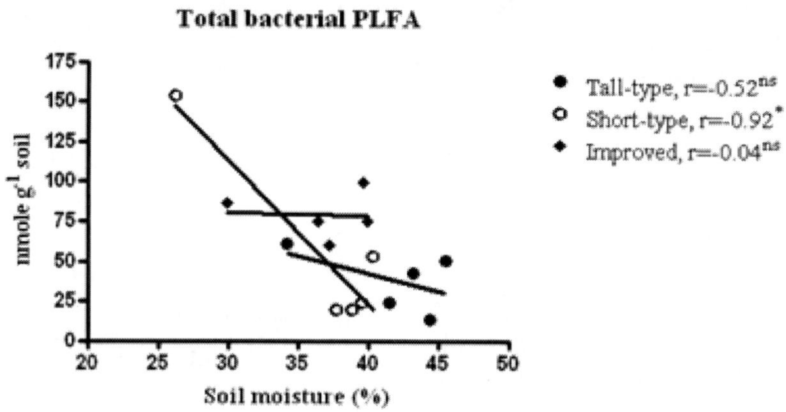

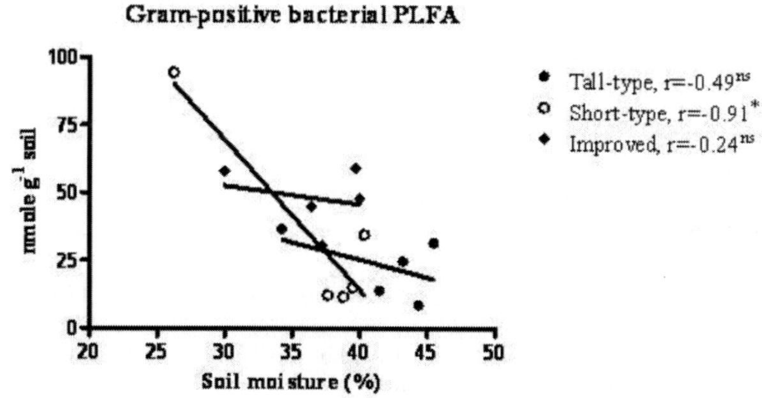

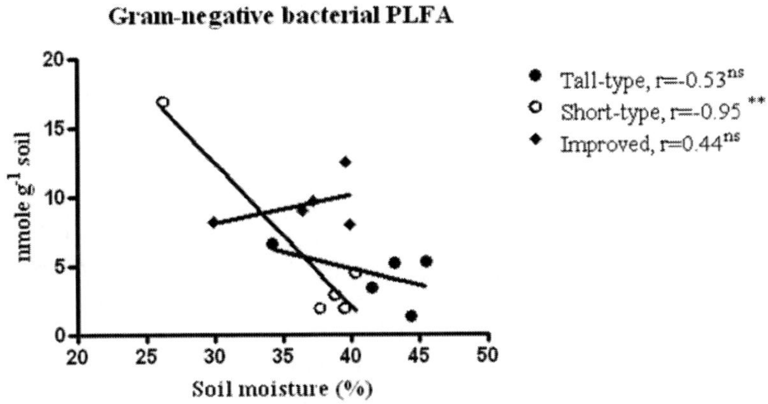

Figure 4. Relationships between microbial PLFA (Phospholipid fatty acids) biomass and the moisture content in soils of the semi-natural Tall-type and Short-type and the improved grasslands of Shiriyazaki, Japan. * and ** indicate significant at $P < 0.05$ and 0.01 levels, respectively. ns = not-significant.

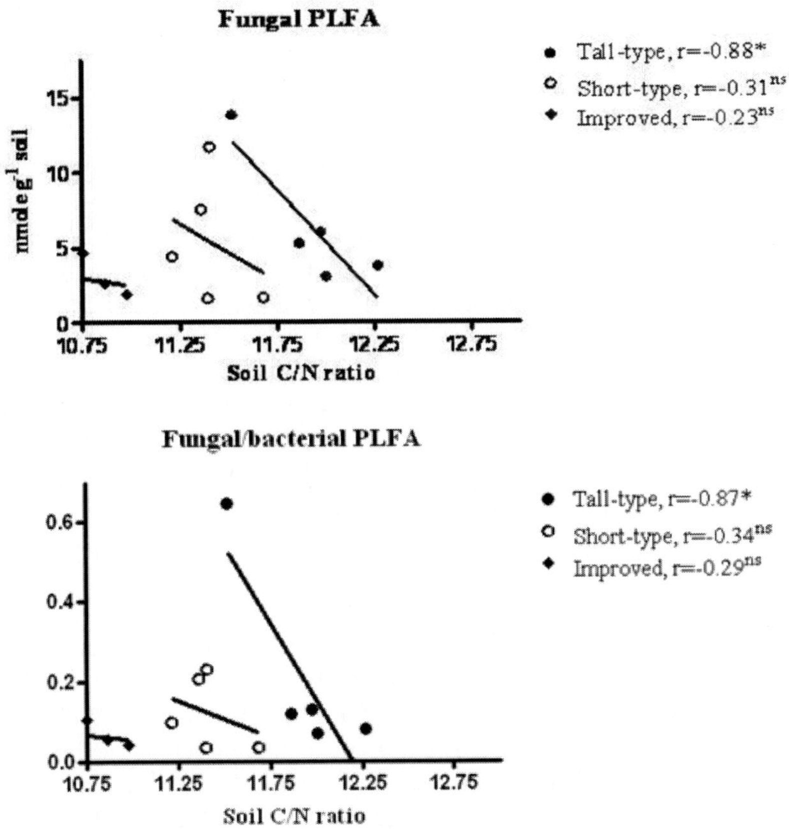

Figure 5. Relationships between soil microbial PLFA (phospholipid fatty acids) biomass and the C/N ratio in soils of the semi-natural tall- and short-type and the improved grasslands of Shiriyazaki, Japan. * indicates significant difference at $P < 0.05$ level. N.S = not significant.

Discussion

This study demonstrated significant seasonal variability of soil microbial communities as evaluated by PLFA profiling in the grasslands of Shiriyazaki in the northern Honshu island (Japan). Results of the study revealed that seasonal changes of soil microbes varied with the type of microbial communities and grasslands and also correlated with soil physico-chemical properties such as soil moisture content and C, N or C/N ratio. Among the microbial groups, total bacterial, Gram-positive and Gram-negative bacterial biomass showed significant correlation with soil moisture content only in the short-type grasslands indicating that different mechanisms were involved in seasonal variability of the soil microbial communities in the three grassland types studied. Seasonal variability in total PLFA, total bacterial PLFA and Gram-positive bacterial PLFA that correlated negatively with soil moisture content in the short-type grassland indicated that seasonal fluctuations of soil moisture can lead to seasonal variation in microbial communities. The seasonal variation in soil moisture content in short-type grassland might be influenced by the effects of direct sunlight, air temperature and wind due to the short stature of plants compared to those in the tall-type and the improved grasslands (Hossain et al., 2007). Relatively high atmospheric temperature and a suitable soil moisture content in September might have favored total microbial, total bacterial and Gram-positive bacterial biomass in the short-type grassland (Curtin et al., 2012). Relatively taller vegetation in tall-type and improved grasslands might have played role in influencing microbial biomass by reducing the seasonal fluctuation of temperature in the soil of these two grasslands compared to those in the short-type grasslands. Although improved grassland showed a significant seasonal variation in soil total bacterial biomass, it did not correlate with any of the soil physico-chemical properties (e.g. pH, moisture, N, P, and C) measured in this study indicating that other factors like fertilization, root exudates and disturbances caused by grazers (Chaparro et al., 2013, Gallardo and Schlesinger 1995, Kuzyakov 2002, Wang et al., 2006) or a complex of several factors might be responsible for the seasonal variation. On the other hand, soil fungal biomass and fungal to bacterial biomass ratio were significantly and negatively correlated with soil C:N ratio indicating that quality of substrates in soil might be more important for the seasonal changes of the fungal community in the tall-type grassland. A peak in fungal biomass was observed in June in each grassland type and this corresponded to the beginning of the growing season in the Shiriyazaki area due to favorable temperature and long day length and to the maximum release of plant root-exudates (Chaparro et al., 2013).

Responses of soil microbial communities to soil properties have been variable (Gordon et al., 2008, Griffiths et al., 2003, Hattori 1988). While Steinberger et al. (1999) reported that soil moisture content and bacterial biomass were positively correlated with each other, Hamel et al. (2006) showed the opposite results. Griffiths et al. (2003), on the other hand, found that soil bacterial communities were resistant to water stress and they explained this result as the adaptation to variation in soil moisture content by regulation of cellular activity. Significant negative correlations between total microbial and bacterial biomass with soil moisture contents across the grassland types as well as in the short-type grasslands observed in the present study were consistent with the results of Hamel et al. (2006) who reported higher microbial biomass under conditions of low moisture content. Although several studies (e.g. Pennanen et al., 1998) reported great influence of soil pH on bacterial PLFA, the results of this chapter did not find any correlation between soil pH and bacterial biomass across seasons and grassland types, indicating that soil pH does not affect microbial communities in the Shiriyazaki grasslands. Fungal biomass in soil is influenced by a number of factors including disturbance from management activities and other soil properties such as soil C and C/N ratio (Hattori 1988, Shi et al., 2013, Bardgett et al., 1999, Bardgett et al., 2001, de Vries et al., 2006, Bardgett and McAlister 1999, Gordon et al., 2008). As demonstrated by other authors (Shi et al., 2013, Hossain and Sugiyama 2011), fungal PLFA was positively correlated with soil total carbon content. Further, bacteria were more influenced by season than grassland and fungi were more influenced by grassland than season indicated that bacteria were more sensitive to seasonal perturbations (Table 2). This result was in agreement with the results of other studies (e.g. Kennedy et al., 2005) suggesting that fungi are more influenced by plants and bacteria are more affected by soil physico-chemical properties.

Significantly greater amount of fungal biomass and fungal to bacterial biomass ratio in the semi-natural tall-type grasslands than the improved grasslands observed in the present study were consistent with the results of the previous studies carried out in the Shiriyazaki area in northern Japan (Hossain et al., 2007). Increased productivity or intensity of human management in grasslands were responsible for transforming a fungal-dominated community to a bacterial-dominated community (Bardgett et al., 1999) and such shift in soil microbial community composition might be related with the long-term history of different management intensities between the grassland types. The grazing is less frequent in the tall-type semi-natural grassland compared to the improved grassland. Further, there are application of chemical fertilizers and artificial sowing of pasture species such as *Trifolium repens* and *Dactylis glomerata* in the improved grassland. Increased grazing intensity has also been found to increase soil microbial biomass through incorporating organic matter, while trampling and adding labile inputs in the form of urine and feces stimulates new root biomass, which exudates labile C in soil (Wang et al., 2006). de Vries et al. (2006) showed a negative correlation between N addition and fungal/bacterial ratio related to the effect of N fertilization on the structure of soil microbial communities. Similarly, Bardgett et al. (2001) reported differences in microbial community structure with higher fungal abundance in less disturbed sites compared to those in highly disturbed sites (Bardgett et al., 1999, Bardgett et al., 2001, de Vries et al., 2006, Bardgett and McAlister 1999, Gordon et al., 2008). Similarly, previous studies reported higher fungal abundance in the semi-natural tall-type grassland than in the improved grasslands in Shiriyazaki area (Hossain et al., 2007). de Vries et al. (2011) also reported higher fungal biomass in unimproved grasslands associated with higher N retention in soils. Results of the present study thus imply that seasonal variability in the soil fungal abundance can influence seasonal fluctuation of N availability in soil in the semi-natural tall-type grassland. Our results highlight higher microbial biomass and bacterial-dominated communities in improved grasslands than in semi-natural grasslands (tall- and short-type). Because bacterial-dominated grasslands are linked with higher N leaching (de Vries et al., 2011), we suggest that improved grasslands would be considered as a non-sustainable system with higher N losses. These findings have important implications for the ways grasslands are managed and suggest that tall-type grassland would be a more sustainable system than improved grasslands.

Microbial communities differ in influencing carbon and N pool in soils with fungal-dominated communities retaining more N (Gordon et al., 2008, de Vries et al., 2011) and showing higher rates of mineralization (Wardle et al., 2004) than bacterial-dominated communities. Further, soil microbial biomass can be used as an indicator of soil quality and management (Anderson 2003, Balota et al., 2003). These results, therefore, indicate that any change in soil microbial communities is likely to alter carbon and N pool in soil. In our study we showed that semi-natural grasslands contained higher soil fungal biomass than improved grasslands reflecting higher N retention. While a large part of Shiriyazaki grasslands has been converted into improved grasslands for economical reasons, our results suggest that this type of management is not sustainable. Moreover, fungal biomass was found maximum in June probably due to maximum root exudation by the plants at their growing stage indicating the higher N retention ability in soil during this season. Therefore, intensive grazing and fertilization should be removed at this period of the year to improve the functioning of the Shiriyazaki grassland ecosystems. Since knowledge about the soil quality is prerequisite for the management of any agricultural systems, better understanding about the seasonal variability of soil microbial communities is relevant for the management and conservation of the grassland ecosystems.

Acknowledgments

The authors are thankful to the Ministry of Education, Culture, Sports, Science and Technology, Japan for financial assistance in conducting this research.

References

Adams, J.M., Faure, H., Faure-Denard, L., McGlade, J.M. and Woodward, F.I. (1990) Increase in terrestrial carbon storage from the last glacial maximum to the present. *Nature*. 348, 711-714.

Anderson, T. (2003) Microbial eco-physiological indicators to assess soil quality. *Agric. Ecosyst. Environ.* 98, 285-293.

André, H.M., Ducarme, X., Lebrun, P. (2002) Soil biodiversity: myth, reality or conning? *Oikos*. 96, 3-24.

Bååth, E. and Anderson, T.-H. (2003) Comparison of soil fungal/bacterial ratios in a pH gradient using physiological and PLFA-based techniques. *Soil Biol. Biochem.* 35, 955-963.

Balota, E.L., Filho, A.C., Andrade, D.S., Dick, R.P. (2003) Microbial biomass in soils under different tillage and crop rotation systems. *Biol. Fertil. Soils*. 38, 15-20.

Bardgett, R.D. (2005) *The Biology of Soil: A Community and Ecosystem Approach*. Oxford University Press, Oxford.

Bardgett, R.D., Bowman, W.D., Kaufmann, R., Schmidt, S.K. (2005) A temporal approach to linking aboveground and belowground ecology. *Trends Ecol. Evol.* 20, 634-641.

Bardgett, R.D. and Cook, R. (1998) Functional aspects of soil animal diversity in agricultural grasslands. *Appl. Soil Ecol.* 10, 263-276.

Bardgett, R.D., Jones, A.C., Jones, D.L., Kemmitt, S.J., Cook, R., Hobbs, P.J. (2001) Soil microbial community patterns related to the history and intensity of grazing in sub-montane ecosystems. *Soil Biol. Biochem.* 33, 1653-1664.

Bardgett, R.D., Lovell, R.D., Hobbs, P.J. and Jarvis, S.C. (1999) Seasonal changes in soil microbial communities along a fertility gradient of temperate grasslands. *Soil Biol. Biochem.* 31, 1021-1030.

Bardgett, R.D. and McAlister, E. (1999) The measurement of soil fungal:bacterial biomass ratios as an indicator of ecosystem self-regulation in temperate meadow grasslands. *Biol. Fertil. Soils.* 29, 282–290.

Bardgett, R.D., Streeter, T.C. and Bol, R. (2003) Soil microbes compete effectively with plants for organic-nitrogen inputs to temperate grasslands. *Ecology,* 84, 1277-1287.

Black, C.A. (1965) *Methods of soil and plant analysis.* Part I and II. American Society of Agronomy.

Bossio, D.A. and Scow, K.M. (1998) Impacts of carbon and flooding on soil microbial communities: Phospholipid fatty acid profiles and substrate utilization pattern. *Microbial Ecol.* 35, 265-278.

Brookes, P.C. (1995) The use of microbial parameters in monitoring soil pollution by heavy metals. *Biol. Fertil. Soils.* 19, 269-279.

Castro, H.F., Classen, A.T., Austin, E.E., Norby, R.J. and Schadt, C.W. (2010) Soil microbial community responses to multiple experimental climate change drivers. *Appl. Environ. Microbiol.* 76, 999-1007.

Chaparro, J.M., Dayakar, V., Badri, D.V., Bakker, M.G., Sugiyama, A., Manter, D.K. and Vivanco, J.M. (2013) Root exudation of phytochemicals in *Arabidopsis* follows specific patterns that are developmentally programmed and correlate with soil microbial functions. *PLOS One.* 8, 1-10.

Curtin, D., Beare, M.H. and Hernandez-Ramirez, G. (2012) Temperature and moisture effects on microbial biomass and soil organic matter mineralization. *Soil Sci. Soc. Am. J.* 76, 2055-2067.

Davet, P. (2004) *Microbial Ecology of the Soil and Plant Growth.* Science Publishers, Inc., New Hampshire.

Defrieri, R.L., Sarti, G., Tortarolo, M.F., Escobar–Ortega, J., García de Salamone, I., Auria, F.D. and Effron, D. (2011) Biochemical and microbiological properties of Argentinean Patagonia soil with implanted forest species. *J. Soil Sci. Plant Nut.* 11, 111-124.

de Vries, F.T., Hofflanda, E., van Eekerenc, N., Brussaarda, L. and Bloem, J. (2006) Fungal/bacterial ratios in grasslands with contrasting nitrogen management. *Soil Biol. Biochem.* 38, 2092-2103.

de Vries, F.T., van Groenigen, J.W., Hoffland, E. and Bloem, J. (2011) Nitrogen losses from two grassland soils with different fungal biomass. *Soil Biol. Biochem.* 43, 997-1005.

Dick, R.P. (1992) A review: long-term effects of agricultural systems on soil biochemical and microbial parameters. *Agric. Ecosyst. Environ.* 40, 25-36.

Flanagan, L.B., Wever, L.A. and Carlson, P.J. (2002) Seasonal and interannual variation in carbon dioxide exchange and carbon balance in a northern temperate grassland. *Global Change Biol.* 8, 599-615.

Frostegård, A., Tunlid, A. and Bååth, E. (1993) Phospholipid fatty acid composition, biomass, and activity of microbial communities from two soil types experimentally exposed to different heavy metals. *Appl. Environ. Microbiol.* 59, 3605-3617.

Gallardo, A. and Schlesinger, W.H. (1995) Factors determining soil microbial biomass and nutrient immobilization in desert soils. *Biogeochemistry.* 28, 55-68.

Gordona, H., Haygarth, PM., and Bardgett, R.D. (2008) Drying and rewetting effects on soil microbial community composition and nutrient leaching. *Soil Biol. Biochem.* 40, 302–311.

Gray, S.B., A.T., Classen, P., Kardol, Z., Yermakov and Miller, R.M. (2011) Multiple climate change factors interact to alter soil microbial community structure in an old-field ecosystem. *Soil Sci. Soc. America J.* 75, 1-10.

Grayston, S.J., Griffith, G.S., Mawdsley, J.L., Campbell, C.D. and Bardgett, R.D. (2001) Accounting for variability in soil microbial communities of temperate upland grassland ecosystems. *Soil Biol. Biochem.* 33, 533-551.

Griffiths, R.I., Whiteley, A.S., O'Donnell, A.G., Bailey, M.J. (2003) Physiological and community responses of established grassland bacterial populations to water stress. *Appl. Environ. Microbiol.* 69, 6961-6968.

Hamel, C., Hanson, K., Selles, F., Cruz, A.F., Lemke, R., McConkey, B. and Zentner, R. (2006) Seasonal and long-term resource-related variation in soil microbial communities in wheat-based rotations of the Canadian prairie. *Soil Biol. Biochem.* 38, 2104-2116.

Hattori, T. (1988) Soil aggregates in microhabitats of microorganisms. Institute of Agricultural Research, Tohoku University. 37, 23-36.

Høgh-Jensen, H. and Schjoerring, J.K. (2001) Rhizodeposition of nitrogen by red clover, white clover and ryegrass leys. *Soil Biol. Biochem.* 33, 439-448.

Hossain, M.Z., Okubo, A. and Sugiyama, S. (2007) Local variation in soil microbial community structure in artificial and semi-natural grasslands. *Grassland Sci.* 53, 165-171.

Hossain, M.Z., Okubo, A. and Sugiyama, S. (2010) Effects of grassland species on decomposition of litter and soil microbial communities. *Ecol. Res.* 25, 255-261.

Hossain, Z. and Sugiyama, S. (2011) Geographical structure of soil microbial communities in northern Japan: effects of distance, land use type and soil properties. *Eur. J. Soil Biol.* 47, 88-94.

Hossain, Z. and Sugiyama, S. (2013) Differences in microbial functions in soils of semi-natural and improved grasslands. *Natural Selection: Biological processes, Theory and Role in Evolution* (eds J. R. Lynch and D. T. Williamson), pp. 317-327. Nova Science Publishers, Inc. New York.

Kaye, J.P. and Hart, S.C. (1997) Competition for nitrogen between plants and soil microorganisms. *Trends Ecol. Evol.* 12, 139-143.

Kennedy, N., Connolly, J. and Clipson, N. (2005) Impact of lime, nitrogen, and plant species on fungal community structure in grassland microcosms. *Environ. Microbiol.* 7, 780-788.

Kuzyakov, Y. (2002) Review: Factors affecting rhizosphere priming effects. *J. Plant Nutr. Soil Sci.* 165, 382 – 396.

Lipson, D.A., Schadt, C.W., Monson, R.K. and Schmidt, S.K. (2002) Changes in microbial community structure and function following snow melt in an alpine soil. *Microbial Ecol.* 43, 307-314.

Lipson, D.A., Schmidt, S.K. and Monson, R.K. (1999) Links between microbial population dynamics and plant N availability in an alpine ecosystem. *Ecology.* 80, 1623-1631.

Merila, P., Strommer, R. and Fritze, H. (2002) Soil microbial activity and community structure along a primary succession transect on the land-uplift coast in western Finland. *Soil Biol. Biochem.* 34, 1647-1654.

Miller, A.E., Schimel, J.P., Meixner, T., Sickman, J.O. and Melack, J.M. (2005) Episodic rewetting enhances carbon and nitrogen release from chaparral soils. *Soil Biol Biochem.* 37, 2195-2204.

Mutoh, N., Kimura, M., Oshima, Y. and Iwaki, H. (1985) Species diversity and primary productivity in *Miscanthus sinensis* grasslands; 1. Diversity in relation to stand structure and dominance. *Bot. Mag. Tokyo.* 98, 159-170.

Orwin, K.H., Wardle, D.A. and Greenfield, L. (2006) Ecological consequences of carbon substrate identity and diversity in a laboratory study. *Ecology.* 87, 580-593.

Peeters, A. (2009) Importance, evolution, environmental impact and future challenges of grasslands and grassland-based systems in Europe. *Grassland Sci.* 55, 113–125.

Pennanen, T., Fritze, H., Vanhala, P., Kiikkila, O., Neuvonen, S. and Bååth, E. (1998) Structure of a microbial community in soil after prolonged addition of low levels of simulated acid rain. *Appl. Environ. Microbiol.* 64. 2173-2180.

Powlson, D.S., Brookes, P.C. and Christensen, B.T. (1987) Measurement of soil microbial biomass provides an early indication of changes in the total soil organic matter due to straw incorporation. *Soil Biol. Biochem.* 19, 159-164.

Schlesinger, W.H. (1997) Biogeochemistry: an Analysis of Global Change. Academic Press Inc., San Diego, CA.

Shi, Y., Lalande, R., Hamel, C., Ziadi, N., Gagnon, B. and Hu, Z. (2013) Seasonal variation of microbial biomass, activity, and community structure in soil under different tillage and phosphorus management practices. *Biol. Fertil. Soils.* 49, 803-818.

Spedding, T.A., Hamel, C., Mehuys, G.R. and Madramootoo, C.A. (2004) Soil microbial dynamics in maize-growing soil under different tillage and residue management systems. *Soil Biol. Biochem.* 36, 499-512.

Steinberger, Y., Zelles, L., Bai, Q.Y., von Lutzow, M. and Munch, J.C. (1999) Phospholipid fatty acid profiles as indicators for the microbial community structure in soils along a climatic transect in the Judean desert. *Biol. Fertil. Soils.* 28, 292-300.

Sugiyama, S., Hossain, M.Z. and Atsushi, O. (2008) Relationships between soil microbial diversity and plant community structure in seminatural grasslands. *Grassland Sci.* 54, 117-124.

Valášková, V. and Baldrian, P. (2009) Denaturing gradient gel electrophoresis as a fingerprinting for the analysis of soil microbial communities. *Plant Soil Environ.* 55, 413–423.

Waldrop, M.P., Balser, T.C. and Firestone, M.K. (2000) Linking microbial community composition to function in a tropical soil. *Soil Biol. Biochem.* 32, 1837-1846.

Wang, K.-H., McSorley, R., Bohlen, P. and Gathumbi, S.M. (2006) Cattle grazing increases microbial biomass and alters soil nematode communities in subtropical pastures. *Soil Biol. Biochem.* 38, 1956-1965.

Wardle, D.A. (2002) *Communities and Ecosystems: Linking the Aboveground and Belowground Components.* Princeton University Press, Princeton.

Wardle, D.A., Bardgett, R.D., Klironomos, J.N., Setälä, H., Van der Putten, W.H., and Wall, D.H. (2004) Ecological linkages between aboveground and belowground biota. *Science.* 304, 1629-1633.

Zelles, L. (1999) Fatty acid patterns of phospholipids and lipopolysaccharides in the characterization of microbial communities in soils: a review. *Biol. Fertil. Soils.* 9, 111-129.

Yazaki, Y., Mariko, S. and Koizumi, H., 2004. Carbon dynamics and budget in a *Miscanthus sinensis* grassland in Japan. *Ecol. Res.* 19, 511-520.

In: Grassland Biodiversity and Conservation …
Editors: Pierre Mariotte and Paul Kardol

ISBN: 978-1-61761-619-8

Chapter 4

RESTORATION MANAGEMENT FOR SPATIALLY AND TEMPORALLY COMPLEX CALIFORNIAN GRASSLAND

Sheri Spiegal*, Loralee Larios, James W. Bartolome and Katharine N. Suding

Department of Environmental Science, Policy and Management
University of California Berkeley, Berkeley, CA, US

ABSTRACT

Introduced species from the Mediterranean dominate plant cover of the Californian grassland, but more than one thousand native species persist at low abundance or may be locally absent. Efforts to successfully increase native abundance are complicated by the spatial and temporal complexity of the system. Highly variable rainfall, topography, and soils result in large differences in species composition across space and time. Managers must deal with this variability to carry out effective restoration. We present a conceptual management toolkit containing five steps to better organize variability, predict suitable restoration sites, and select and time treatments. The toolkit relies on the key concepts of ecological site classification, state-and-transition models, and opportunistic adaptive management to help managers achieve their restoration goals.

INTRODUCTION

California's grassy hills and plains have long captured imaginations with their beauty and promise of economic opportunity. These grasslands provide forage for livestock and wildlife (Jackson & Bartolome 2007), habitat for 40% of the state's native plant species (Schiffman 2007), and carbon sequestration (Booker et al., 2013). Despite the ecosystem services they provide, grasslands are among the most fragmented and developed natural systems in California (Jantz et al., 2007). A dramatic invasion of species from the Old World has also greatly impacted Californian grassland. Restoration must address historical sources of

* Corresponding author: Email: spiegal@berkeley.edu.

degradation (e.g. species invasion, past cultivation) along with new ones (e.g. fragmentation due to increasing urbanization; FRAP 2010).

Managers are grappling with the critical decision of how to best allocate their restoration resources; here we argue that the spatial and temporal variability of these grasslands should be a central consideration in the decision-making process (Landres et al., 1999; Bartolome et al., 2009; McBride et al., 2010). We present a conceptual toolkit that can help managers understand the variability and incorporate their understanding into restoration management. We begin by outlining the drivers of grassland species composition, using examples from central and southern California. Next we describe some of the restoration constraints of this system. We then present the five-step toolkit, first by explaining its conceptual framework and second by outlining step-by-step instructions for its use. We conclude with a case study describing the application of the toolkit in southern California. While our focus is on California's Mediterranean-climate grasslands, this toolkit can be applied to other terrestrial ecosystems with high spatial and temporal variability across the state and world.

Controls over Species Composition in Californian Grassland

The Californian grassland discontinuously covers about 11% of the state (Davis et al., 1998), spanning a significant precipitation gradient and a variety of topographic and edaphic features (Huenneke 1989). This extensive spatial distribution contributes to regional differences in species composition, and three major grassland types have been identified: Coastal Prairie, Coast Range Grassland, and Valley Grassland (see Barbour et al., 2007; Jackson & Bartolome 2002 for more detailed information). Here we focus on the drivers of community composition within Valley Grassland, the most extensive grassland type in California.

The majority of Valley Grassland sits on the alluvium and foothills on the edges of California's Great Central Valley, a sediment basin circumscribed by the Sierra Nevada to the east and the California coast ranges to the west (Figure 1; Heady 1977; Heady et al., 1991). Valley grasslands are in California's Mediterranean climate zone of hot dry summers and wet cool winters. Like grasslands in other Mediterranean climate zones worldwide (Dallman 1998), valley grasslands are dominated by annual species. The most abundant valley grassland annuals, however, are not native to California. A widespread type conversion from native species to Mediterranean species occurred in the nineteenth century (Jackson 1985; Baker 1989). A handful of exotic annual species are the overall dominants; however, native species, especially native annual forbs, often contribute to the high species richness found in valley grasslands (Gennet 2007; Fernandez-Going et al., 2012). Also, though they are relatively rare, valley grasslands with extensive native communities still exist (Kimball & Schiffman 2003; Lulow & Young 2011).

Valley grasslands span a regional-scale latitudinal climatic gradient, from the wetter north to the drier south. This north-south gradient is modified by the rain shadow of the coast ranges, with the coastal foothills and valleys in the west more mesic than the xeric interior Central Valley and Sierra Nevada foothills to the east. The longitudinal precipitation gradient is reflected in a parallel gradient in species composition in which perennial bunchgrasses are more common in the more mesic valley grasslands nearer to the coast and less common in the xeric interior valley grasslands (Burcham 1975; Heady 1977; Bartolome et al., 2007a).

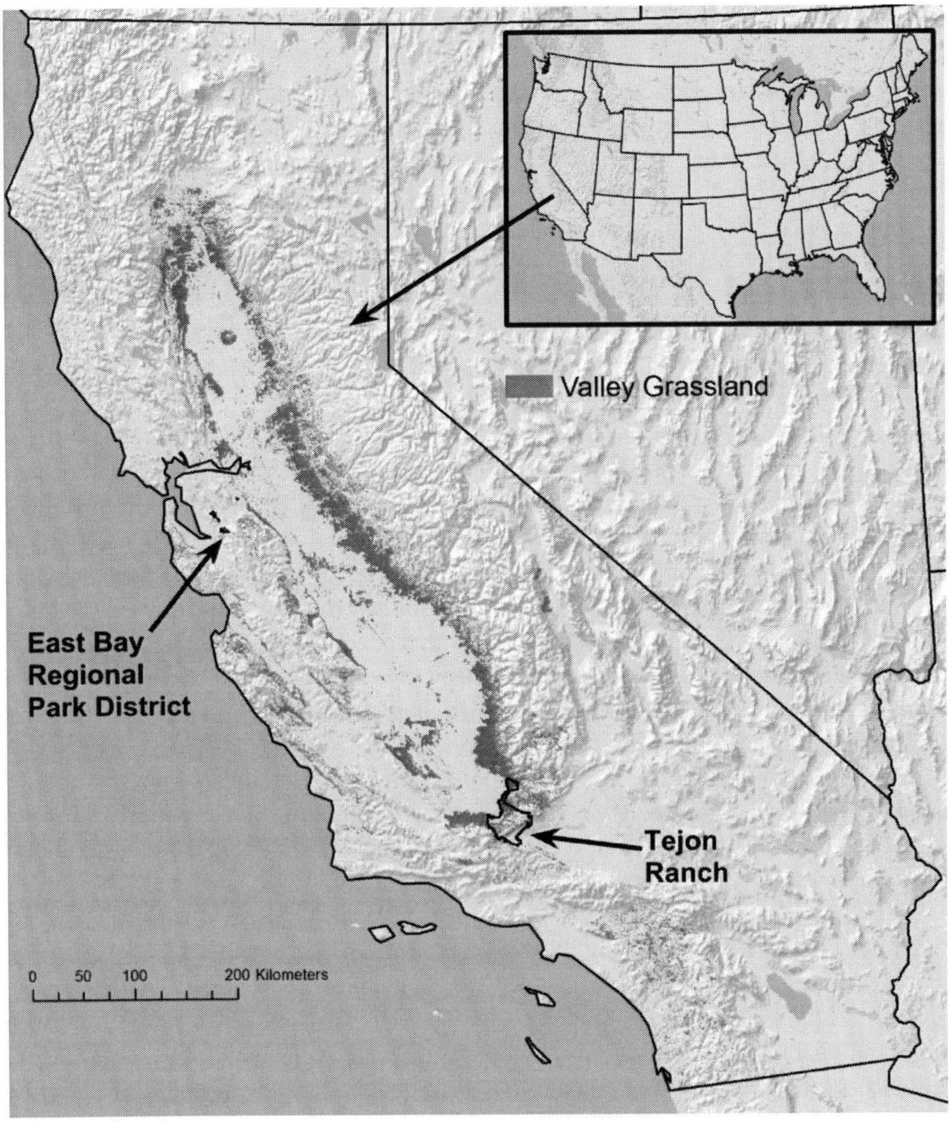

Figure 1. Distribution of the Valley Grassland vegetation type in California, USA (USFS 2010). The managed landscapes featured in the two case studies, the East Bay Regional Park District and Tejon Ranch, are also shown.

In addition to this spatial variation, the climate of valley grasslands also varies temporally: intra-annually due to the Mediterranean seasons, and inter-annually due to largely stochastic rainfall timing and amount. In a recent data synthesis, Hallett et al. (*in press)* found that compared to more mesic grasslands like the tall grass prairie in Kansas or old fields in Michigan, California grasslands have greater inter-annual precipitation variability. The coefficient of variation of annual precipitation is 20-25 in Kansas and Michigan but 31-37 in California. This temporal variation results in strong inter- and intra-annual differences in species composition (Heady 1958; Pitt & Heady 1978; Bartolome 1989). A classic example of how this temporal variation affects species composition is "grass" vs. "filaree" years. After the initial germinating rains of a given year, a period of drought is associated with a higher abundance of filaree (*Erodium* spp.), while frequent precipitation after the initial rains results in a higher abundance of annual grasses like wild oats (*Avena fatua*) and soft chess (*Bromus hordeaceus*). The degree of temporal variability of species composition at a site is modified by the site's productivity, topography, and soils (Bartolome et al., 2007b; Gennet 2007; Fernandez-Going et al., 2012).

Within zones of a singular climate, topographic and soil properties such as elevation, slope, and soil type are fundamental controls on valley grassland species distribution (Burcham 1975; Evans et al., 1975; Evans & Young 1989; Eviner & Firestone 2007). This control was apparently in place in the past, as well. For instance, there is strong evidence that before the major invasion of exotic species, when annual forbs were more extensive than grasses (Schiffman 2007; Minnich 2008), grasses were restricted to sites like stream terraces (Evett & Bartolome 2013).

Contrasting findings on the particular soil nutrients that control species abundance and distribution in valley grasslands underscore the need to consider spatial heterogeneity when conducting restoration. For example, there is a general trend that lower fertility sites are more likely to support native perennial bunchgrasses, but the roles of specific nutrients can vary among sites in a landscape. In the highly invaded valley grasslands of the Diablo Range of central California, for instance, one study documented that low-nitrogen sites support the native perennial bunchgrass *Stipa pulchra* (Robertson 2004), whereas another study in neighboring grasslands found that the species is more often found in low-phosphorus sites (Gea-Izquierdo et al., 2007).

A useful concept for understanding controls on species composition in rangelands (grasslands, shrublands, savannas) is a theoretical continuum with equilibrium and non-equilibrium community models at its extremes (Wiens 1984; Vetter 2005). In equilibrium models, dominant controls over the structure and function of communities are biotic interactions like plant competition and herbivory. In contrast, in non-equilibrium models, communities are structured more by external factors such as climate (Wiens 1984; Ellis & Swift 1988), topography (Milchunas et al., 1989), and abiotic soil properties (Booker et al., 2012). The spatial scale at which rangeland is observed affects whether equilibrium or non-equilibrium dynamics are detected (Wiens 1989), but there is a general trend that rangelands with highly variable abiotic environments are best described using non-equilibrium models (Ellis & Swift 1988; Briske et al., 2003). A working hypothesis is that non-equilibrium models are appropriate for rangelands with a coefficient of variation of inter-annual rainfall exceeding 33 (von Wehrden et al., 2012).

As one might expect given their variability, including their high coefficient of variation of inter-annual rainfall, California's valley grasslands exhibit non-equilibrium dynamics on multiple spatial scales (Jackson & Bartolome 2002). Biotic interactions (e.g. cattle grazing, plant litter on germinating seeds) certainly affect community dynamics (Huntsinger et al., 2007), but overall, abiotic factors appear to be stronger controls (Jackson & Bartolome 2007). Californian grassland can also be strongly affected by long term processes that cannot be categorized as biotic or abiotic, such as altered fire regimes, novel species invasions, and anthropogenic nitrogen enrichment. Establishing the system's placement on the theoretical continuum from equilibrium to non-equilibrium can help managers make predictions about the impacts of these processes.

CASE STUDY 1: SPATIAL AND TEMPORAL VARIATION IN VALLEY GRASSLAND

To illustrate spatial and temporal species composition dynamics, we present data from valley grasslands managed by the East Bay Regional Park District (EBRPD) and the Tejon Ranch Conservancy (Tejon). EBRPD grasslands lie in the Diablo Range east of San Francisco, California. The Tejon Ranch, a vestige of the Mexican-California Rancho land ownership system, spans 1093 km^2 (Figure 1). It is the largest privately owned contiguous property in California. The EBRPD and Tejon grasslands are largely dominated by naturalized exotic annual grasses but maintain high native species richness (Gennet 2007). At Tejon, native grassland communities are extensive on some sites (see Case Study 2).

Here we compare grasslands at three EBRPD regional parks (Pleasanton Ridge Regional Park, Sunol Regional Wilderness-Ohlone Regional Wilderness, and Vasco Caves Regional Preserve) with grasslands of the northern half of Tejon Ranch. Tejon's northern grasslands cover 22,000 ha on four distinct geophysical areas: the San Joaquin Valley (the southern portion of the Great Central Valley, SJV), southernmost Sierra Nevada mountains (SN), Tejon Hills (TH), and Tehachapi Mountains (TM).

In order to guide long term management efforts, grasslands at EBRPD and Tejon were sampled for species composition over multiple years, 2005-2012 at EBRPD and 2010-2012 at Tejon. At EBRPD, six 500 m^2 circular plots were sampled at each park. Within each plot, four 17 m transects were laid in the cardinal directions and identity of the first species hit by a dropped pin was recorded every 0.5 m along each transect. At Tejon, different numbers of plots were sampled in each geophysical area (SJV $n = 9$, SN $n = 3$, TH $n = 8$, TM $n = 9$). Like the plots at EBRPD, the plots at Tejon comprised four transects, each pointing in a cardinal direction, and the first species hit by a dropped pin was recorded every 0.5 m on each transect. Tejon plots are larger than EBRPD plots, with 25 m transects instead of 17 m transects. All plots at Tejon and EBRPD were lightly to moderately grazed by cattle or sheep (~0.4 Animal Unit Months/ha; McDougald et al., 1991) throughout sampling, except for three plots in Pleasanton Ridge Regional Park that were not grazed for two years.

Figure 2 shows the precipitation (Figure 2a), species richness (Figure 2b), and temporal and spatial beta-diversity (Figure 2c) of EBRPD and Tejon. The precipitation curves represent the average of the precipitation recorded at each EBRPD park and each Tejon geophysical area. The species richness and beta-diversity curves (Figures 2b and 2c, respectively) represent twice-averaged values. First, plot values were averaged within each park and geophysical area to yield park-scale and geophysical area-scale values. Second, the park-scale values were averaged together to yield EBRPD landscape-scale values (circles) and the geophysical area-scale values were averaged together to yield Tejon landscape-scale values (squares).

Precipitation differed between the EBRPD and Tejon landscapes in 2010-2012 (Figure 2a); however, both typically experience high variability in inter-annual rainfall (PRISM Climate Group 2013). This variability supports the working hypothesis that these grasslands should be placed on the non-equilibrium end of the theoretical continuum discussed above.

Species richness of the two landscapes is similar (Figure 2b). It fluctuated slightly over time in EBRPD and decreased in 2011 for both EBRPD and Tejon when precipitation was high.

In addition to changes in species richness, variation in species composition can also be measured as the turnover of species (beta-diversity) in a plot from one year to the next or among plots within any given year (Whittaker 1960, Anderson et al., 2006). We used a dissimilarity metric (Bray-Curtis), in which values range from 0 (similar) to 1 (dissimilar), to estimate beta-diversity. This analysis revealed that the variation in community structure differs slightly between EBRPD and Tejon, with Tejon having greater turnover (Figure 2c). The variation of the two landscapes, however, fluctuated similarly over time (i.e. the Tejon and EBRPD curves track each other from 2010-2012). The slight increase in temporal variability in species composition between 2010 and 2011 in both landscapes likely reflects changes in composition that resulted from the high rainfall across California in 2011. Temporal turnover (open symbols) was lower compared to spatial turnover (filled symbols) suggesting that community structure changes less from one year to the next at a given plot than it does from one plot to the next in a given year. This temporal similarity in species composition supports the previously observed pattern of temporal nestedness within California grasslands, where interannual rainfall patterns similarly influence most species within a community (Elmendorf & Harrison 2009).

Variability in species composition among plots at EBRPD and Tejon was relatively consistent over the years (filled symbols, Figure 2c), suggesting that the site-specific constraints on species composition are maintained over time. Strong site effects on species abundance patterns were also observed for EBRPD by Gea-Izquierdo et al. (2007), who found that the abundance of *Stipa pulchra* was strongly regulated by the phosphorus levels and sand content of the soil. Both Gea-Izquierdo et al. (2007) and Gennet (2007) found minimal impacts of livestock grazing on vegetation structure in EBRPD.

The range of spatial variability in species composition observed in these grasslands (0.5-0.7) is similar to that which has been observed in grasslands in other dry climates. Using a similar dissimilarity metric (Jaccard's), Chalcraft et al. (2004) found that a xeric grassland in southern New Mexico had variation in community structure that ranged from 0.4-0.75 compared to a mesic grassland in northeastern Kansas where variability did not exceed 0.45.

Teasing apart the effects of temporal and spatial variability on species composition is an important consideration when developing management strategies for Californian grassland. At the central and southern California valley grasslands showcased here, species composition varied more strongly across space within each managed landscape than it did across time. This finding reveals the strong influence that site conditions have on species composition. Accordingly, site conditions should be strongly considered during restoration planning.

Management of Californian Grassland

Grassland restoration in California typically involves reducing the abundance of exotics while increasing the abundance of natives (Seabloom et al., 2003a; Bartolome et al., 2004; Stromberg et al., 2007; Lulow 2008), but completely eradicating the naturalized exotic species is usually not feasible (Bartolome et al., 2007a). Many restoration projects are geared toward re-introducing or augmenting populations of native perennial bunchgrasses using agronomic methods (Stromberg et al., 2007). Livestock is the major economic product of the Californian grassland, and though restoration is largely the domain of conservation organizations, public agencies, and land trusts, common objectives of livestock operators can be synergistic with objectives of restoration managers. Such objectives include increasing forage productivity, improving soil stability, and providing wildlife habitat (Huntsinger et al., 2007).

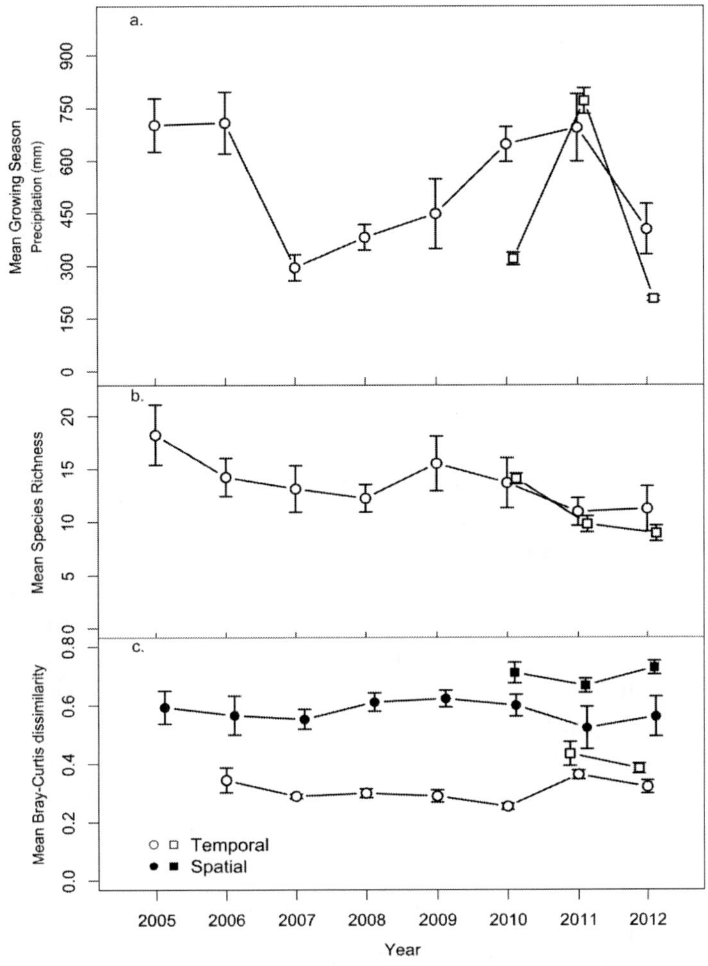

Figure 2. Mean growing season precipitation (a), species richness (b), and temporal and spatial beta-diversity (c) in the valley grasslands of the East Bay Regional Park District (circles) and Tejon Ranch (squares). Growing season precipitation was defined as rainfall from October to June (e.g. 2006 growing season was from October 2005 to June 2006). EBRPD species richness and beta-diversity curves represent values from 18 plots across 3 parks, averaged within and then across the parks. Tejon species richness and beta-diversity curves represent values from 30 plots across 4 geophysical areas, averaged within and then across the geophysical areas. Species richness was estimated as the total number of unique species hit along four transects in each sampling plot. Temporal beta-diversity is estimated as the pairwise comparison of the previous and current year species composition of a plot. Spatial beta-diversity is estimated as the mean pairwise difference in species composition of the plots sampled in a given year. Precipitation data for three EBRPD parks was downloaded from the nearest weather stations of California Department of Water Resources and any missing data was supplemented by weather stations from California Irrigation Management Information Services; precipitation data for four Tejon geophysical areas was downloaded from PRISM (PRISM Climate Group 2013). Error bars represent ± 1standard error.

Restoration efforts have been met with mixed success (Stromberg et al., 2007; D'Antonio et al., 2002). A vexing problem for assessing the success of any California grassland restoration effort is that the original vegetation is relatively unknown, resulting in the lack of a suitable reference system (Bartolome et al., 2009). Restoration success is also thwarted by the inability to replicate management efforts across California due to the strong effects of spatial and temporal variability on vegetation structure and productivity (Bartolome et al., 2007b & 2009; Wilson et al., 2011) as well as the variability of past human activity among sites (Stromberg & Griffin 1996; Hamilton et al., 2002).

Land use legacies, invasive species, and biotic interactions may significantly impede efforts to re-establish native species (Stromberg & Griffin 1996; Orrock et al., 2008). Old fields experienced large-scale soil disturbances such as plowing and disking that permanently altered soil structure and soil microbial communities (Steenwerth et al., 2002), creating dynamics that favor exotics (Stromberg & Griffin 1996). Invasive species are major obstacles to grassland restoration in California. Long-term invader dominance of an area can greatly reduce the native seed bank, preventing native recovery even once the invader has been removed (Cox & Allen 2002; Seastedt et al., 2008). Additionally, some invaders create soil legacies by altering the soil biota, releasing allelopathic chemicals, or creating litter that detrimentally affects native recruitment (Hawkes et al., 2005; Reinhart & Callaway 2006; Vogelsang & Bever 2009). The small mammals of Californian grasslands can also thwart native recovery. Gopher activity can often favor exotic annual grasses (Hobbs & Mooney 1995; Stromberg & Griffin 1996; Seabloom 2003b) and preferential seed predation of native species can also limit the success of restoration management (Orrock et al., 2009). Understanding the degree to which these factors constrain recovery can guide the selection of management treatments.

Another recovery constraint is financial. Improving the efficacy of restoration management requires descriptive and predictive models that can account for the spatial and temporal dynamics of vegetation as well as any potential site-specific recovery constraints (Hobbs 2007; Bartolome et al., 2009; Suding & Hobbs 2009). The science-based toolkit we present in the next section requires a site-specific and data-intensive approach, which can be time-consuming and therefore expensive. We contend, however, that while the cost of a science-based approach is high, it reduces the uncertainty that inevitably complicates Californian grassland restoration planning (McBride et al., 2010; Wilson et al., 2011). More certainty should translate into better chances for using limited financial resources on successfully achieving restoration objectives (Landres et al., 1999).

A Management Toolkit for Spatially and Temporally Complex Grassland

The variability and restoration constraints outlined above can be daunting obstacles to the conservation and enhancement of native vegetation in California's valley grasslands. One response to these obstacles is a heavy-handed approach, in which natives, often perennial bunchgrass species, are cultivated using agronomic methods to overcome potential constraints imposed by soils or rainfall signatures (Stromberg et al., 2007). We suggest an alternative approach in which spatiotemporal variability is actively embraced (Eviner & Hawkes 2008) and used as a guide in restoration planning (Landres et al., 1999), to match species palettes and restoration techniques with suitable sites and auspicious timing (White & Walker 1997).

This alternative approach is a science-based strategy founded on ecological site, state-and transition, and auxiliary models and is appropriate in a framework of adaptive and opportunistic management. Our approach is appropriate for managers with extensive land and multiple conservation goals but limited funding — a common scenario in California's valley grasslands (Huntsinger et al., 2007).

Ecological site and state-and-transition models are useful tools for describing and understanding terrestrial landscapes with high spatiotemporal variability. These conceptual tools are important elements of our five-step toolkit (Figure 3). An ecological site is an assemblage of landscape units, or sites, that have the same potential vegetation and respond similarly to management (Bestelmeyer et al., 2009). The basic idea that geologic, topographic, and soil (topo-edaphic) properties are the primary governing agents of a site's potential vegetation and response to management is increasingly accepted for terrestrial landscapes, having been verified through observation and manipulative experimentation (Grigal et al., 1999; Bestelmeyer et al., 2009; Brown 2010). Accordingly, the sites in ecological site classifications are best defined by their topo-edaphic properties (Bestelmeyer et al., 2009). Given the strong influence of abiotic factors on vegetation dynamics in non-equilibrium systems, topo-edaphic control of site potential may be especially applicable to these systems (Booker et al., 2012). The sites constituting an ecological site are not necessarily contiguous, but instead are typically arrayed across the landscape in a mosaic pattern (Bestelmeyer et al., 2009). The sites are irregularly sized, but they are larger than plot or patch and smaller than landscape, at approximately 5-50 ha (10^4 - 10^5 m^2; Bestelmeyer et al., 2011a).

A state-and-transition model is a description of the spatial (Bestelmeyer et al., 2011b) or temporal vegetation change at an ecological site, in which a change from one vegetative state to another is called a transition (Westoby et al., 1989). By pairing each ecological site with a state-and-transition model, the grassland manager can understand the vegetation dynamics occurring on that ecological site. This understanding can help the manager predict future dynamics and select suitable management objectives and tactics for the ecological site (Herrick et al., 2006; Bestelmeyer et al., 2009).

The theory underlying ecological site and state-and-transition modeling has important implications for restoration. An ecological site can contain reference areas with desirable states as well as degraded areas with undesirable states. The reference and degraded areas are part of the same ecological site, so they theoretically have the same potential vegetation. As a result, the management team can have some certainty that restoration treatments on the degraded areas can foster transitions to the desirable states seen on the reference areas (White and Walker 1997; Bestelmeyer et al., 2011b).

Due to the demonstrated utility of ecological sites and state-and-transition models, United States (U.S.) land management agencies have formally adopted their use and have specified official protocols for their creation (Caudle et al., 2013). Our approach to building and presenting these models differs from the government's in a few ways, which we describe in our explanation of Step 2, below.

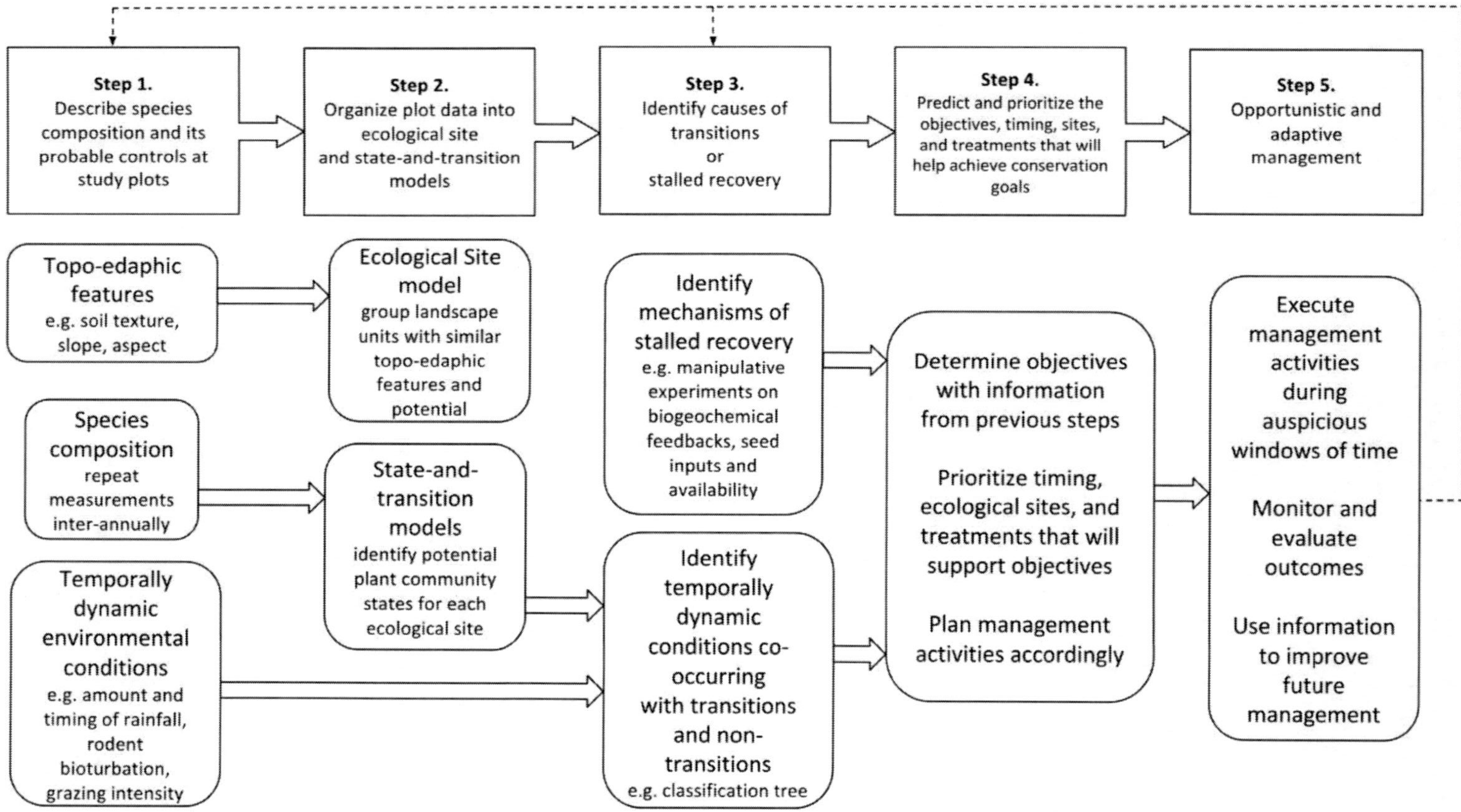

Figure 3. Framework of the management toolkit for spatially and temporally complex grassland. Tools in the steps help managers describe and understand the variation of the managed landscape and use their understanding to guide management decisions. The dashed arrow represents the adaptation of understanding about the landscape (developed in Steps 1-3), which may be necessary if monitoring indicates that the execution of management activities (Step 5) did not achieve objectives (defined in Step 4).

An assumption underlying the toolkit is that the management team will specify conservation and restoration goals before taking any of the toolkit steps. Goal setting is especially important when multiple stakeholders are involved. Because goals, which represent a future desired state or condition, are often general or idealistic, they should be paired with specific and practical objectives that specify how the goals will be achieved (Bush 2006). In Step 4, management teams will use their understanding gleaned from Steps 1-3 to specify objectives and determine the tactics that will support the realization of those objectives. The plans made in Step 4 will be executed in Step 5, opportunistic and adaptive management. Pairing adaptive management (Holling 1978) with opportunistic management (Westoby et al., 1989) can translate into higher chances of achieving conservation goals in terrestrial systems with high spatiotemporal variation.

A second assumption is that a science-based approach – in which ecosystem description leads to an understanding that can inform sound predictions and ultimately enhance the ability to control the ecosystem (Begon et al., 1996; Bartolome et al., 2009) – is more effective than a non-scientific approach. We refer the reader to a complementary framework for science-based management of rangelands put forth by Herrick and others (2006), who advise that management teams define the ecological potential of, and management strategies for, an ecosystem using ecological site and state-and-transition models; assess the functional status of the ecosystem with respect to its potential; and monitor how that functional status changes over time with and without management inputs (Herrick et al., 2006; Karl & Herrick 2010). Sheley and others (2010) also offer a framework for model-based management of rangelands, with a focus on invasive plants. While our understanding of rangelands has been improved by the work of these colleagues, our framework differs from theirs. Next we describe the tools of our toolkit as a series of five steps that evolve from description to understanding to prediction to control.

Step 1 - Describe Species Composition and Its Probable Controls at Study Plots

The models in Steps 2 and 3 that form the backbone of the toolkit are built upon plot-based measurements of topo-edaphic features, species composition, and temporally-dynamic environmental conditions. Using hierarchy theory, we make the assumption that linkages between the environment and species observed at 10^2 to 10^3 m^2 study plots also occur at the extent of the 10^4 to 10^5 m^2 sites that constitute an ecological site. We assume this ability to scale up is strengthened by randomly siting plots but stratifying them across multiple zones in order to embody gradients in topography, soil properties, and vegetation "nativeness". Sampling the plots repeatedly through time should also help justify scaling up (Bartolome et al., 2009).

Random sampling may not be appropriate for all projects. For example, if the management goal is the restoration of a single species (e.g. increase abundance and distribution of *Stipa pulchra*), investigators may prefer to measure ecosystem characteristics only in areas known to support that single species. For goals that are not tied to particular species but instead represent broader ideals such as "enhancing native biodiversity," random sampling can protect management teams from reliance on preconceived notions about restoration potential that tend to be strengthened by subjectively placing plots only in areas that reinforce those preconceived notions.

Within a valley grassland of uniform climate, we predict that topo-edaphic characteristics will be the principal controls on plant community structure, and accordingly, we recommend the ecological site classification model in Step 2 be populated with topo-edaphic data. Deciding the particular topo-edaphic features to measure is a site-specific process, with local expertise the best guide. In valley grasslands, ideal sampling would cover the chemistry, texture, and depth of the soils as well as the factors that impact soil formation, such as slope, aspect, climate, and site history (Jenny 1941). When site history cannot be reconstructed, empirical soil characteristics can act as proxies (Steenwerth et al., 2002). Because many soil properties change diurnally, seasonally, annually, or on longer timescales, site-specific considerations must dictate whether one-time sampling of topo-edaphic properties is sufficient for the ecological site model (Grigal et al., 1999).

Sampling the entire soil profile will reveal information about depth trends, pedogenesis, and taxonomic classification of the soil (Brady & Weil 2002; Soil Survey Staff 2010), but we recommend that sampling depth be determined by the rooting depths of the majority of the extant and target species. We predict that for California grasslands, in which most plants have the majority of roots at 60 cm or shallower (Holmes & Rice 1996; Hunter & Wu 2005), assaying soil taken from a sub-meter depth and measuring depth to restrictive layer is sufficient to understand the most informative plant-soil relationships.

Deciphering how strongly the topo-edaphic properties being measured are structuring the plant community can indicate whether an ecological site classification based on those properties will, in fact, efficaciously inform predictions. A modeling technique useful for this purpose is BIO-ENV (Clarke & Ainsworth 1993), which relates community and environmental data by correlating community structure with each possible combination of environmental variables. Model output provides a list of correlations, which offers a sense of how well the full set of environmental variables explains community structure, as well as which environmental variables best explain community structure.

A correlation lower than 0.3 between the species and the "best" subset of topo-edaphic variables is unexpected in non-equilibrium ecosystems. A management team faced with such a low correlation may elect to sample a different set of topo-edaphic factors. Alternatively, they may conclude that defining site potential using topo-edaphic properties is not appropriate and may classify landscape units using recovery constraints such as land use legacies and invasive species (the toolkit does not cover this approach). The BIO-ENV modeling exercise is also useful for streamlining future sampling efforts. Because the variables that are not part of the "best" subset may be redundant, they can be omitted from future sampling with some certainty that no critical information will be lost.

Repeatedly measuring species composition at numerous plots, though expensive, would allow the management team to build temporal state-and-transition models. The transitions in such data-driven models would symbolize inter-annual changes observed during the study; however, we hypothesize that these would be predictive with respect to inter-annual as well as decadal time scales. In valley grasslands with few perennials, three or more inter-annual samples of species composition taken during springtime peak growing season can inform predictive state-and-transition models. Systems with extensive perennials may require more sampling years than a system dominated by annuals.

Measuring temporally-dynamic environmental conditions concurrently with replicated species composition samples allows the management team to isolate the temporal conditions associated with plant community transitions and stalled recovery. Determining the suite of potential influences on temporal species composition dynamics is ultimately a local process, but likely factors in valley grasslands include weather (e.g. monthly rainfall totals, Julian date of first significant rainfall, monthly average coldest temperature), seedbank dynamics, herbivory and disturbance by livestock and wildlife, and plant litter. Measuring some potential influences on species composition in the spring and then other potential influences in the fall when only residual dry vegetative matter (RDM) remains can be instructive (Heady 1956; Bartolome et al., 2002). Isolating the set of variables may be best achieved through small-scale experiments, such as livestock grazing trials with paired grazed and ungrazed plots (Kettenring & Adams 2011).

Step 2 - Organize Plot Data into Ecological Site and State-and-Transition Models

If the management team is satisfied with the correlation of species composition and environmental properties measured, they can proceed to Step 2: using the data to build ecological site and state-and-transition models. The models are analytically straightforward, but local, objective expertise is critical to interpreting the models and communicating results to stakeholders.

The ecological site model is borne out of describing and understanding the topographic and edaphic environment of the focal area. Once decisions have been made about the set of variables to measure and data have been collected at the plots (Step 1), hierarchical cluster analysis can parse the plots into ecological site groupings. Several auxiliary analyses are available to ensure accurate and unbiased pruning of the cluster dendrogram (e.g. Mantel test; Borcard et al., 2011); to identify the key variables that are differentiating the clusters (e.g. conditional box plots; Verfaillie et al., 2009); and to verify that the clusters are statistically significantly different from each other with respect to the key variables (e.g. non-parametric Kruskal-Wallis test for differences between groups; Whitlock & Schluter 2009). Local decisions should be made about whether all plots in the topo-edaphic dataset should be included in a global cluster analysis, or whether plots should be modeled in separate groups. A possible reason for splitting the plots *a priori* would be a distinct climatic difference among two groups of plots that is recognized by stakeholders but would not be recognized by a global cluster analysis.

For managers to effectively use ecological site classifications, they need the ability to stand in any area of the management landscape and determine its ecological site. Managers adopting our approach can approximate the ecological site class of an area by consulting geology and soils maps; using visual cues of topography and elevation; and informally determining surface soil texture. In order to absolutely verify that the soil of the area falls within the ecological site, the manager must determine depth to restrictive layer and assay soil taken from a depth of 15 to 60 cm, to be determined by management goals. They would not be required to identify diagnostic soil horizons to taxonomically classify the soil.

If the budget allows, we recommend replicating species composition data collection inter-annually to build a state-and-transition model for each ecological site. A cluster analysis and/or ordination on the community data matrix (McCune & Grace 2002; Borcard et al., 2011) can reveal states and transitions. For this type of analysis, we create a community data matrix with each row representing a particular plot's species during a particular sampling year (hereafter "plot × year"). Transitions are detected by the plots' movement through ordination or cluster space. For example, if a particular study plot ("Plot A"), for a particular year ("Plot A × 2007"), is in a cluster with plot × years dominated by native annual forbs, but for a subsequent year, the plot ("Plot A × 2009") is in a cluster with plot × years dominated by exotic annual grass, a transition has occurred. It is critical to demarcate ordination groupings and prune cluster diagrams objectively. Indicator Species Analysis, which isolates the species that are driving the differences between the cluster groupings (Dufrene & Legendre 1997), can be used for this purpose when used in concert with a randomization procedure (McCune & Grace 2002).

Statistical analyses for the state-and-transition and ecological site models may not be sensitive enough to pick up dynamics that are observed on the ground, and it may be necessary to alter model results *a posteriori*. Such alterations may include assigning a study plot to a different ecological site or re-assigning plots to an extra state that is known to occur on the ecological site and is important to stakeholders, but that was non-existent during sampling. In addition, if a conservation objective is enhancing a particular functional group (e.g. native annual forb, native perennial bunchgrass), the management team may choose to lump the species composition clusters into states defined by functional groups. Similarly, the team may lump multiple site clusters into a single ecological site. A general rule of thumb when altering the models *a posteriori* is remaining as objective as possible and favoring empirical information instead of preconceived notions.

Presenting the models clearly will be critical to communication among stakeholders. State-and-transition models are commonly presented as box and arrow diagrams showing states and transitions, with a catalogue of states and a catalogue of transitions that includes hypotheses about why the transitions occurred (Westoby et al., 1989). Ecological sites can be displayed with stylized diagrams (see Case Study 2) or maps with plot symbols color-coded to represent the ecological site groupings. Ecological site names that incorporate the local vernacular can help facilitate communication.

As mentioned above, three U.S. government land management agencies formally adopted ecological site and state-and-transition modeling to inform management (Caudle et al., 2013). Each Ecological Site Description in the Ecological Site Information System (NRCS 2013) includes a wealth of useful information. Our toolkit has been informed and enhanced by the agencies' work, but the two approaches differ in a few key ways. One difference is the degree to which plot-based data "drives" modeling. The agencies first create ecological site and state-and-transition concepts using existing records and expert knowledge and secondly collect field data in plots that are randomly and subjectively placed to test the concepts and describe and define the ecological sites and their states. In our alternative approach, stratifying study plots across gradients discussed in Step 1 requires conceptualization, but we emphasize the use of plot data as the primary foundation for defining and describing the ecological sites and their states. A second difference between the official and alternative approaches is the role of maps. The U.S. government has created soil maps for the majority of the country. Soil map polygons typically contain multiple soil types classified per U.S. soil taxonomy and *in situ* conditions (Soil Survey Staff 1999). Each soil type corresponds with one official ecological site. Because the soil types are undifferentiated on soil maps, multiple ecological sites can exist in one soil map unit (Soil Survey Staff 2013). Accordingly, ecological sites exist at spatial scales smaller than the scale of the basic map unit. A potential problem is misinformed management planning founded on a false sense of the scale of the ecological site. Researchers are working towards mapping soils at higher resolutions to mitigate this problem (Duniway et al., 2010), but such maps are not yet widespread. In contrast with the governmental approach, we do not formally link our classification with maps. We believe this helps to mitigate possible confusion. A third difference between the official and alternative approach is the lexicon used to describe vegetation dynamics. Official state-and-transition models label temporal changes in grassland species assemblages "community pathways" within a grass state, even when the change is from exotic to native dominance or vice versa (NRCS 2013). The change from exotic to native dominance is a basic goal of California grassland restoration, and the "community pathway" nomenclature may underestimate the value that society places on a change from exotic to native dominance. In our approach, such nomenclature is not rigid. For instance, in the case study below, we refer to an inter-annual change in the grassland community as a transition between states. This sort of flexibility can help preserve the widespread utility of state-and-transition models (Jackson et al., 2002).

Step 3 - Identify Causes of Transitions or Stalled Recovery

In the toolkit framework, restoration can be considered a transition from an undesirable to a desirable state. The persistence of an undesirable state can indicate a recovery constraint: processes and/or factors that contribute to the resistance of degraded communities to restoration management. Knowledge about the temporal conditions that coincide with transitions, as well as the recovery constraints working against desired transitions, is invaluable for management prioritization and planning (Step 4).

Pairing the results from the state-and-transition modeling with data on temporally-dynamic environmental conditions can isolate the circumstances that are prerequisites for community succession on each ecological site. A classification tree (Breiman et al., 1984) can be used for this purpose. In the model, the dependent categorical variable represents the transition or lack of transition (hereafter "non-transition") that occurred during a given transition year at a given plot. Independent variables represent the temporal conditions that occurred at each plot during the year the transition was detected (Jackson & Bartolome 2002). The classification tree will indicate the relative importance of the independent variables (conditions) to the variability of the categorical dependent variable (transitions/non-transitions). Conditions higher on the classification tree have stronger control over temporal dynamics of the ecological site than those lower on the tree, with the top, "root," node having the greatest influence (De'ath & Fabricius 2000; Zuur et al., 2007). Relationships between the transitions and conditions will necessarily be constrained to the spatial and temporal scales of the data collection, but we hypothesize that the relationships can be predictive, as well.

Management teams faced with long-term persistence of unwanted states may require additional information on recovery constraints, which may not be isolated by the classification tree analysis. Recovery constraints can range from local-scale interactions such as trophic interactions and exotic species impacts on soil properties and nutrient cycling to long-term, regional-scale processes such as changes in disturbance regimes and availability of seeds due to landscape connectivity. Management intervention may be necessary to overcome recovery constraints that are forcing states to persist. Expanding understanding beyond plot-based observations of temporal conditions, to the mechanisms underlying the persistence of an undesirable state, can be achieved through small-scale sampling or manipulative experiments. For example, a land manager hoping to foster a transition from high exotic cover to high native forb cover can assess the potential for natural recruitment of native species with seed bank soil samples. In another instance, if a land manager is concerned about the potential recolonization of exotic species in a restored area, the land manager can assess potential seed inputs by clearing a few small areas, ideally in a gradient from a nearby uncleared site, and observe which species establish.

Step 4 - Predict and Prioritize the Objectives, Timing, Sites, and Treatments That Will Help Achieve Conservation Goals

As stated above, we assume the management team will set general, idealistic conservation and restoration goals before undertaking the steps of the toolkit. An example of a goal is "enhance native forb abundance". A complementary objective would specify how that goal could be achieved, for instance, "decrease exotic annual grass biomass by 50% during March, April, and May, to promote ample sunlight and moisture for native forbs" (Bush 2006). Results from the modeling in Step 3 can help the management team determine specific objectives and predict which timing, sites, and treatments have the highest potential for accomplishing the objectives. The team can plan its management activities using those predictions.

In non-equilibrium valley grasslands, it can be expected that management actions will be constrained by the timing and amount of rainfall. Managers must be prepared to implement treatments opportunistically, during the unique windows of time when conditions are auspicious for meeting restoration targets (Westoby et al., 1989; Holmgren & Scheffer 2001). Identifying the temporal conditions that co-occur with desired transitions (Step 3) can inform sets of plans to be deployed when proper conditions arise.

In addition to temporal constraints, managers of Californian grassland also typically have extensive land, but limited financial resources (Huntsinger et al., 2007), and therefore must prioritize sites and restoration treatments (Hopkinson et al., 2009; Meinke et al., 2009; Noss et al., 2009). Predicting the sites with the highest potential for successful treatments can guide site prioritization during the planning process. For instance, in valley grasslands, it may be hypothesized that on a particular ecological site, low levels of RDM left on the ground at the end of the dry season may be prerequisite to a desired transition detected during the subsequent rainy season. The management team could elect to use prescriptive livestock grazing or mowing to achieve low levels of RDM in order to create conditions that are auspicious for the desired transition. On a different ecological site, rodent bioturbation could be the prevalent temporally dynamic condition coinciding with the stability of a single, unwanted state. A possible management action for this ecological site would be eradicating the rodents, which would likely be very difficult as well as controversial among stakeholders. The management team may decide that the rodent recovery constraint is not worth mitigating and instead turn their attention to the ecological site without the rodent problem.

Another example illustrates how the models from Step 3 can inform predictions about, and prioritization of, treatments at a particular ecological site. If classification tree modeling reveals that light but increasing precipitation in October, November, and December predicates a desirable transition on a given ecological site, the management team will be faced with a conundrum. The Mediterranean climate requires that seeds of desirable species be planted in October, when light and soil temperature conditions are most favorable. Unfortunately, even if the rain in October is matching the weather signature isolated as ideal by the classification tree, the team will not know with certainty that November and December will follow suit. Faced with this uncertainty, the team may elect to focus its efforts on treatments other than seeding at the ecological site.

Assessment of recovery constraints on an ecological site may also influence its prioritization. A site with a single constraint may require less effort and be prioritized over a site with multiple constraints (Suding et al., 2004). To ensure achieving the desired transition of a site with multiple constraints, the management team should attempt to tackle those constraints in one effort. The management team can use augmentative restoration, a restoration strategy that is synergistic with tackling multiple constraints, to guide the development and subsequent execution of any management plans. Augmentative restoration aims to enhance ecological processes that are necessary for the recovery of target plant communities but are not operating at adequate levels (i.e. recovery constraints). Under this approach practitioners have developed a suite of potential factors that can be modified to improve plant establishment (Bard et al., 2004). Along with predicting and prioritizing temporal windows, sites, and treatments that will foster the achievement of goals and objectives, the science-based management team should also consult restoration literature and other managers for best practices. Using site-specific data to inform management decisions is an intuitive process that is gaining a new foothold with web-based databases. For instance, the Nature Reserve of Orange County (http://www.naturereserveoc.org/), which coordinates the land management activities of governmental and private landowners in a southern California land reserve system, has built a web-based database through which landowners can access useful ecological data, learn what other landowners are doing, and share information about their own management.

Step 5 - Opportunistic and Adaptive Management

Adaptive management is the process through which management is initiated, evaluated, and refined (Holling 1978). It differs from traditional management approaches by recognizing and preparing for the uncertainty that inevitably complicates resource management decisions. A typical formal adaptive management process consists of a multi-step cycle (Nyberg 1999; Murray & Marmoneck 2003; Reever-Morghan et al., 2006). The cycle promoted by the Forest Service of British Columbia, Canada consists of six steps (Table 1; Nyberg 1999; Murray & Marmoneck 2003). The first step, assessing the problem, includes setting goals and objectives, identifying possible actions that could be taken to achieve the goals, and identifying key uncertainties about the managed system. The second step is designing a plan that will foster the achievement of goals and test hypotheses about the system; the third step is implementing the plan. Next, in the fourth step, the management team monitors the implementation and effectiveness of the plan with attention to the key uncertainties identified in the first step. In the fifth step of the adaptive management cycle, the team evaluates the plan's outcomes. If the evaluation reveals that the desired outcome was not reached, managers move on to the sixth step and adjust their hypotheses and plan by returning back to the beginning of the six-step cycle. Table 1 contextualizes our toolkit with this six-step adaptive management cycle.

Table 1. A comparison of the toolkit with a 6-step adaptive management cycle

Adaptive management	Toolkit for science-based restoration of California grasslands
1. Assess problem	1. Describe species composition and its probable controls at study plots 2. Organize plot data into ecological site and state-and-transition models 3. Identify causes of transitions or stalled recovery
2. Design a plan	4. Predict and prioritize the objectives, timing, sites, and treatments that will help achieve conservation goals
3. Implement the plan	5. Opportunistic and adaptive management
4. Monitor plan implementation and effectiveness	5. Opportunistic and adaptive management
5. Evaluate the results	5. Opportunistic and adaptive management
6. Adjust plans and practices based on what was learned. Return to 1. Assess problem	Return to 3. Identify causes of transitions or stalled recovery; or 1. Describe species composition and its probable controls at study plots

Six-step adaptive management cycle per Nyberg 1999; Murray & Marmoneck 2003.

We do not discuss monitoring or evaluation in detail here, so we refer the reader to the Landscape Toolbox at http://www.landscapetoolbox.org/. This website is a valuable repository for field and remote sensing monitoring methods designed to inform managers' method selection, with attention to appropriate scale (Karl et al., 2011). The conceptual foundation of the Landscape Toolbox is consistent with the integrated framework for science-based management of arid lands put forth by Herrick and others (2006), discussed above. Herrick et al., (2012) expand this work by integrating ecological site-based monitoring and assessment into an adaptive management strategy for rangelands. Like teams managing rangelands worldwide, teams managing California grasslands can benefit from the tools offered in the Landscape Toolbox as well as the restoration toolkit we propose here. We believe that our toolkit allows for substantial improvement in effectively achieving conservation goals as we show in Case Study 2.

Case Study 2: Use of the Toolkit in Southern California Valley Grassland

Here we illustrate the use of the toolkit in the valley grasslands of the Tejon Ranch, California (Figure 1). We elected to use Tejon Ranch as an example in favor of the East Bay Regional Park District (Case Study 1) for two main reasons. Primarily, valley grasslands in the southern part of the state have been underrepresented in the scientific literature in comparison with their more northerly counterparts. Secondly, Tejon Ranch is an ideal laboratory for our toolkit due to the exceptional areal extent of its grasslands (i.e. 48,000 ha), the variation in its topography and soils, and the high abundance of natives compared with grasslands statewide (Hopkinson et al., 2009).

The mission of the Tejon Ranch Conservancy (hereafter, Conservancy) is to preserve, enhance and restore native species biodiversity on Tejon Ranch. The Conservancy is using the toolkit in a framework of adaptive, opportunistic management (Tejon Ranch Conservancy 2013). Cattle grazing is the Conservancy's primary on-the-ground management tool. Livestock have grazed the ranch since at least its founding in 1843, and the current cattle grazing regime consists of light to moderate stocking rates with minimal rotation between large, heterogeneous pastures (see Case Study 1 for additional information about the study area).

Step 1 - Describe Species Composition and Its Probable Controls at Study Plots

The Conservancy is using the toolkit approach to manage grasslands across the entire ranch, but we constrain our example to the grasslands on the ranch's northern side. The southern side will need to be modeled separately because the Tehachapi Mountains bifurcate the ranch, and the grasslands south of the mountains are in a different regional climate zone than the grasslands north of the mountains (Bailey 1995). In total, 57 permanent plots have been established across 48,000 ha of grasslands ranch-wide, but the data populating the models in this case study are from only 30 plots established across 22,000 ha of Valley Grassland on the ranch's north side.

Study plots were randomly sited to ensure unbiased sampling. To capture the variation in soils and grassland "nativeness," the plots were stratified by the major geophysical areas on the north side of the ranch: San Joaquin Valley, Sierra Nevada, Tejon Hills, and Tehachapi Mountains. Sampling was conducted during the spring peak growing season in 2010, 2011, and 2012.

Each 3024 m^2 elliptical plot comprises four 25 m transects radiating in the cardinal directions from the corners of a 100 m^2 *relevé*. We sampled topographic factors at the spatial scale of the entire plot, soils at the corners of the *relevé*, and species on the transects. We assume relationships observed between the environment and species observed at the 3×10^3 m^2 plots scale up to the extent of the landscape units of 10^4-10^5 m^2 (Bestelmeyer et al., 2011a) that aggregate into ecological sites. Here we describe the sampling of probable controls on species composition; sampling of the species composition itself is described in Case Study 1.

As this was the first effort to describe environmental and species composition patterns on the ranch's grasslands, we sampled numerous environmental variables (34 in total). Future sampling efforts can be streamlined to include only the variables deemed essential by this first effort. The environmental data fall into two major subsets. The first subset contains variables measured once at each plot and treated as "constant/spatial" factors for the ecological site classification (Step 2). Topographic variables in this subset are slope, aspect, and elevation. Edaphic variables are phytolith content, depth to restrictive layer, bulk density, and select properties of soil samples augered from depths of 0-15 cm and 30-45 cm [percent sand, silt, and clay; pH; C_{total}:N_{total}; plant-available sulfate (SO_4-S); plant-available phosphate (PO_4-P); and exchangeable cations Ca^{2+}, K^+, Mg^{2+}, and Na^+]. Geology underlying each plot was obtained from maps by the late geologist David Dibblee. Data is also collected for a second, "temporal" subset, to be used to isolate conditions associated with transitions/non-transitions (Step 3). Field-based variables include cover of bare ground, litter, gravel, cobbles, and boulders at time of sampling and estimated use by cattle, rodents, wild pigs, and humans since the first germinating rains of the growing season. This subset also includes monthly weather for each plot obtained from the PRISM climate model (PRISM Climate Group 2013). Seven of the 34 environmental variables, including geology, were useful for system description and understanding, but not for quantitative analyses. We excluded those seven descriptive variables as well as ten variables that were highly correlated with others, leaving seventeen remaining for quantitative analyses (bottom row of Table 2).

Table 2. Correlations between topo-edaphic properties and 2010 species composition

Set of topo-edaphic variables	Correlation
Slope	0.2590
Elev Slope	0.3816
Elev Heat Slope	0.4463
C:N Elev Heat Slope	0.4100
C:N Elev Heat Slope Depth	0.1740
Ca^{2+} C:N Elev Heat Slope Depth	*0.5474*
B.D. Ca^{2+} C:N Elev Heat Slope Depth	0.5437
B.D. Ca^{2+} C:N Elev Heat pH Slope Depth	0.5455
B.D. C:N Elev K^+ pH PO_4 Sand Slope Depth	0.5461
B.D. C:N Elev Heat K^+ pH PO_4 Sand Slope Depth	0.5432
B.D. Ca^{2+} C:N Elev Heat K^+ pH **PO_4** Sand Slope Depth	0.5427
B.D. Ca^{2+} C:N Elev Heat K^+ pH PO_4 **PO_4** Sand Slope Depth	0.5341
B.D. Ca^{2+} C:N Elev Heat K^+ pH PO_4 **PO_4** Sand Slope SO_4 Depth	0.5294
B.D. Ca^{2+} C:N **C:N** Elev Heat K^+ pH PO_4 **PO_4** Sand Slope SO_4 Depth	0.4991
B.D. Ca^{2+} C:N **C:N** Elev Heat K^+ Na^+ pH PO_4 **PO_4** Sand Slope SO_4 Depth	0.4644
B.D. Ca^{2+} C:N **C:N** Elev Heat K^+ Mg^{2+} Na^+ pH PO_4 **PO_4** Sand Slope SO_4 Depth	0.4356
*B.D. Ca^{2+} C:N **C:N** Elev Heat K^+ Mg^{2+} Na^+ pH PO_4 **PO_4** Sand Slope SO_4 **SO_4** Depth*	*0.4015*

The full set of topo-edaphic variables correlated with the species data at 0.4015, and the optimal subset correlated at 0.5474 per the BIO-ENV routine (Clarke & Ainsworth 1993). Soil properties in bold were from soils sampled at 30-45 cm depth; all other soil properties were from 0-15 cm depth samples. "B.D." denotes bulk density; "Depth" denotes depth to restrictive soil layer; "Heat" represents heat load, a proxy for aspect; and "Sand" denotes percentage of sand in the soil. Additional details about the variables can be found in the text.

Given the hypothetical placement of Tejon Ranch's grasslands on the non-equilibrium end of the theoretical continuum, as well as the general support for the idea that topo-edaphic factors control species composition, we hypothesized that our topo-edaphic metrics would "explain" plant community structure. We verified this hypothesis using a BIO-ENV modeling approach (Table 2; Spearman correlation, Sørensen dissimilarity for the community data matrix, Euclidean distance for the log-transformed topo-edaphic data matrix, 53 species after omission of species encountered in only a single plot). We found that the full suite of topo-edaphic variables (n = 17) was correlated with the species composition data at 0.4015. We perceived this correlation as sound rationale to proceed with classifying landscape units based on their topo-edaphic properties. The subset (n = 6) with the highest correlation, 0.5474, contained soil exchangeable Ca^{2+} (0-15 cm depth sample), soil C:N (0-15 cm depth sample), elevation, heat load (a proxy for aspect; McCune and Keon 2002), slope, and soil depth. Subsequent sampling efforts can be streamlined using this subset as a guide.

Step 2 - Organize Plot Data into Ecological Site and State-and-Transition Models

The "constant/spatial" topo-edaphic data collected at the 30 study plots resolved into five ecological sites per hierarchical cluster analysis (Figure 4; Euclidean distance measure, Ward's linkage, 17 continuous log-transformed variables). A Mantel test run on the cluster dendrogram (Borcard et al., 2011) identified five as the optimal number of clusters (r = 0.49). Conditional box plots with the five clusters as the categorical independent variable and 17 continuous environmental variables as the conditional variables revealed the key variables driving the differences between ecological sites (Verfaillie et al., 2009). We used a non-parametric Kruskal-Wallis analysis of variance to test for differences between groups (Whitlock & Schluter 2009) with respect to the key variables. We included geologic characteristics in ecological site names for ease of communication among stakeholders, even though geologic variables were not in the quantitative cluster analysis.

The 'Flat sands and loamy sands on Holocene surficial sediments' ecological site comprises sites that lie at low elevations of the ranch (~150-450 m). Its surface soils (i.e. soils sampled at 0-15 cm and 30-45 cm) have low SO_4-S compared with the surface soils of the other ecological sites. Generally situated at elevations slightly higher than the previously described ecological site, the 'Gently sloping sandy loams on dissected alluvial terraces' ecological site has lower surface soil C:N than the other ecological sites. Sites composing the 'Sandy clay loams on steeply-sloped, low-relief hills of Miocene sediments' ecological site lie on marine and nonmarine sediments deposited during the Miocene when the Great Central Valley was an inland sea. In comparison with the surface soils of the other ecological sites, surface soils of this ecological site have low PO_4-P, high SO_4-S, and high exchangeable Ca^{2+} and Na^+. The 'Moderate slopes on Pleistocene landslide debris' and 'Steep slopes on Sierran plutonic basement complex' ecological sites lie on mountainous terrain at a wide range of elevations.

For the remainder of the case study, we use the 'Flat sands and loamy sands on Holocene surficial sediments' ecological site, (hereafter "Flat sands ecological site") to illuminate our approach. The Conservancy has prioritized this ecological site for management because of its demonstrated potential to support desired native plant communities as well as special status wildlife species (Tejon Ranch Conservancy 2013). Several objectives aimed at maintaining and enhancing the ecological site's native vegetation are synergistic with objectives for its wildlife, but here we focus only on the vegetation.

We offer an ecological site classification, not an ecological site map; however, we used existing maps of local geology, soils, and elevation; visual cues of slope; and informal texture-by feel estimates to approximate the areal extent of the Flat sands ecological site. We estimate it covers approximately 3600 ha (3.6×10^7 m^2), not necessarily contiguously, on the western edges of the ranch north of the Tehachapi Mountains.

Within those estimated 3600 ha, eight U.S. soil map units (soil map polygons) occur once or repeatedly. The eight map units comprise 17 undifferentiated soil types, two of which correlate with official ecological sites. Within the estimated area, the Arvin soil correlates with 'R017XF072CA Dry Coarse Loamy' and the Steuber soil correlates with 'R018XE032CA Coarse Loamy Fan' and 'R018XE006CA Fine Loamy' (Soil Survey Staff 2013). Future work could include intensively relating our Flat sands ecological site with these three officially-described ecological sites. As of the writing of this chapter, Ecological Site Descriptions for the three ecological sites were not published in the Ecological Site Information System (NRCS 2013).

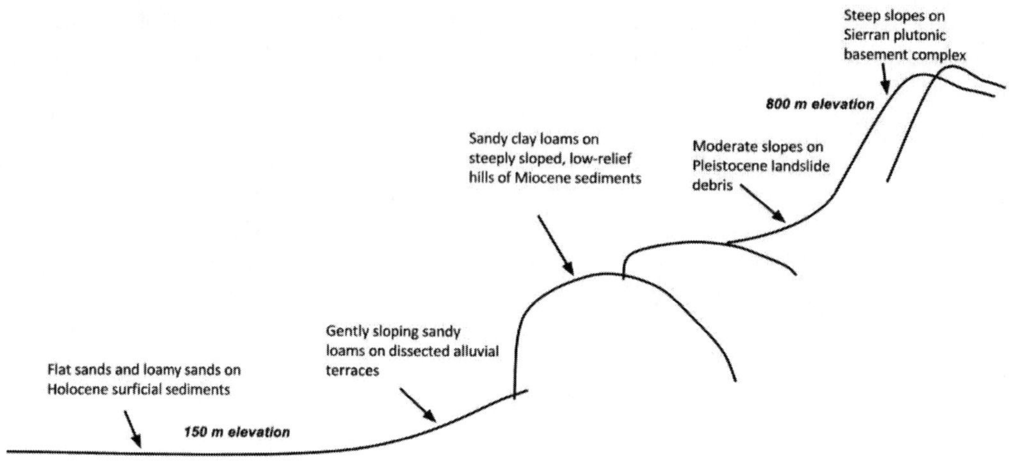

Figure 4. Ecological sites on a schematic cross-section of the characteristic landforms underlying the grasslands on the northern half of Tejon Ranch. The ecological sites were classified using topo-edaphic data collected at 30 randomly located study plots across 22,000 ha of valley grasslands. We assume data collected at the plots (3.6×10^3 m^2) scales up to landscape units of 10^4 - 10^5 m^2, and we estimate that the landscape units aggregate into ecological sites that cover $\geq 10^7$ m^2. The ecological sites are arrayed in a mosaic across the landscape, but are depicted here along a simple elevational gradient to illustrate general associations among topography, geology, and ecological sites.

To identify states and transitions (Figure 5), we ran hierarchical cluster analysis (Sørensen dissimilarity; flexible beta linkage set at -0.25) on 21 plot × years (7 plots in the ecological site × 3 years of vegetation sampling). The dataset comprised thirty-five species (species encountered in more than one plot × year, square-root-transformed relative cover). We objectively cut the dendrogram to nine clusters using a randomization procedure with Indicator Species Analysis (Dufrene & Legendre 1997; McCune & Grace 2002). Next we modified the nine species composition clusters *a posteriori* based on Conservancy goals and on-the-ground knowledge not identified by the model. Because the enhancement of native forb communities is among Conservancy objectives, clusters that had functional overlap in species composition were combined into states (two clusters were lumped into the 'Native annual forbs characteristically abundant state'; five clusters were lumped into the 'Exotic annual grass dominant' state). In addition, the composition at two plot × years was very distinct (i.e. low cover of live vegetation paired with either high litter cover or high litter and bare ground cover), but because we input relative cover, the cluster model was not sensitive enough to pick up this distinction. We removed those two plot × years from their clusters, and designated each as a state ('Sparse live vegetation, litter dominant'; 'Sparse live vegetation, litter and bare dominant'). 'Dominance' and 'characteristic abundance' in state names were defined with respect to mean relative cover across all plot × years in the state (species or functional groups with ≥40% mean relative cover were classified as dominant; those with >40% mean relative cover were classified as "characteristically abundant").

The resulting state-and-transition model for the Flat sands ecological site (Figure 5) consists of six states (boxes), five transitions (straight arrows), and two non-transitions (curved arrows). The catalogue of states adjacent to the boxes and arrows describes species composition of each state. The catalogue of transitions lists hypotheses about why the transitions and non-transitions occurred and can be used to predict future dynamics. Transitions and non-transitions symbolized with solid arrows were linked to temporally-dynamic environmental conditions by a classification tree (Step 3; Figure 6); those with dashed arrows were linked to temporally-dynamic conditions by local knowledge. While the transitions and non-transitions occurred from 2010 to 2011, 2011 to 2012, or 2010 to 2012, we hypothesize that these are predictive with respect to inter-annual through decadal time scales.

Step 3 - Identify Causes of Transitions or Stalled Recovery

We used a classification tree to identify the temporal conditions co-occurring with state transitions and stability at the Flat sands ecological site during 2010-2012 (Figure 6). The dependent, categorical variable represented the seven transitions and non-transitions symbolized by the arrows in Figure 5. The independent variables were temporally-dynamic conditions that we hypothesize to be influences on vegetation succession: September-March monthly rainfall totals, litter cover, categorical cattle grazing severity, and rodent bioturbation. In the data matrix, row headings were "plot × transition years," with transition year being the year the transition was detected. Each plot × transition year was assigned a category representing the transition or non-transition detected at the plot during the year specified (Y variable) along with a unique set of temporal conditions that preceded or were concurrent with the detection of the transition or non-transition (X variables). Monthly rainfall totals are specific to each plot due to our use of the PRISM model (PRISM Climate Group 2013). The resulting data matrix had 14 rows (7 plots × 2 possible transition years 2011 and 2012), one Y-variable "transition" column, and several X-variable "temporal condition" columns. The classification tree isolated influences on four of the seven transitions/non-transitions that occurred (solid arrows in Figure 5; terminal node circles in Figure 6). The root node of the classification tree, representing the condition with the strongest overall effect, was December precipitation during the growing season in which the transition/non-transition was detected. Rainfall, an abiotic factor, had more control over community temporal dynamics than did the biotic interactions of litter cover, cattle grazing, and rodent bioturbation. This finding lends strength to the hypothesis that Tejon Ranch's grasslands should be placed on the non-equilibrium end of the theoretical continuum from equilibrium to non-equilibrium. Local knowledge about conditions on the ground was used to hypothesize causes of the remaining three transitions not identified by the classification tree (dashed arrows in Figure 5). Data-driven and local knowledge are recorded in the catalogue of transitions adjacent to Figure 5.

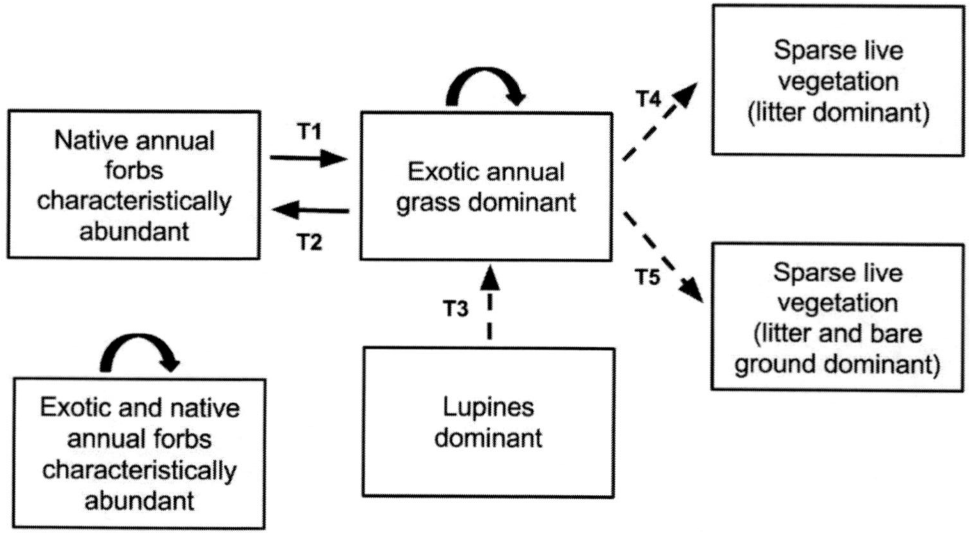

Figure 5. State-and-transition model for the 'Flat, low elevation, sands and loamy sands on Holocene surficial sediments' ecological site of Tejon Ranch. Transitions (straight arrows) and non-transitions (curved arrows) occurred during 2010-2012, but we hypothesize that they are predictive on inter-annual as well as decadal time scales. Transitions/non-transitions symbolized with solid arrows were linked to temporally-dynamic environmental conditions by the classification tree in Figure 6; those with dashed arrows were linked to such conditions by local knowledge (see Catalogue of transitions).

Catalogue of states:

- Native annual forbs characteristically abundant (*Plagiobothrys* spp., *Trifolium* spp.)
- Exotic annual grass dominant (*Avena barbata, Bromus diandrus,* or *Vulpia myuros*)
- Lupines dominant (native *Lupinus* spp.)
- Sparse live vegetation, litter dominant
- Sparse live vegetation, litter and bare dominant
- Native and exotic annual forbs characteristically abundant (*Dicholostemma capitatum* (native) and *Erodium botrys* (exotic). Native forb cover is, on average, less than that of 'Native annual forbs characteristically abundant'.)

Catalogue of transitions:

(Temporal conditions and transition/non-transition occurred during the same September-May growing season.)

Transition 1 (T1): Native annual forbs characteristically abundant → Exotic annual grass dominant

This is an undesirable transition in relation to Conservancy goals.

Temporal conditions (classification tree): December rainfall greater than or equal to 120 mm.

Transition 2 (T2): Exotic annual grass dominant → Native annual forbs characteristically abundant

This is a desirable transition in relation to Conservancy goals.

Temporal conditions (classification tree): December rainfall less than 120 mm and litter covering less than 5% of the ground.

Transition 3 (T3): Lupines dominant → Exotic annual grass dominant

This is an undesirable transition in relation to Conservancy goals.

Temporal conditions (local knowledge): This transition was likely linked to high December rainfall.

Transition 4 (T4): Exotic annual grass dominant → Sparse live vegetation (litter dominant)

Neutral in relation to Conservancy goals, but succession should be monitored

Temporal conditions (local knowledge): This transition occurred between 2011 and 2012. It was likely linked to the heavy production of grass in the 2010-2011 growing season and the lack of fire and livestock grazing in 2011-2012.

Transition 5 (T5): Exotic annual grass dominant → Sparse live vegetation (litter and bare ground dominant)

Neutral in relation to Conservancy goals, but succession should be monitored

Temporal conditions (local knowledge): This transition was likely linked to a fire that burned through part of the Ecological Site in September 2011.

Non-transition: 'Exotic annual grass dominant' persisted

Undesirable in relation to Conservancy goals

Temporal conditions (classification tree): December precipitation below 120 mm, litter covering greater than 5% of the ground, and September precipitation exceeding 2 mm. If this state persists for the long-term, recovery constraints should be investigated using manipulative experiments.

Non-transition: 'Exotic and native annual forbs characteristically abundant' persisted

Neutral in relation to Conservancy goals, but succession should be monitored

Temporal conditions (classification tree): December precipitation below 120 mm, litter covering greater than 5% of the ground, and September precipitation below 2mm.

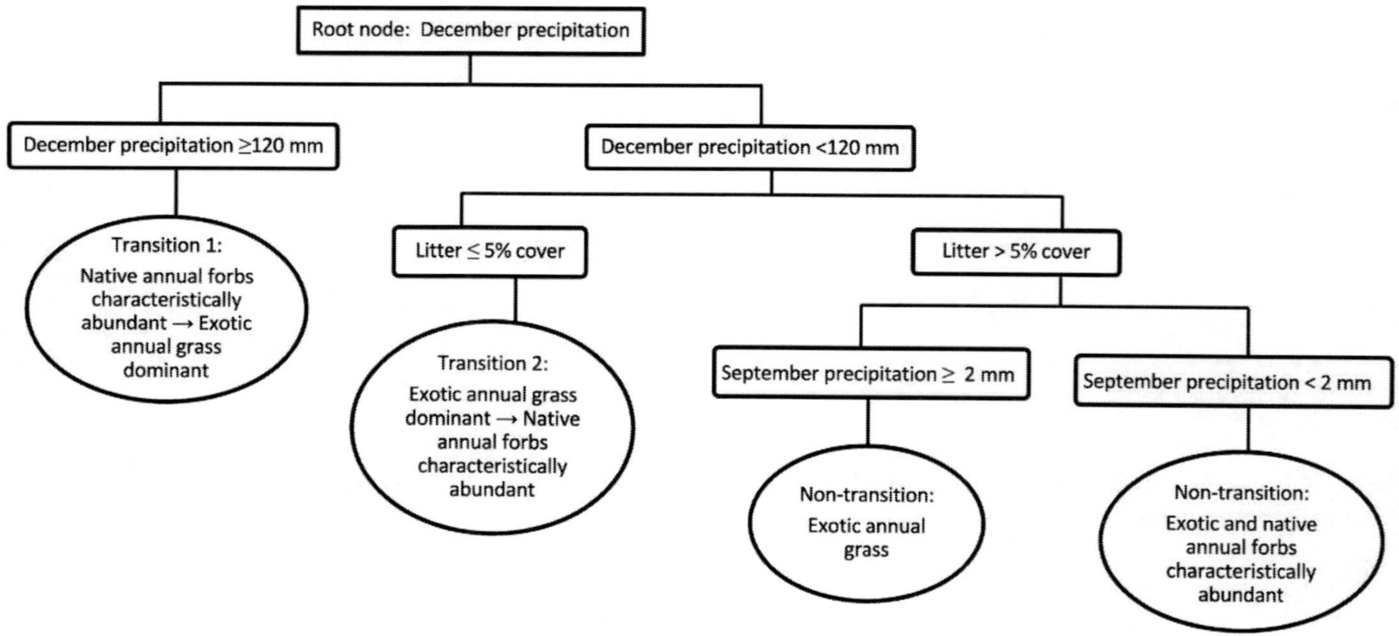

Figure 6. Classification tree identifying temporal environmental conditions predicating transitions/non-transitions at the Flat sands ecological site. The tree identified conditions for four of the seven transitions/non-transitions observed in 2010-2012. Terminal nodes (circles) represent the transitions/non-transitions that the tree linked to conditions; conditions are in the squares preceding the terminal nodes. For example, rainfall greater than or equal to 120 mm in December was the prerequisite of T1: 'Native annual forbs characteristically abundant → Exotic annual grass dominant'. The December rainfall and detection of T1 occurred during the same September-May growing season. Conditions higher on the classification tree have stronger control over community dynamics on the Flat sands ecological site than those lower on the tree.

Step 4 - Predict and Prioritize the Objectives, Timing, Sites, and Treatments That Will Help Achieve Conservation Goals

In light of its management goal to "Preserve, enhance and restore native biodiversity," the Conservancy may prioritize objectives, timing, and treatments that facilitate T2: Exotic annual grass dominant → Native annual forbs characteristically abundant (Figures 5 and 6). The classification tree analysis in Step 3 specified that the prerequisite to this transition is plant litter at <5% ground cover in the spring following a December with low rainfall (< 120 mm). Accordingly, the Conservancy would specify an objective such as "maintain <5% ground cover in the spring following a December with low rainfall to facilitate T2". The team would elect cattle grazing as the treatment type because it is the most powerful on-the-ground treatment option. To achieve its objective, the Conservancy must opportunistically capitalize on low December rainfall and may be challenged to find a livestock operator who can manage with this degree of flexibility. An alternative tactic that would not require such flexibility would be consistently maintaining springtime litter ground cover levels at 5% or less on the areas of the ecological site with the 'Exotic annual grass dominant' state, thereby creating conditions auspicious for the desired transition if a dry December does occur.

The classification tree indicates that minimal springtime plant litter is prerequisite to T2, but it does not stipulate the role of litter in T1 (Native annual forbs characteristically abundant → Exotic annual grass dominant). The classification tree links T1 only to a wet December. Because the model indicates that T1 is beyond human control, the Conservancy may elect to turn its attention to different transitions, especially those that the classification tree links with manipulatable conditions.

Other management teams, however, may not have the luxury to turn their focus elsewhere. A hypothetical scenario involves a team that is managing the ecological site having a mandate to spend restoration funding by the financial year-end (i.e. the June following the wet December). A possible restoration strategy would be using livestock on the areas of the ecological site dominated by exotic annual grass to reduce grass growth, cover, and seed production while not reducing native annual forb populations (but see Kimball & Schiffman 2003). The management team would work with the livestock operator to turn the cattle out during the wet December but remove them when desirable native species begin to set seed. Because the team would be adopting a strategy that is not explicitly prescribed by the data, we would recommend incorporating timely effectiveness monitoring.

Step 5 - Opportunistic and Adaptive Management

Upon completion of the treatment (using cattle to maintain springtime litter at <5% ground cover during a growing season with December rainfall < 120 mm) to foster T2 (Exotic annual grass dominant → Native annual forbs characteristically abundant), the Conservancy would monitor and evaluate the results of the treatment. If the desired litter levels were achieved within the specified weather signature, but the desired transition did not occur, the Conservancy would adapt its hypotheses about the causes of T2 by returning to toolkit Step 3. If a repeated return to Step 3 does not result in the achievement of restoration goals, the Conservancy may elect to return to Step 1 to re-assess the probable controls on species composition of the grasslands of Tejon Ranch.

CONCLUSION

California's grassland restoration managers bear the important responsibility of preserving and restoring native plant communities. They face many challenges, including widespread exotic species, land use changes, and extreme spatial and temporal variability. We propose a set of conceptual tools to help them deal with these challenges.

Our proposed toolkit uses a general framework proposed by Begon et al., (1996) that the scientific knowledge base for ecosystems proceeds through description to understanding into prediction. For managers predictions are in the end ideally tied into control of the system to achieve management goals (Bartolome et al., 2009). Adaptive management has been lauded as an effective management approach, but the expenses of detailed monitoring and long-term assessment mean that it is rarely executed. Westoby et al., (1989) proposed the state-and-transition approach as a compromise allowing for selection and adoption of management controls without waiting for complete understanding and prediction of system responses. The ecological site concept is also helpful to structure information about spatial and temporal variations and accordingly enhance ecosystem understanding, prediction, and ultimately control.

Rigorous adherence to the toolkit will not be appropriate for all management teams at all times, and teams can take a less structured approach by continually fine-tuning plans as monitoring reveals useful information and/or conditions change. For instance, a management team with a limited science budget may elect to sample a small set of environmental variables to develop an ecological site classification and fine-tune the model as money is available by sampling additional variables isolated through monitoring. If only a single year of species sampling can be conducted, the team could assign single-year vegetation states to the ecological sites. On the other hand, if monitoring reveals that extreme environmental changes (e.g., atmospheric nitrogen deposition, fire regimes) are impacting plant community structure, the management team may elect to shift its focus from the topo-edaphic and climatic controls on species composition emphasized by the toolkit to controls imposed by the extreme environmental changes.

We illustrated the use of the toolkit in the most extensive grassland type in California, Valley Grassland, but it is suitable for other spatially and temporally complex terrestrial ecosystems. This toolkit can help managers describe and understand the complexity and use their understanding to achieve their restoration goals.

ACKNOWLEDGMENTS

We thank L.M. Hallett, M.W. White, F.P. Ratcliff, and the editors for their helpful comments on earlier versions of this chapter. We also thank Michele Hammond, Peter Hopkinson, and the many other talented people who have participated in the East Bay Regional Park District and Tejon Ranch studies. Thanks to the East Bay Regional Park District and Tejon Ranch Conservancy for providing funding, field research sites, and excellent examples of grassland management. This material was supported by the NSF Postdoctoral Research Fellowship in Biology Award #1309014 to L. Larios.

References

Anderson, M. J., Ellingsen, K. E. & McArdle, B. H. (2006). Multivariate dispersion as a measure of beta diversity. *Ecology Letters*, *9*, 683–693.

Bailey, R. G. (1995). *Description of the Ecoregions of the United States*, 2nd ed. United States Department of Agriculture - Forest Service, Miscellaneous Publication 1391, Washington, D.C.

Barbour, M., Keeler-Wolf, T. & Schoenherr, A. A. (2007). *Terrestrial Vegetation of California*, 3rd ed. University of California Press, Berkeley.

Bard, E. C., Sheley, R. L., Jacobsen, J. S. & Borkowski, J. J. (2004). Using ecological theory to guide the implementation of augmentative restoration. *Weed Technology*, *18*, 1246–1249.

Bartolome, J.W. (1989). Local temporal and spatial structure. *Grassland Structure and Function: California Annual Grassland* Tasks for Vegetation Science 20. (eds L.F. Huenneke & H. A. Mooney), 73–80. Kluwer Academic Publishers, Dordrecht.

Bartolome, J. W., Barry, W. J., Griggs, T. & Hopkinson, P. (2007a). Valley Grassland. *Terrestrial Vegetation of California*, 3rd ed (eds M. Barbour, T. Keeler-Wolf & A.A. Schoenherr), 367–393. University of California Press, Berkeley.

Bartolome, J. W., Fehmi, J. S., Jackson, R. D. & Allen-Diaz, B. (2004). Response of a native perennial grass stand to disturbance in California's Coast Range Grassland. *Restoration Ecology*, *12*, 279–289.

Bartolome, J. W., Frost, W. E., McDougald, N. K. & Connor, J. M. (2002). *California Guidelines for Residual Dry Matter (RDM) Management on Coastal and Foothill Annual Rangelands*. University of California - Division of Agriculture and Natural Resources, Rangeland Management Series Publication 8092, Oakland.

Bartolome, J. W., Jackson, R. D. & Allen-Diaz, B. (2009). Developing data-driven descriptive models for Californian grasslands. *New Models for Ecosystem Dynamics and Restoration* (eds R.J. Hobbs & K.N. Suding), 124–138. Island Press, Washington, D.C.

Bartolome, J. W., Jackson, R. D., Betts, A. D. K., Connor, J. M., Nader, G. A. & Tate, K. W. (2007b). Effects of residual dry matter on net primary production and plant functional groups in Californian annual grasslands. *Grass and Forage Science*, *62*, 445–452.

Begon, M., Harper, J. L. & Townsend, C. R. (1996). *Ecology: Individuals, Populations, and Communities*, 3rd ed. Blackwell Science, Oxford.

Bestelmeyer, B., Brown, J., Fuhlendorf, S., Fults, G. & Wu, X. B. (2011a). A landscape approach to rangeland conservation practices. *Conservation Benefits of Rangeland Practices: Assessment, Recommendations, and Knowledge Gaps*. (ed D.D. Briske), 337–370. United States Department of Agriculture - Natural Resources Conservation Service.

Bestelmeyer, B. T., Goolsby, D. P. & Archer, S. R. (2011b). Spatial perspectives in state-and-transition models: a missing link to land management? *Journal of Applied Ecology*, *48*, 746–757.

Bestelmeyer, B. T., Tugel, A. J., Peacock, Jr. G. L., Robinett, D. G., Shaver, P. L., Brown, J. R., Herrick, J. E., Sanchez, H. & Havstad, K. M. (2009). State-and-transition models for heterogeneous landscapes: a strategy for development and application. *Rangeland Ecology & Management*, *62*, 1–15.

Booker, K., Huntsinger, L., Bartolome, J. W., Sayre, N. F. & Stewart, W. (2012). What can ecological science tell us about opportunities for carbon sequestration on arid rangelands in the United States? *Global Environmental Change*, *23*, 240–251.

Borcard, D., Gillet, F. & Legendre, P. (2011). *Numerical Ecology with R*. Springer Verlag, New York.

Brady, N. C. & Weil, R. R. (2002). *The Nature and Properties of Soils*, 13th ed. Pearson Education, Upper Saddle River, New Jersey.

Breiman, L., Friedman, J. H., Olshen, R. A. & Stone, C. J. (1984). *Classification and Regression Trees*. Wadsworth & Brooks, Belmont, California.

Briske, D. D., Fuhlendorf, S. D. & Smeins, F. E. (2003). Vegetation dynamics on rangelands: a critique of the current paradigms. *Journal of Applied Ecology*, *40*, 601–614.

Brown, J. R. (2010). Ecological sites: their history, status, and future. *Rangelands*, *32*, 5–8.

Burcham, L. T. (1957). *California Range Land: An Historico-Ecological Study of the Range Resource of California*. State of California - Deptartment of Natural Resources, Division of Forestry, Sacramento.

Burcham, L. T. (1975). Climate, structure, and history of Californa's annual grassland ecosystem. *The California Annual Grassland Ecosystem* (ed R.M. Love), 16–34. University of California, Davis - Institute of Ecology, Davis.

Bush, L. (2006). *Grazing Handbook: A Guide for Resource Managers in Coastal California*. Sotoyome Resource Conservation District, Santa Rosa.

Caudle, D., DiBenedetto, J., Karl, M., Sanchez, H. & Talbot, C. (2013). *Interagency Ecological Site Handbook for Rangelands*. United States Department of the Interior - Bureau of Land Management, United States Department of Agriculture - Forest Service, and United States Department of Agriculture - Natural Resources Conservation Service.

Chalcraft, D. R., Williams, J. W., Smith, M. D. & Willig, M. R. (2004). Scale dependence in the species-richness-productivity relationship: the role of species turnover. *Ecology*, *85*, 2701–2708.

Clarke, K. R. & Ainsworth, M. (1993). A method of linking multivariate community structure to environmental variables. *Marine Ecology Progress Series*, *92*, 205–219.

Cox, R. D. & Allen, E. B. (2008). Composition of soil seed banks in southern California coastal sage scrub and adjacent exotic grassland. *Plant Ecology*, *198*, 37–46.

D'Antonio, C., Bainbridge, S., Kennedy, C., Bartolome, J. & Reynolds, S. (2002). *Ecology and Restoration of California Grasslands with Special Emphasis on the Influence of Fire and Grazing on Native Grassland Species*. The Packard Foundation.

Dallman, P. R. (1998). *Plant Life in the World's Mediterranean Climates: California, Chile, South Africa, Australia, and the Mediterranean Basin*. University of California Press, Berkeley.

Davis, F. W., Stoms, D. M., Hollander, A. D., Thomas, K. A., Stine, P. A., Odion, D., Borchert, M. I., Thorne, J. H., Gray, M. V., Walker, R. E., Warner, K. & Graae, J. (1998). *The California Gap Analysis Project - Final Report*. University of California, Santa Barbara, Santa Barbara.

De'ath, G. & Fabricius, K. E. (2000). Classification and regression trees: a powerful yet simple technique for ecological data analysis. *Ecology*, *81*, 3178–3192.

Dufrene, M. & Legendre, P. (1997). Species assemblages and indicator species: the need for a flexible asymmetrical approach. *Ecological Monographs*, *67*, 345–366.

Duniway, M. C., Bestelmeyer, B. T. & Tugel, A. (2010). Soil processes and properties that distinguish ecological sites and states. *Rangelands*, *32*, 9–15.

Ellis, J. E. & Swift, D. M. (1988). Stability of African pastoral ecosystems: alternate paradigms and implications for development. *Journal of Range Management*, *41*, 450–459.

Elmendorf, S. & Harrison, S. (2009). Temporal variability and nestedness in California grassland species composition. *Ecology*, *90*, 1492–1497.

Evans, R. A., Kay, B. L. & Young, J. A. (1975). Microenvironment of a dynamic annual community in relation to range improvement. *Hilgardia*, *43*, 78–102.

Evans, R. A. & Young, J. A. (1989). Characterization and analysis of abiotic factors and their influences on vegetation. *Grassland Structure and Function: California Annual Grassland Tasks for Vegetation Science*, *20*, 13–28. Kluwer Academic Publishers, Dordrecht.

Evett, R. R. & Bartolome, J. W. (2013). Phytolith evidence for the extent and nature of prehistoric Californian grasslands. *The Holocene*, *23*, 1644–1649.

Eviner, V. T. & Firestone, M. K. (2007). Mechanisms determining patterns of nutrient dynamics. *California Grasslands: Ecology and Management* (eds M.R. Stromberg, J.D. Corbin & C.M. D'Antonio), 94–106. University of California Press, Berkeley.

Eviner, V. T. & Hawkes, C. V. (2008). Embracing variability in the application of plant-soil interactions to the restoration of communities and ecosystems. *Restoration Ecology*, *16*, 713–729.

Fernandez-Going, B. M., Anacker, B. L. & Harrison, S. P. (2012). Temporal variability in California grasslands: soil type and species functional traits mediate response to precipitation. *Ecology*, *93*, 2104–2114.

Fire and Resource Assessment Program (FRAP). (2010). *California's Forests and Rangelands: 2010 Assessment*. California Department of Forestry and Fire Protection - Fire and Resource Assessment Program, Sacramento.

Gea-Izquierdo, G., Gennet, S. & Bartomole, J. W. (2007). Assessing plant-nutrient relationships in highly invaded Californian grasslands using non-normal probability distributions. *Applied Vegetation Science*, *10*, 343–350.

Gennet, S. (2007). Environmental Determinants of Plant Community Composition and Songbird Abundance in a California Coast Range Grassland. PhD thesis, University of California, Berkeley, Berkeley.

Grigal, D. F., Bell, J. C., Ahrens, R. J., Boone, R. D., Kelly, E. F., Monger, H. C. & Sollins, P. (1999). Site and landscape characterization for ecological studies. *Standard Soil Methods for Long-Term Ecological Research* (eds G.P. Robertson, D.C. Coleman & C.S. Bledsoe), 29–51. Oxford University Press, Cary, North Carolina.

Hallett, L. M., Hsu, J. S., Cleland, E. E., Collins, S. L., Dickson, T. L., Farrer, E. C., Gherardi, L. A., Gross, K. L., Hobbs, R. J., Turnbull, L. & Suding, K. N. (in press). Biotic mechanisms of community stability shift along a precipitation gradient. *Ecology*.

Hamilton, J. G., Griffin, J. R. & Stromberg, M. R. (2002). Long-term population dynamics of native *Nassella* (Poaceae) bunchgrasses in Central California. *Madroño*, *49*, 274–284.

Hawkes, C. V., Wren, I. F., Herman, D. J. & Firestone, M. K. (2005). Plant invasion alters nitrogen cycling by modifying the soil nitrifying community. *Ecology Letters*, *8*, 976–985.

Heady, H. F. (1956). Changes in a California annual plant community induced by manipulation of natural mulch. *Ecology*, *37*, 798–812.

Heady, H. F. (1958). Vegetational changes in the California annual type. *Ecology*, *39*, 402–416.

Heady, H. F. (1977). Valley Grassland. *Terrestrial Vegetation of California* (eds M.G. Barbour & J. Major), 491–514. Wiley, New York.

Heady, H. F., Bartolome, J. W., Pitt, M. D., Stroud, M. G. & Savelle, G. D. (1991). California prairie. *Natural Grasslands* Ecosystems of the World 8A. (ed R.T. Coupland), 313–335. Elsevier Science Publishing Company, Amsterdam.

Herrick, J. E., Bestelmeyer, B. T., Archer, S., Tugel, A. J. & Brown, J. R. (2006). An integrated framework for science-based arid land management. *Journal of Arid Environments*, *65*, 319–335.

Herrick, J. E., Duniway, M. C., Pyke, D. A., Bestelmeyer, B. T., Wills, S. A., Brown, J. R., Karl, J. W. & Havstad, K. M. (2012). A holistic strategy for adaptive land management. *Journal of Soil and Water Conservation*, *67*, 105–113.

Hobbs, R. J. (2007). Setting effective and realistic restoration goals: key directions for research. *Restoration Ecology*, *15*, 354–357.

Hobbs, R. J. & Mooney, H. A. (1995). Spatial and temporal variability in California annual grassland - results from a long-term study. *Journal of Vegetation Science*, *6*, 43–56.

Holling, C. S. (1978). *Adaptive Environmental Assessment and Management*. John Wiley & Sons, London.

Holmes, T. H. & Rice, K. J. (1996). Patterns of growth and soil-water utilization in some exotic annuals and native perennial bunchgrasses of California. *Annals of Botany*, *78*, 233–243.

Holmgren, M. & Scheffer, M. (2001). El Niño as a window of opportunity for the restoration of degraded arid ecosystems. *Ecosystems*, *4*, 151–159.

Hopkinson, P., Hammond, M., Spiegal, S. & Bartolome, J. W. (2009). *Quantitative Assessment and Characterization of Selected State Park Grasslands*. California Department of Parks and Recreation - Grassland Assessment and Prioritization Project.

Huenneke, L. F. (1989). Distribution and regional patterns of Californian grasslands. *Grassland Structure and Function: California Annual Grassland* Tasks for Vegetation Science 20. (eds L.F. Huenneke & H.A. Mooney), 1–12. Kluwer Academic Publishers, Dordrecht.

Hunter, K. A. M. & Wu, L. (2005). Morphological and physiological response of five California native grass species to moderate salt spray: implications for landscape irrigation with reclaimed water. *Journal of Plant Nutrition*, *28*, 247–270.

Huntsinger, L., Bartolome, J. W. & D'Antonio, C. (2007). Grazing management on California's Mediterranean grasslands. *California Grasslands: Ecology and Management* (eds M.R. Stromberg, J.D. Corbin & C. D'Antonio), 233–253. University of California Press, Berkeley.

Jackson, L. E. (1985). Ecological origins of California's Mediterranean grasses. *Journal of Biogeography*, *12*, 349–361.

Jackson, R. D. & Bartolome, J. W. (2002). A state-transition approach to understanding nonequilibrium plant community dynamics in Californian grasslands. *Plant Ecology*, *162*, 49–65.

Jackson, R. D. & Bartolome, J. W. (2007). Grazing ecology of California grasslands. *California Grasslands: Ecology and Management* (eds M.R. Stromberg, J.D. Corbin & C. D'Antonio), 197–206. University of California Press, Berkeley.

Jackson, R. D., Bartolome, J. W. & Allen-Diaz, B. (2002). State and transition models: response to an ESA symposium. *Bulletin of the Ecological Society of America, 83*, 194–196.

Jantz, P. A., Preusser, B. F. L., Fujikawa, J. K., Kuhn, J. A., Bersbach, C. J., Gelbard, J. L. & Davis, F. W. (2007). Regulatory protection and conservation. *California Grasslands: Ecology and Management* (eds M.R. Stromberg, J.D. Corbin & C.M. D'Antonio), 297–318. University of California Press, Berkeley.

Jenny, H. (1941). *Factors of Soil Formation*. McGraw-Hill Book Company, New York.

Karl, J. W., Colson, K. & Swartz, H. (2011). Rangeland Assessment and Monitoring Methods Guide: an interactive tool for selecting methods for assessment and monitoring. *Rangelands, 33*, 48–54.

Karl, J. W. & Herrick, J. E. (2010). Monitoring and assessment based on ecological sites. *Rangelands, 32*, 60–64.

Kettenring, K. M. & Adams, C. R. (2011). Lessons learned from invasive plant control experiments: a systematic review and meta-analysis. *Journal of Applied Ecology, 48*, 970–979.

Kimball, S. & Schiffman, P. M. (2003). Differing effects of cattle grazing on native and alien plants. *Conservation Biology, 17*, 1681–1693.

Landres, P. B., Morgan, P. & Swanson, F. J. (1999). Overview of the use of natural variability concepts in managing ecological systems. *Ecological Applications, 9*, 1179–1188.

Lulow, M. E. (2008). Restoration of California native grasses and clovers: the roles of clipping, broadleaf herbicide, and native grass density. *Restoration Ecology, 16*, 584–593.

Lulow, M. E. & Young, T. P. (2011). Is there still native diversity in California's grasslands? *Fremontia, 39*, 6–11.

McBride, M. F., Wilson, K. A., Burger, J., Fang, Y.-C., Lulow, M., Olson, D., O'Connell, M. & Possingham, H. P. (2010). Mathematical problem definition for ecological restoration planning. *Ecological Modelling, 221*, 2243–2250.

McCune, B. & Grace, J. B. (2002). *Analysis of Ecological Communities*. MjM Software Design, Gleneden Beach, Oregon.

McCune, B. & Keon, D. (2002). Equations for potential annual direct incident radiation and heat load. *Journal of Vegetation Science, 13*, 603–606.

McDougald, N. K., Clawson, W. J., Bartolome, J. W. & Frost, W. E. (1991). *Estimating Livestock Grazing Capacity on California Annual Rangeland*. University of California, Davis Range Science Report 29.

Meinke, C. W., Knick, S. T. & Pyke, D. A. (2009). A spatial model to prioritize sagebrush landscapes in the Intermountain West (USA) for restoration. *Restoration Ecology, 17*, 652–659.

Milchunas, D. G., Lauenroth, W. K., Chapman, P. L. & Kazempour, M. K. (1989). Effects of grazing, topography, and precipitation on the structure of a semiarid grassland. *Vegetatio, 80*, 11–23.

Minnich, R. A. (2008). *California's Fading Wildflowers: Lost Legacy and Biological Invasions*. University of California Press, Berkeley.

Murray, C. & Marmorek, D. (2003). Adaptive management and ecological restoration. *Ecological Restoration of Southwestern Ponderosa Pine Forests* (ed P. Freiderici), 417–428. Island Press, Washington, D.C.

Natural Resources Conservation Service (NRCS). (2013). Ecological Site Information System (ESIS). United States Department of Agriculture. URL https://esis.sc.egov.usda.gov [accessed 21 December 2013]

Noss, R., Nielsen, S. & Vance-Borland, K. (2009). Prioritizing ecosystems, species, and sites for restoration. *Spatial Conservation Prioritization: Quantitative Methods and Computational Tools* Oxford biology. (eds A. Moilanen, K.A. Wilson & H.P. Possingham), 158–171. Oxford University Press, Oxford.

Nyberg, B. (1999). *An Introductory Guide to Adaptive Management for Project Leaders and Participants*. British Columbia Forest Service - Forest Practices Branch, Victoria, B.C.

Orrock, J. L., Witter, M. S. & Reichman, O. J. (2008). Apparent competition with an exotic plant reduces native plant establishment. *Ecology*, *89*, 1168–1174.

Orrock, J. L., Witter, M. S. & Reichman, O. J. (2009). Native consumers and seed limitation constrain the restoration of a native perennial grass in exotic habitats. *Restoration Ecology*, *17*, 148–157.

Pitt, M. D. & Heady, H. F. (1978). Responses of annual vegetation to temperature and rainfall patterns in northern California. *Ecology*, 336–350.

PRISM Climate Group. (2013). Gridded climate data. URL http://prism.oregonstate.edu [accessed 1 March 2013]

Reever Morghan, K. J., Sheley, R. L. & Svejcar, T. J. (2006). Successful adaptive management - the integration of research and management. *Rangeland Ecology & Management*, *59*, 216–219.

Reinhart, K. O. & Callaway, R. M. (2006). Soil biota and invasive plants. *New Phytologist*, *170*, 445–457.

Robertson, D. M. (2004). Relationships between Historic Land Use, Plant Species Composition and Environmental Factors in the Foothills South of Mt Diablo, California. Masters Thesis, University of California, Berkeley, Berkeley.

Schiffman, P. M. (2007). Species composition at the time of first European settlement. *California Grasslands: Ecology and Management* (eds M.R. Stromberg, J.D. Corbin & C.M. D'Antonio), 52–56. University of California Press, Berkeley.

Seabloom, E. W., Borer, E. T., Boucher, V. L., Burton, R. S., Cottingham, K. L., Goldwasser, L., Gram, W. K., Kendall, B. E. & Micheli, F. (2003a). Competition, seed limitation, disturbance, and reestablishment of California native annual forbs. *Ecological Applications*, *13*, 575–592.

Seabloom, E. W., Harpole, W. S., Reichman, O. J. & Tilman, D. (2003b). Invasion, competitive dominance, and resource use by exotic and native California grassland species. *Proceedings of the National Academy of Sciences*, *100*, 13384–13389.

Seastedt, T. R., Hobbs, R. J. & Suding, K. N. (2008). Management of novel ecosystems: are novel approaches required? *Frontiers in Ecology and the Environment*, *6*, 547–553.

Sheley, R., James, J., Smith, B. & Vasquez, E. (2010). Applying ecologically based invasive-plant management. *Rangeland Ecology & Management*, *63*, 605–613.

Soil Survey Staff. (1999). *Soil Taxonomy: A Basic System of Soil Classification for Making and Interpreting Soil Surveys*, 2nd ed. United States Department of Agriculture - Natural Resources Conservation Service, Washington, D.C.

Soil Survey Staff. (2010). *Keys to Soil Taxonomy*, 11th ed. United States Department of Agriculture - Natural Resources Conservation Service, Washington, D.C.

Soil Survey Staff. (2013). Web Soil Survey. United States Department of Agriculture - Natural Resources Conservation Service. URL http://websoilsurvey.nrcs.usda.gov [accessed 21 December 2013]

Steenwerth, K. L., Jackson, L. E., Calderon, F. J., Stromberg, M. R. & Scow, K. M. (2002). Soil microbial community composition and land use history in cultivated and grassland ecosystems of coastal California. *Soil Biology and Biochemistry*, *34*, 1599–1611.

Stromberg, M. R., D'Antonio, C. M., Young, T. P., Wirka, J. & Kephart, P. R. (2007). California grassland restoration. *California Grasslands: Ecology and Management* (eds M.R. Stromberg, J.D. Corbin & C. D'Antonio), 254–280. University of California Press, Berkeley.

Stromberg, M. R. & Griffin, J. R. (1996). Long-term patterns in coastal California grasslands in relation to cultivation, gophers, and grazing. *Ecological Applications*, *6*, 1189–1211.

Suding, K. N., Gross, K. L. & Houseman, G. R. (2004). Alternative states and positive feedbacks in restoration ecology. *Trends in Ecology & Evolution*, *19*, 46–53.

Suding, K. N. & Hobbs, R. J. (2009). Threshold models in restoration and conservation: a developing framework. *Trends in Ecology & Evolution*, *24*, 271–279.

Tejon Ranch Conservancy. (2013). *Ranch-Wide Management Plan, Volume 2: Conservation Activities and Best Management Practices*. Tejon Ranch, California.

United States Forest Service (USFS). (2010). Calveg existing vegetation tiles. United States Department of Agriculture - Forest Service, Pacific Southwest Region Remote Sensing Lab. Vector digital data. URL http://www.fs.fed.us/r5/rsl/projects/mapping [accessed 1 March 2013]

Verfaillie, E., Degraer, S., Schelfaut, K., Willems, W. & Van Lancker, V. (2009). A protocol for classifying ecologically relevant marine zones, a statistical approach. *Estuarine Coastal and Shelf Science*, *83*, 175–185.

Vetter, S. (2005). Rangelands at equilibrium and non-equilibrium: recent developments in the debate. *Journal of Arid Environments*, *62*, 321–341.

Vogelsang, K. M. & Bever, J. D. (2009). Mycorrhizal densities decline in association with nonnative plants and contribute to plant invasion. *Ecology*, *90*, 399–407.

Von Wehrden, H., Hanspach, J., Kaczensky, P., Fischer, J. & Wesche, K. (2012). Global assessment of the non-equilibrium concept in rangelands. *Ecological Applications*, *22*, 393–399.

Westoby, M., Walker, B. & Noy-Meir, I. (1989). Opportunistic management for rangelands not at equilibrium. *Journal of Range Management*, 266–274.

White, P. S. & Walker, J. L. (1997). Approximating nature's variation: selecting and using reference information in restoration ecology. *Restoration Ecology*, *5*, 338–349.

Whitlock, M. C. & Schluter, D. (2009). *The Analysis of Biological Data*. Roberts & Company Publishers, Greenwood Village, Colorado.

Whittaker, R. H. (1960). Vegetation of the Siskiyou Mountains, Oregon and California. *Ecological Monographs*, *30*, 279–338.

Wiens, J. A. (1984). On understanding a non-equilibrium world: myth and reality in community patterns and processes. *Ecological Communities: Conceptual Issues and the Evidence* (eds D.R. Strong, D. Simberloff, L.G. Abele & A.B. Thistle), 439–458. Princeton University Press, Princeton.

Wiens, J.A. (1989) Spatial scaling in ecology. *Functional Ecology*, 3, 385–397.

Wilson, K. A., Lulow, M., Burger, J., Fang, Y. C., Andersen, C., Olson, D., O'Connell, M. & McBride, M. F. (2011). Optimal restoration: accounting for space, time and uncertainty. *Journal of Applied Ecology*, *48*, 715–725.

Zuur, A. F., Ieno, E. N. & Smith, G. M. (2007). *Analysing Ecological Data*. Springer Verlag, New York.

In: Grassland Biodiversity and Conservation ...
Editors: Pierre Mariotte and Paul Kardol
ISBN: 978-1-61761-619-8

Chapter 5

BIODIVERSITY RESTORATION AND MANAGEMENT OF DEGRADED GRASSLAND IN SEMIARID AREAS: RESULTS FROM LONG-TERM EXPERIMENTS

Zhaobin Jing*[*1], Jimin Cheng[*1,2] and Xin Zhang[3]
[1]College of Animal Science and Technology, Northwest A&F University, China
[2]State Key Laboratory of Soil Erosion and Dryland Farming in the Loess Plateau, Institute of Soil and Water Conservation,
Chinese Academy of Science/Northwest A&F University, China
[3]Management Bureau of Yunwushan Grassland Natural Reserve, China

ABSTRACT

Grasslands are important terrestrial ecosystems, covering 20% of the terrestrial surface of the Earth. Globally, in arid and semiarid areas, they offer important benefits such as water and soil conservation and livestock forage. In arid and semiarid areas in China, however, grasslands have experienced rapid degradation, over the last few decades, resulting from climate change, overgrazing, and other anthropogenic impacts. The grassland degradation is defined as a decrease in plant species diversity, plant height, vegetation cover and plant productivity. Restoring degraded grassland ecosystems is critical to the ecological and economical sustainability of these systems. The Chinese Loess Plateau is a geological formation featuring extensive areas of arid and semiarid grasslands, but these systems are ecologically vulnerable. Several practices can be used to maintain grassland biodiversity and enhance productivity in this region. Using a 30-year data set representing grasslands typical of semiarid areas of China, we investigated the effects of these practices on grassland biodiversity and productivity. We also determined the effects of these activities as they interact with the climate. First, we showed that grazing exclusion is an effective method to restore degraded grasslands. Plant diversity and productivity reached peak values during the mid stages of restoration; however, long-term grazing exclusion may have a negative influence on species renewal and community stability. Second, we showed that the intensity of grazing and mowing indicated that light

[*] Corresponding author: Zhaobin Jing, email: zxxjzb520@gmail.com.
[*] Corresponding author: Jimin Cheng, email: gyzcjm@ms.iswc.ac.cn.

grazing (2 sheep ha^{-1}) and mowing once every two years would maintain the highest diversity and productivity. The impacts of grazing and mowing management on plant diversity followed the 'intermediate disturbance hypothesis'. Burning also influenced plant diversity, but is not a commonly used management strategy in maintaining grassland biodiversity in China. Third, climate change, grassland use and management practices all have an important influence on the genetic structure of dominant populations. Long-term grazing exclusion decreased the genetic diversity of *Stipa bungeana* populations, with grazing and mowing enhancing the genetic diversity. Given the important role of grasslands in semiarid areas, we should use the most effective and appropriate methods - grazing and mowing - to maintain plant species and genetic diversity under current and future climatic conditions.

Keywords: Biodiversity restoration, grassland management, grazing exclusion, grazing, mowing, fire, semiarid areas

INTRODUCTION

Semiarid and arid ecosystems, also known as drylands, cover about 41% of the earth's terrestrial surface. Over 38% of the global population depends on drylands for ecosystem services (Qiu et al., 2013; Maestre et al., 2012). Grasslands play a critical role in dryland region by maintaining biodiversity and productivity through the provision of important ecosystem services, such as water retention and soil conservation (Cheng et al., 2012; Zavaleta et al., 2010). However, due to global climate change, human disturbance and desertification, grasslands have been degraded. The grassland degradation is defined as a decrease in plant species diversity, plant height and plant productivity (Han et al., 2008) and mainly as a reduction in vegetation cover and soil erosion (Sivakumar et al., 2007; Slimani et al., 2010). This degradation influences ecological functions and socio-economic issues worldwide (Zhang et al., 2011). In addition, reduction of vegetation cover and soil erosion impact species diversity, above- and below-ground productivity, soil organic carbon, and soil total N and P (Wu et al., 2010; Wang et al., 2009). Biodiversity decline, in turn, causes direct consequences for grassland ecosystem functioning and services (Hooper et al., 2005). Specifically, decline of biodiversity alters ecosystem processes and decreases the resilience of ecosystems to environmental change (Chapin et al., 2000). Therefore, it is important to restore biodiversity of degraded grasslands in semiarid and arid ecosystems.

Grassland management can help maintain biodiversity in these ecosystems. Here, we distinguish six main aspects of grassland management and conservation, five practices as well as one grassland characteristic (hereafter 1 to 6). By reviewing past studies, we show these practices have a wide variety of effects on grassland ecosystems, depending on many factors, including study location and scale. 1) *Passive restoration.* Passive restoration, defined as grazing exclusion, is considered as an effective restoration approach for degraded grasslands (Jing et al., 2013; Golodets et al., 2010; Wu et al., 2010; Spooner and Briggs 2008). Grazing exclusion can change the vegetation structure and composition; could improve the vegetation coverage, biomass, species diversity and richness; and can enhance soil properties of grasslands (Jing et al., 2013; Qiu et al., 2013; Mayer et al., 2009). 2) *Grazing.* Previous studies have shown that grazing has the potential to: (a) increase or decrease plant coverage, species richness, biodiversity and productivity (Georgia et al., 2004; Gillen and Sims 2004;

Zhao et al., 2004; Tsutsumi et al., 2003); (b) change the species composition of grasslands (Cooper et al., 2005; Thomas et al., 2003); (c) promote soil C and N sequestration in grasslands (Han et al., 2008; He et al., 2008; Steffens et al., 2008; Su et al., 2005); and (d) affect the spatial distribution of vegetation and soil fertility (Lin et al., 2010). 3) *Mowing.* Mowing is a widely practiced human disturbance and pasture management practice in grassland ecosystems. Mowing has a significant impact on plant species composition and diversity (Reisch et al., 2009). For example, Kolos and Banaszuk (2013) reported that the annual practice of mowing during June and July significantly affected meadow vegetation, leading to a substantial increase in species diversity and richness. Mowing can cause substantial changes in grassland canopy structure, species composition and functional groups (Poptcheva et al., 2009); lead to changes in the variability of environmental resources in a grassland ecosystem (Chen et al., 2004); and indirectly increase the soil temperature (Klein et al., 2005; Wan et al., 2002). 4) *Fire.* Burning management plays a key role in the ecology of many grassland systems. This practice has been shown to: (a) control undesirable plants and pathogens (Holdo et al., 2009; Higgins et al., 2007; DiTomaso et al., 2006; Bond 2005; Stolle et al., 2003). (b) increase the production of forage grasses (Masunga et al., 2013; Collins and Calabrese 2012); (c) increase nitrogen, calcium, magnesium, sodium, potassium, and phosphorus levels in plants and in the soil (Holdo 2013; Lavoie et al., 2010; Brockway et al., 2002; Wan et al., 2001; Moe and Wegge 1997); and (c) increase soil microbial biomass C and N (Liu et al., 2007). 5) *Fertilization.* Nutrient enrichment can also impact the species richness and community structure of grassland ecosystems. For example, several studies have shown that fertilization or nitrogen deposition can cause a decline in species richness (e.g., Crawley et al., 2005; Gough et al., 2000), which can recover, at least partially, after an initial diversity decline (Pierik et al., 2011). 6) *Genetic diversity.* Plant genetic diversity is a prerequisite for species conservation and an important characteristic of any species or population for environmental adaptation (Lande 1988; Hummer and Sugar 1998; Jing et al., 2013). Because it has been shown that overgrazing and mowing can affect the genetic diversity among populations (Jing et al., 2013; Wu et al., 2010; Matlaga and Kleijn and Steinger 2002), it is important to analyze the genetic diversity of grassland populations under different management practices.

Stipa species dominate large parts of the Eurasian zonal vegetation, with *Stipa* steppe considered to be an important model system in arid and semiarid ecosystems (Hassan et al., 2009). *Stipa bungeana* is widely distributed in the typical steppe of semiarid areas and plays an important role in soil and water maintenance in China (Jing et al., 2013; Cheng et al., 2012). Previous studies have indicated that disturbance using grazing, mowing and fertilization can affect the genetic structure of a population (Jing et al., 2013; Reisch et al., 2009; Kolliker et al., 1998). Little is known, however, about the impact of disturbance on genetic variation in *Stipa* steppe. Here, we compared the genetic structure of *S. bungeana* under different disturbance regimes in semiarid grassland. With this chapter, we have two goals.: (i) evaluate changes in biodiversity and productivity under different grassland management practices including grazing exclusion, grazing, mowing, fertilization and fire; and (ii) analyze the genetic diversity and structure in dominant populations under those management practices. The goal is to provide a solid basis for biodiversity restoration and management of degraded grassland in semiarid areas of the Loess Plateau in China.

METHODS

Study Area

The study area was located in typical steppe grasslands of the Loess Plateau at Yunwu Mountain (106°21′–106°27′E, 36°10′–36°17′N) in Ningxia Hui Autonomous Region, China. Yunwu Mountain was given protection status as a long-term monitoring site in 1982. Elevation ranges from 1800 to 2100 m, and the area covers a total area of 6660 hectares. The grassland protection zone includes a core conservation area (1000 hectares), a buffer conservation area (1300 hectares) and an experimental area (4360 hectares; Figure 1). Various management strategies are used in the grassland protection zone. The climate of the study area is semi-arid with mean July temperature of 22–25 °C and mean annual temperature of 7.0 °C. The mean annual rainfall is 425 mm, with approximately 60%–75% of the annual rainfall occurring in July to September. The annual evaporation rate is 1017–1739 mm, with the frost-free season averaging 137 days (from 1982-2013). The vegetation of the grassland protection zone consists of a total of 297 plant species. *Stipa*, *Artemisia* and *Potentilla* are the predominant plant genera, with the main dominant species including *Stipa bungeana*, *Stipa grandis*, *Thymus mongolicus*, *Artemisia sacrorum*, *Potentilla acaulis*, and *Androsace erecta*.

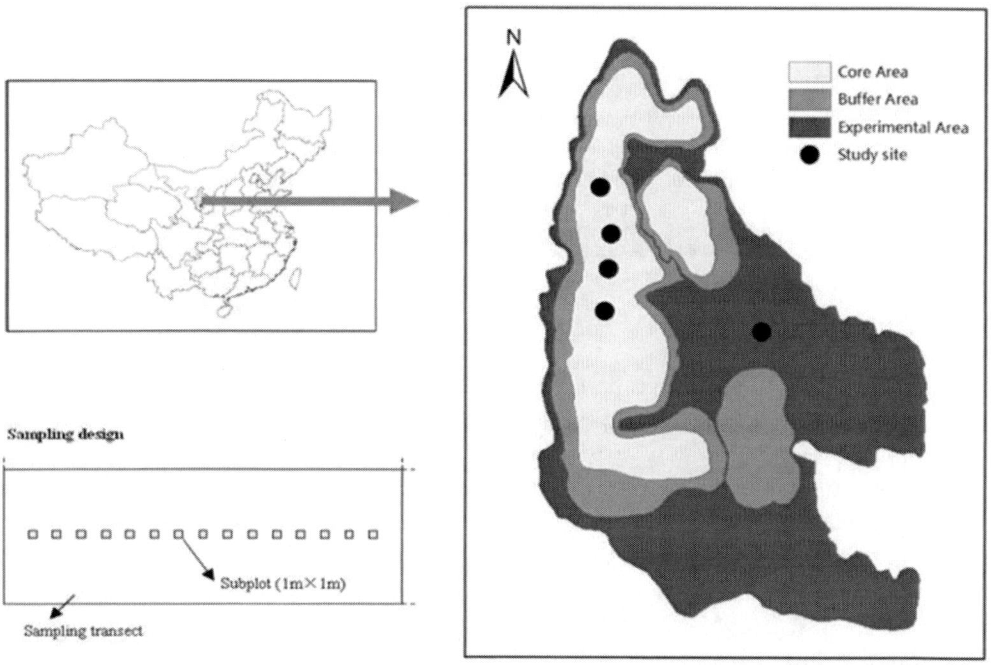

Figure 1. Location of the study area and a schematic overview of the sampling design (example only). The upper left panel shows China. The right panel shows the grassland protection zone at the Loess Plateau in Ningxia Hui Autonomous Region.

Experimental Design and Sampling

The experimental sites were located in the largest grassland areas in the Loess Plateau. This grassland has been fenced since 1982, as a permanent monitoring site for grassland restoration practices for the Institute of Soil and Water Conservation, Chinese Academy of Sciences and the Ministry of Water Resources. Before being fenced, the grassland was significantly degraded with reduction of biodiversity (Qiu et al., 2013). The long-term study has consisted of six main aspects of grassland management and conservation: (1) natural restoration; (2) grazing management; (3) mowing management; (4) fire management; (5) fertilization management; (6) genetic diversity analysis of a dominant plant population. The grazing exclusion (A and B), grazing (C), mowing (D), and fire (E and F) areas are shown in Figure 2. Descriptions of the experiments pertaining to aspects 1-6 are as follows:

(1) *Passive restoration by grazing exclusion.* This experiment was divided into two parts: grazing exclusion grassland (GEG) and grazing grassland (GG; control). In 1982, the experimental sites for grazing exclusion were established in the core conservation area. There are three grassland types in the grazing exclusion area, each of which is dominanted by *S. bungeana*, *T. mongolicus*, or *A. sacrorum*. These types of grasslands were located in similar natural grassland landscapes (eg. soils, geology) and the sites were no further than 1 km away from each other. A grazed site was selected as a control in an adjacent area outside the grassland protection zone. This control treatment used a grazing density of 8 sheep ha^{-1} during the year. The dominant plant species at the control site was the perennial bunchgrass *S. bungeana*.

At each site, three equal-sized replicate blocks (300 m × 100 m) were established, with fifteen permanently established sampling quadrats (1 m × 1 m) per block. At the control site, ten random quadrats (1 m × 1 m) were established in a 200 m × 100 m block. Vegetation surveys were carried out annually (1982-2011) in mid-August when standing plant biomass was at its peak. Plant coverage was measured using the projection method, with height measured using a ruler. Species richness and plant density were recorded. The aboveground biomass of each plant species (green parts only) was measured in each sampling quadrat. All aboveground samples were dried at 80°C until constant weight.

(2) *Grazing management.* Grazing experiments were conducted in the core conservation area from 1985 to 2012. The grazing management experiment included four treatments: heavy grazing (HG, 8 sheep ha^{-1}), moderate grazing (MG, 4 sheep ha^{-1}), light grazing (LG, 2 sheep ha^{-1}) and no grazing (NG). The total area per treatment was 1 ha, divided into three equal-sized replicate plots (about 0.33 ha) per treatment. Plots were adjacent and contiguous to each other. The grazing periods per year were divided into three stages: June 15 to 30; August 15 to 30; October 15 to 30. Sheep grazed the plots during the day and were removed at night. During the experimental period, ten random quadrats (1 m × 1 m) were established per plot. In each quadrat, the number of plant species and plant height and coverage were recorded each year (1985-2012) between September 1 and 15.

(3) *Mowing management.* Mowing experiments were carried out in the core conservation area at three grassland sites where grazing was excluded. The mowing experiment ran from 1982 to 2012. The mowing experiment consisted of three treatments: mowing once every two years (September), mowing once a year (September), and mowing twice a year (July and September). Each treatment area was 0.5 ha, divided into two equal-sized replicate plots (0.25 ha). In each 0.25 ha plot fifteen 1 m × 1 m quadrats were randomly allocated. In each quadrat,

the number of plant species, height, and coverage were recorded annually in mid-August from 1982-2012.

(4) *Fire management.* The fire management experiment was carried out in the core conservation area at different sites where grazing was excluded. Prior to burning, these sites were fenced, and the burning was caused accidentally by human activities. The fire experiment consisted of four treatments: burned in 2000, burned in 2009, burned in 2011, and no burn control. The burned areas for each treatment varied because the fire was difficult to control. Fifteen quadrats (1 m×1 m) were randomly allocated per fire treatment. In each quadrat, the number of plant species, height, and coverage were recorded in mid-August in 2012.

(5) *Fertilization.* The fertilization experiment was established in the core conservation area in 2010. Prior to fertilization the area for the fertilizer experiment was fenced. The fertilizers were added once at the start of experiment in 2010. The fertilizer experiment consisted of six treatments in a chronosequence of 5, 9, 15, 22, 25 and 30 years of grazing exclusion. Each treatment area was 0.012 ha, divided into three equal-sized replicate plots (0.004 ha). The fertilizer rate was 600 kg/ha. The fertilizer was Nitrogen-Phosphorus-Potassium mixed fertilizer, with 20.2% N, 9.5% P_2O_5, and 10.5% K_2O.

(6) *Genetic diversity.* The genetic diversity of the dominant plant species was estimated under different management regimes. In May 2013, five sites were selected based on the different management treatments as described above (grazing exclusion, 2 sheep ha^{-1} grazing, mowing once every two years, fire, fertilization). Twenty plants were randomly sampled per management type, with the minimum distance between individual plants higher than 20 m. Fresh leaves were collected from individual *of S. bungeana* plants, immediately stored in zip-lock bags with silica gel, returned to the laboratory and stored at -80°C for DNA extraction. Genomic DNA from silica gel-dried leaves was extracted using a modification of the protocol of standard CTAB methods (Zhao et al., 2006). DNA quality was analyzed by electrophoresis on 0.8% (w/v) agarose gel. DNA concentration was measured using a UV-VIS spectrophotometer and adjusted to 40 ng μL^{-1}, then stored at -20°C for Sequence-Related Amplified Polymorphism (SRAP) analysis. SRAP markers were used to detect the genetic diversity among five *S. bungeana* populations for each of the five management types, according to previously established protocols (Li and Quiros 2001). The primer sequences were synthesized from Beijing Aoke Biological Technology and Service Co. Ltd. Ten primer combinations were selected according to Jing et al. (2013). PCR amplification reactions were carried out in 20 μL volume, containing 40 ng of template DNA, 10 × PCR buffer (100Mm Tris-HC, pH 8.3; 500 mM KCl), 0.18 mM of each dNTP, 0.75 mM of each primer, 1.87 mM of $MgCl_2$, 1 unit of Taq DNA polymerase (TaKaRa Biotechnology Dalian Co., Ltd., China). PCR amplification was performed as follows: initial 5 min at 94 °C, followed by 5 cycles of 1 min at 94 °C, 1 min at 36 °C, and 1min at 72 °C. In the following 30 cycles, the annealing temperature was increased to 51 °C and a final 7 min extension at 72 °C. PCR products were separated on 6% denatured polyacrylamide gels and detected by silver staining. Then clear, reproducible, amplified DNA polymorphic bands between 200 bp and 2000 bp were scored as 1 for presence or 0 for absence to form a binary matrix for further analysis. DL2000 DNA ladder (TaKaRa Biotechnology Dalian Co., Ltd., China) was used as DNA markers (Jing et al., 2013).

Figure 2. Steppe grasslands in the Loess Plateau (Ningxia Hui Autonomous Region, China) under different management regimes. A) Grazing exclusion; B) Permanent fixed monitoring quadrat; C) Grazing; D) Mowing; E) Grassland shortly after burning; F) Grassland four months after burning.

Data Analysis

(1) For the experiments testing different management treatments (grazing exclusion, grazing, mowing, fertilization, fire), species richness (S) was used to describe species diversity. The summed dominance ratio (SDR) was used to analyze species dominance in the

burned grassland treatments. The SDR for a given species is defined as the average of four quantities: relative coverage, relative plant height, relative plant density and relative aboveground biomass of each species to the respective maximum values (Yamamoto et al., 1995).

All data are presented as mean ± standard error. We used regression analyses and one-way ANOVA for comparative analyses among different treatments, and linear regression to examine the correlations between rainfall and species diversity and productivity. Significant differences for all statistical tests were estimated at a significance level of $p < 0.05$. All statistical analyses were performed using SPSS 13.0 (SPSS Inc., Chicago, IL, USA).

(2) For the genetic diversity experiment, the data analyses followed methods as described in Jing et al. (2013). POPGENE version 1.32 (Yeh et al., 2000) was used to compute the number of effective loci, the percentage of polymorphic loci, observed number of alleles (*na*), effective number of alleles (*ne*), Nei's genetic diversity (*h*) (Nei 1973), Shannon's information index (*H*), and Nei's coefficient of genetic differentiation (*Gst*). An analysis of molecular variance (AMOVA) was carried out using AMOVA, version 1.5 (Excoffier et al., 1992). Genetic differentiation among the populations of *S. bungeana* was calculated using AMOVA.

Results and Discussion

Impacts of Grazing Exclusion on Grassland Diversity and Productivity

Plant species richness was significantly higher in grassland where grazing was excluded than in grazed grassland ($P<0.01$; Figure 3A). When grazers were excluded, species richness increased during the first years of restoration, but decreased thereafter, peaking in the 12th year after sheep removal (Figure 3A).

In the grazing exclusion grassland, the aboveground biomass showed a similar change over the 30-year period, with biomass increasing during the early stages, followed by a slight decrease (Figure 3B). Aboveground biomass was significantly higher in the grazing exclusion grassland compared with the grazed grassland ($P < 0.01$; Figure 3B). In the grazing exclusion grassland there was a significant positive correlation between aboveground biomass and species richness ($R = 0.859$, $P < 0.0001$; Figure 4A). For grazed grasslands, the correlation between aboveground biomass and species richness was only marginally significant ($R = 0.402$, $P = 0.021$; Figure 4B).

Figure 5 shows the relationship between rainfall and species diversity over the 30-year period of grassland restoration. Despite previous suggestions that water availability is a limiting factor for vegetation development in the Loess Plateau (Zhang et al., 2006), we did not find any relation between annual rainfall and species richness. Hence, annual rainfall may not be an important factor for plant diversity in the steppe grasslands of our study; instead, soil moisture may be a more important influence. This conjecture is corroborated by Jing et al. (2013), who found that in the same study areas, the coverage, species richness, Shannon-Wiener index, above- and below-ground biomass all significantly increased with increasing annual mean temperature, growing season temperature and soil moisture in each soil layer, but not with rainfall.

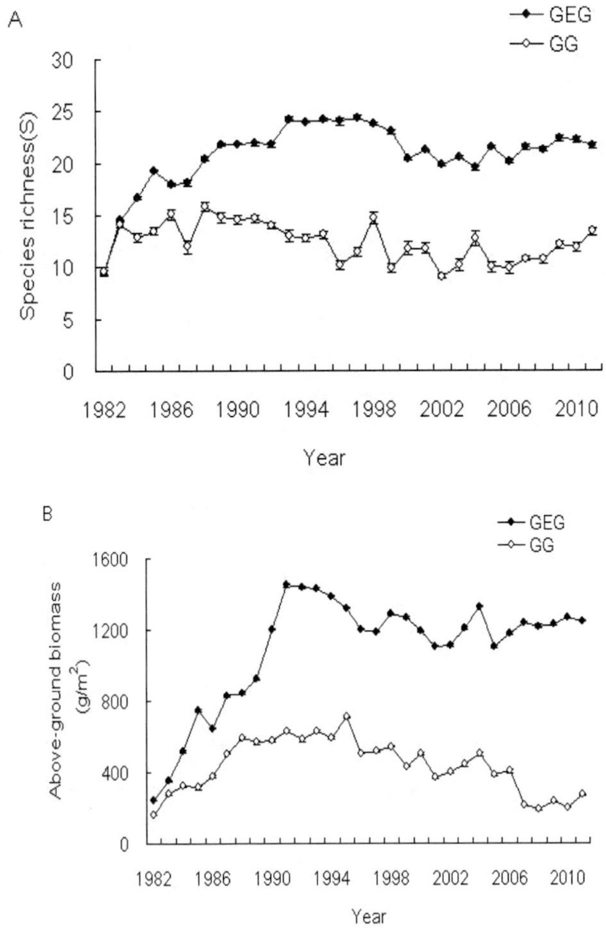

Figure 3. Changes in species richness (A) and aboveground biomass (B) in grazing exclusion grasslands (GEG) and grazed grasslands (GG). The values given are mean ± SE. For grazing exclusion, the mean was calculated from 15 quadrats in each of three grassland types (*S. bungeana*, *T. mongolicus*, *A. sacrorum*.) (N = 45). For grazed grassland, the mean was calculated from 10 quadrats in one grassland type (*S. bungeana*) (N = 10).

The restoration of degraded grassland ecosystems is a complex and long-term ecological process (Zhang et al., 2010). Many studies indicate that the restoration process could be evaluated using vegetation traits and species diversity (Wang et al., 2012; Wilkins et al., 2003). We selected species diversity and aboveground biomass to evaluate the effects of long-term grazing exclusion. Our results showed that when grazing was excluded, the grassland had higher species diversity and aboveground biomass compared to grazed grassland (Figure 3). Our results support the hypothesis that an undisturbed community supports relatively higher species richness, diversity and community productivity compared to a degraded grassland community (e.g. bare soil and soil erosion; Guo 2005). In China, research has shown that vegetation traits, such as species diversity and productivity could be improved using excluding grazers in degraded sandy grasslands, alpine swamp meadows and wetlands (Wu et al., 2010; Yuan et al., 2012; Wang et al., 2012). However, many of these restoration studies are based on relatively short-term experiments (Wu et al., 2010; Wu et al., 2009), and often use space-for-time substitution as an alternative for long-term observations (Wang et al., 2009; Zhang et al., 2011). At the community level, when grazing was excluded, the aboveground biomass reached the highest value in the 10th year and species diversity reached the highest value during the mid- and late stages of restoration. These results support the hypothesis that species diversity and productivity are highest in the mid- and transitional stages of succession in arid and semiarid grasslands (Guo 2005). Similar patterns were found by Wang et al., (2009), who showed that the species diversity, plant density and community coverage increased rapidly with successional time. After excluding grazing for 10–15 years, the species diversity and aboveground biomass decreased probably because of litter accumulation, light and water resource competition. Previous studies have shown a negative relationship between litter accumulation and vegetation structure, species diversity, richness and evenness (Cheng et al., 2012; Liu et al., 2010; Lamb 2008; Guo 2005).

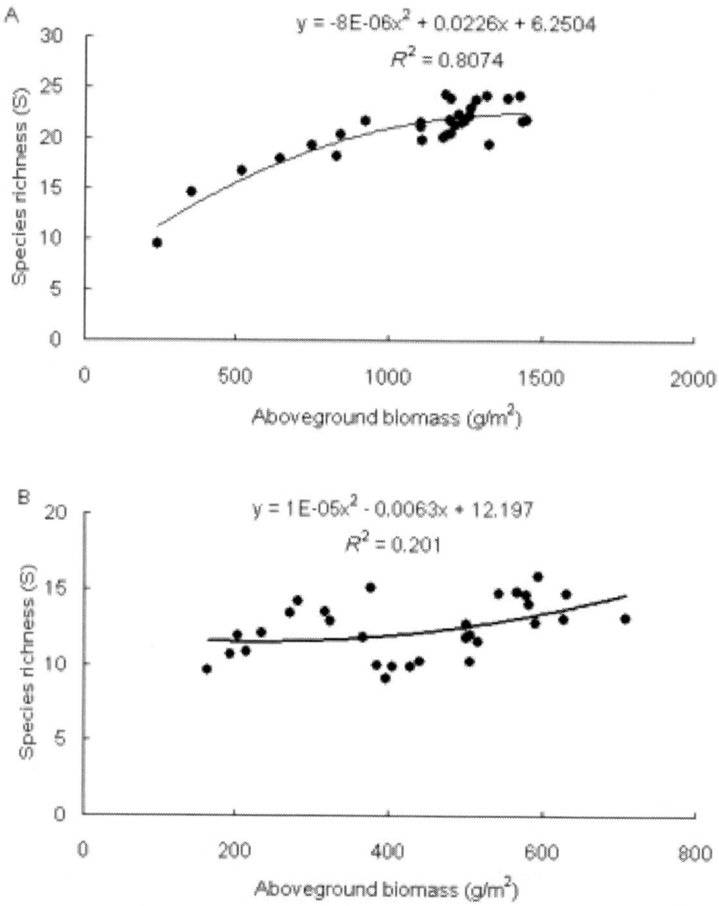

Figure 4. The relationship between species richness and aboveground biomass in A) grazing exclusion grassland and B) grazed grassland. For grazing exclusion grassland, the mean was calculated from 15 quadrats in each of three grassland types (*S. bungeana*, *T. mongolicus*, *A. sacrorum*.). For grazed grassland, the mean was calculated from 10 quadrats in one grassland type (*S. bungeana*).

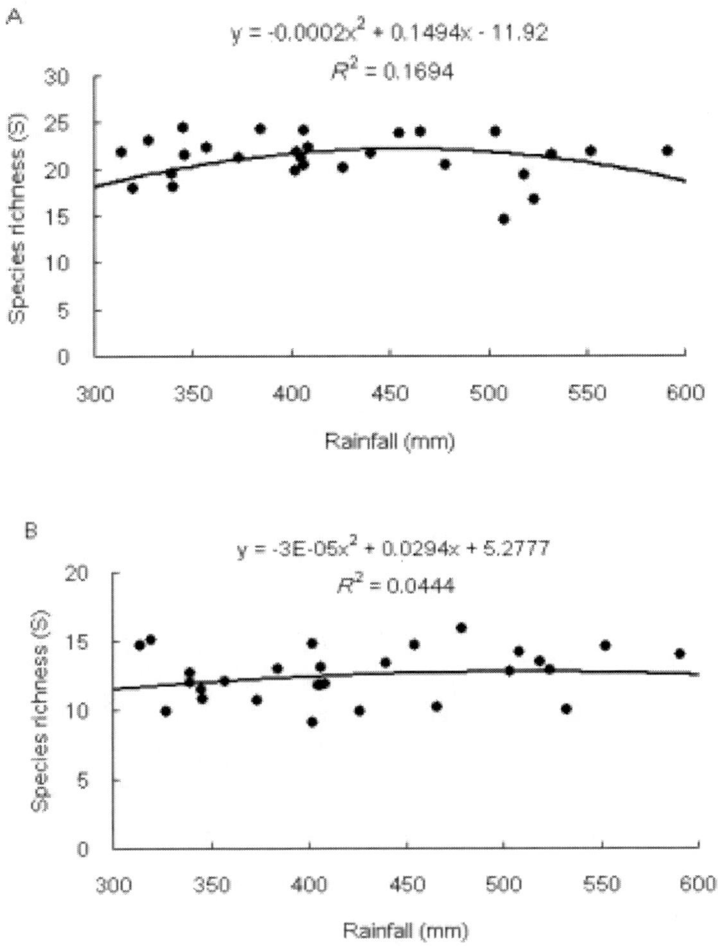

Figure 5. The relationship between rainfall and species richness in A) grazing exclusion grassland and B) grazed grassland. For grazing exclusion grassland, the mean was calculated from 15 quadrats in each of three grassland types (*S. bungeana*, *T. mongolicus*, *A. sacrorum*) (N = 45). For grazed grassland, the mean was calculated from 10 quadrats in one grassland type (*S. bungeana*) (N = 10).

Some studies have shown that climatic factors (e.g., rainfall, temperature) are major drivers of vegetation structure and function of semiarid grasslands (Wu et al., 2009; Akiyama et al., 2007), and many studies have shown positive relationships between rainfall and plant diversity and aboveground biomass (e.g., Malo et a. 1990; Chen and Li 2004). In our study area, the annual mean temperature tended to increase over the 30-year period of our study, whereas the annual rainfall slightly decreased (Figure 6). This points towards a warmer and drier local climate, which may potentially have influenced low species diversity and aboveground biomass in the later stages of community succession (Bai et al., 2004). In our study, rainfall was not the main limiting factor; light and temperature may instead be more important for plant growth (see also Jing et al., 2013).

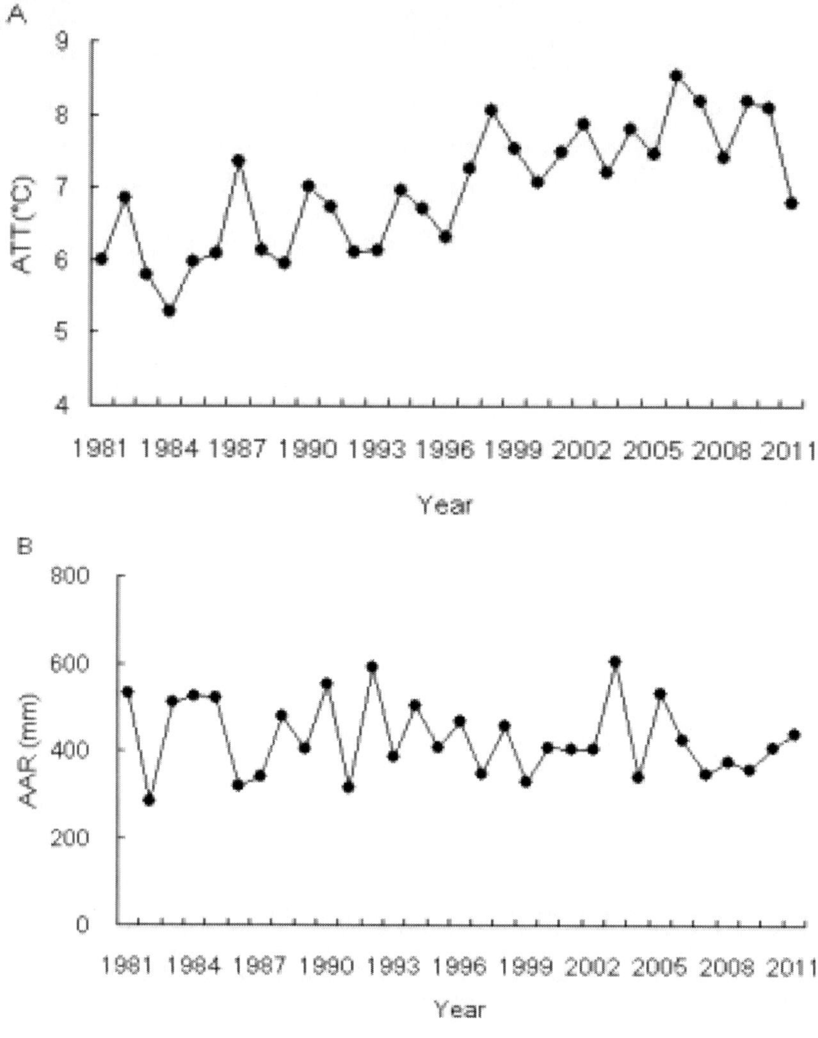

Figure 6. Changes in A) annual average rainfall (AAT) and B) annual average temperature (AAR) from 1982 to 2011 for our study site at the Loess Plateau in Ningxia Hui Autonomous Region, China.

Impacts of Grazing Management on Grassland Diversity and Productivity

Species diversity showed different temporal trends depending on the grazing treatments (Figure 7A). Compared with HG and MD, species diversity increased significantly in NG and LG treatments ($P < 0.05$). During early and later stages of succession, species diversity was higher in LG. Over the period of study, species diversity showed an increasing tendency with grazing time (Figure 7A). These results indicated that light disturbance was beneficial for retaining plant diversity in grasslands. This result may provide evidence for the moderate disturbance hypothesis (Connell 1978), i.e., the highest species richness usually occurs where disturbance is intermediate; disturbance often creates structural and functional heterogeneity in time and space, and promotes the coexistence of species. Positive effects from low-intensity grazing on plant species diversity have been demonstrated before (e.g., Dumont et al., 2009; Hickman et al., 2004; Sarmiento et al., 2004). For example, Hobbs and Huenneke (1992) found that species richness and diversity decreased under high-intensity grazing and no grazing compared to low-intensity grazing. Dumont et al., (2009) concluded that low and intermediate stocking rates were suitable for providing a high diversity of the plants.

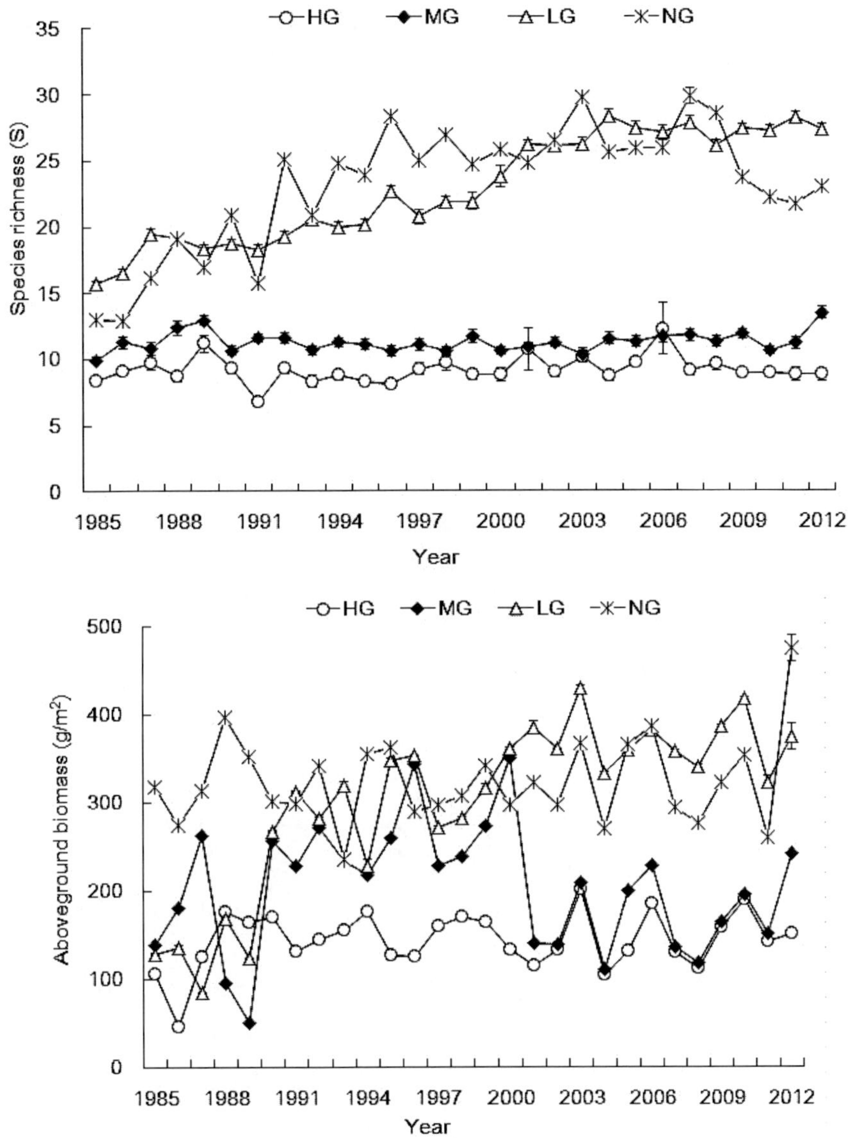

Figure 7. Changes in species richness (A) and aboveground biomass (B) under different grazing intensities: HG (heavy grazing, 8 sheep ha^{-1}), MG (moderate grazing, 4 sheep ha^{-1}), LG (light grazing, 2 sheep ha^{-1}), and NG (no grazing). The values are given as mean ± SE; the mean was calculated from 10 quadrats in each of three plots.

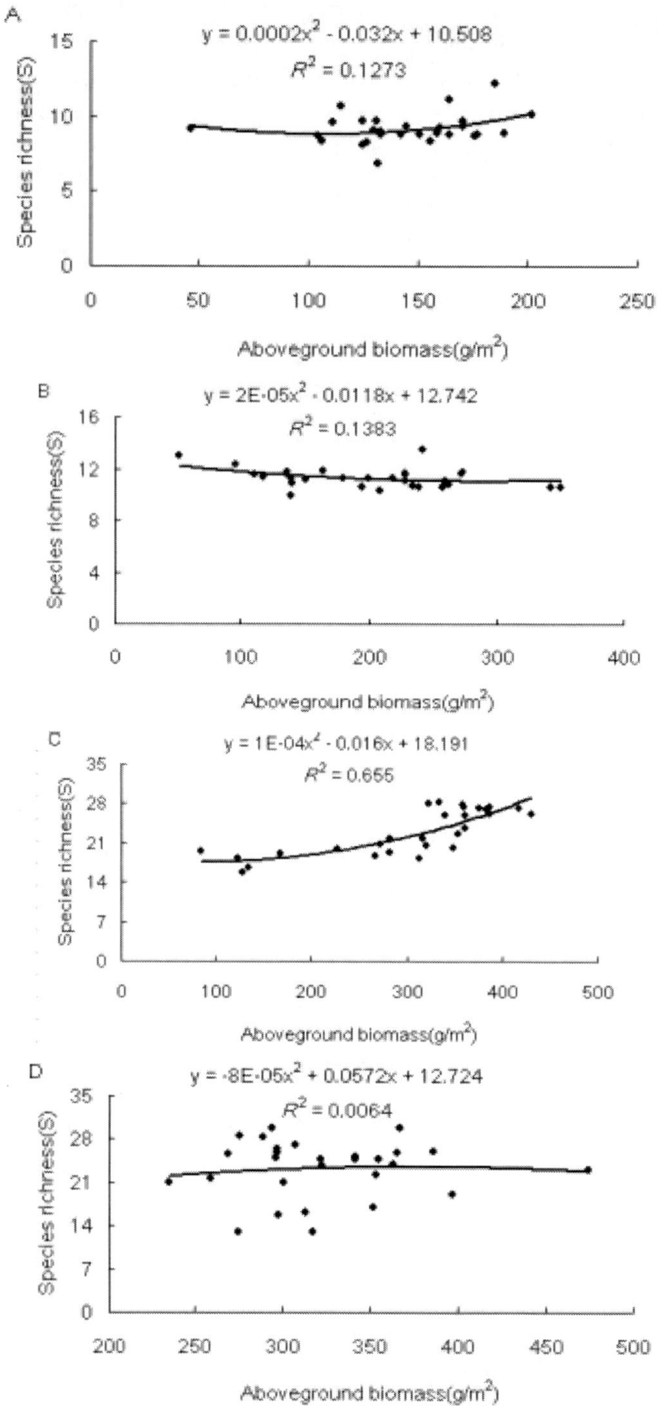

Figure 8. The relationship between species richness and aboveground biomass under different grazing intensities: A (heavy grazing, 8 sheep ha^{-1}); B (moderate grazing, 4 sheep ha^{-1}); C (light grazing, 2 sheep ha^{-1}); D (no grazing).

The aboveground biomass of the plant community showed differences depending on the grazing treatments (Figure 7B). The aboveground biomass was significantly lower in the HG treatment compared with the other treatments ($P < 0.05$). During later stages of succession, compared with the aboveground biomass in the HG, MG and NG treatments, LG had high aboveground biomass. Aboveground biomass decreased with increasing grazing intensity. Our results indicate that light grazing could increase grassland productivity. Light grazing could maintain the typical vegetation composition. In contrast, heavy grazing changes the vegetation composition and can lead to the invasion of undesirable plant species (e.g., Metera et al., 2010). Our results support findings by He et al. (2011) and Zhou et al. (2006) that light grazing stimulates aboveground biomass, and imply that low-intensity grazing can maintain the sustainable use and species diversity of semiarid grasslands. According to the diversity-productivity hypothesis, higher species diversity leads to greater productivity (Tilman et al., 1996). However, there were no significant correlation between species richness and aboveground biomass in all grazing treatments (Figure 8 A-D). There were also no significant correlations between species diversity and annual rainfall (P>0.05).

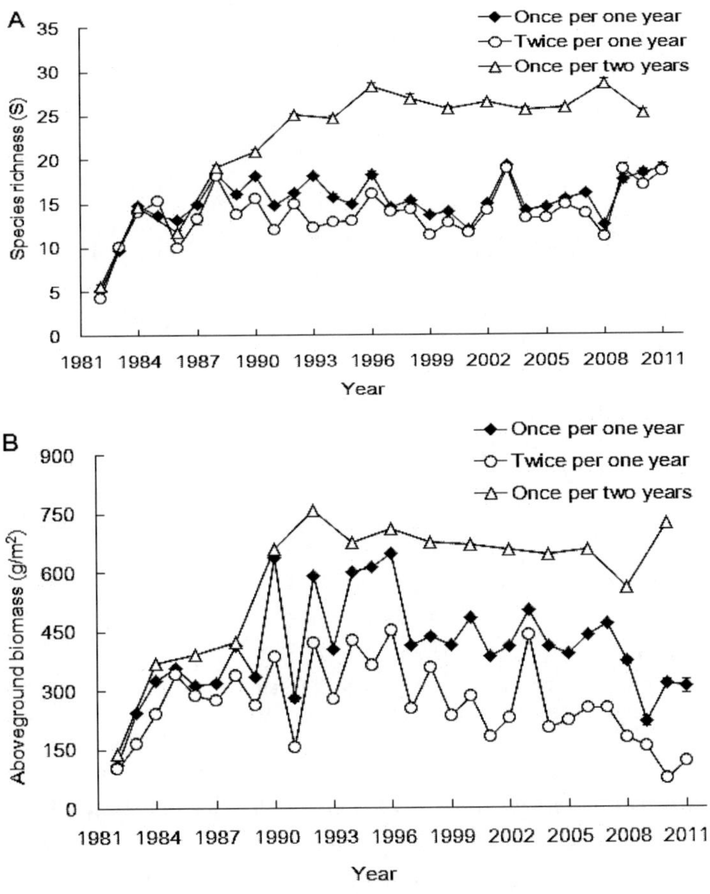

Figure 9. Changes in species richness (A) and aboveground biomass (B) for three mowing treatments: once a year, twice a year, and once every two years. The values are given as mean ± SE; the mean was calculated from 15 quadrats in each of two plots.

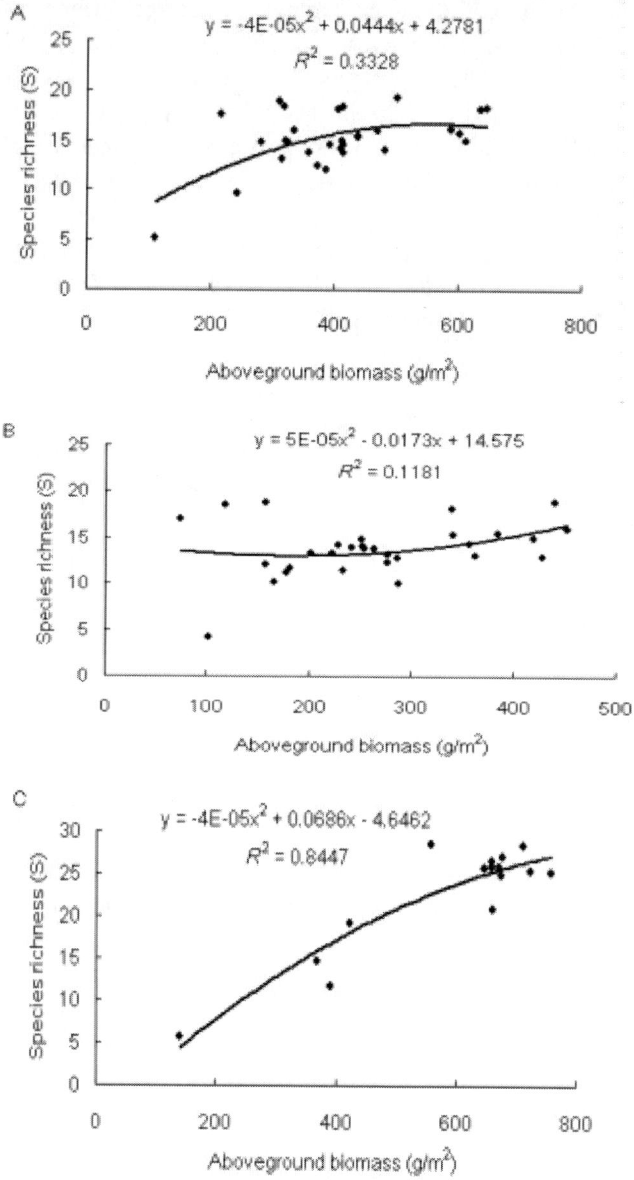

Figure 10. The relationship between species richness and aboveground biomass under three mowing treatments: A) once a year, B) twice a year, and C) once every two years.

Impacts of Mowing Management on Grasslands Diversity and Productivity

Species richness and aboveground biomass of the degraded grassland were very low at the start of the experiment in 1982 (Figure 9). Species diversity of the three mowing treatments showed significant differences during the 30-year period of the experiment ($P <$ 0.0001) (Figure 9A). The species diversity of these three treatments was similar in the first six years (1982-1988), with changes occurring from the 7th year. Van Dyke et al. (2004)

suggested that the mowing of grasslands was an important management practice for biodiversity, which can influence above- and belowground vegetation characteristics and soil properties of grasslands (Schrama et al., 2013; Liira et al., 2009; Hamza and Anderson 2005). We found that the time and intensity of mowing were critical factors for improving grassland diversity and productivity. Mowing once every two years showed the highest species diversity, followed by mowing once a year and mowing twice a year. These results indicate that mowing once every two years was beneficial to natural regeneration and the sustainable use of the grassland. In contrast, mowing once or twice a year may inhibit the natural regeneration of the grassland. There was a significant difference in aboveground biomass among the three mowing treatments ($P < 0.0001$) (Figure 9B). The average aboveground biomass over 30 years was: 581 g/m^2 (once every two years), 405 g/m^2 (once a year), and 264 g/m^2 (twice a year), respectively. In the once-per-year treatment, the aboveground biomass of the community showed a gradual increase over time, with biomass reaching the maximum value in the 8th year. There was a significant positive relationship between diversity and biomass for the once-per-year treatment ($R = 0.496$ $P < 0.001$) and the once-per-every-two-years treatment ($R = 0.750$, $P < 0.001$) (Figure 10 A, C). These results indicated that at low and moderate mowing disturbance levels, the relationship of plant diversity and biomass showed a positive linear pattern. There was no significant relationship for mowing twice a year ($P > 0.05$) (Figure 10B). Turner et al. (1993) reported that mowing could increase aboveground productivity in tall grass prairies. Low intensity moving (once every two years) could provide a way to maintain the existing vegetation, and allows perennial species to allocate the nutrients to their belowground reserves. Results from our long-term experiment, indicate that mowing once every two years could promote the natural renewal and growth of vegetation, increasing the species diversity and biomass in grasslands. These finding are consistent with Lemen et al. (1984), who reported that a cycle of mowing every three to five years may be effective in preventing the degradation of the prairie. Over the 30-year period, there was no significant relationship between annual rainfall and species richness for the different treatments (data not shown).

Impacts of Fire Management on Grassland Diversity and Productivity

The species richness was higher in burned treatments than in the unburned control (Table 1). As such, our results indicated that fire could increase species richness. Compared with the summed dominance ratio (SDR) of different fire treatments, the SDR of *S. bungeana* and *S. grandis* was higher than for other species (Table 1). The Shannon–Wiener diversity index and Richness index showed a significant difference between year 1 and all other years ($p < 0.05$). One year after burning, the Shannon-Wiener diversity index and Richness index was higher than 3 and 11 years after burning and higher than the unburned control treatment. Shannon–Wiener diversity and evenness 3 and 11 years after burning did not differ from the unburned control treatment (Table 2). The aboveground biomass was 242 g/m^2, 244 g/m^2, 294 g/m^2 and 212 g/m^2 1, 3, and 11 years after burning and in the control treatment, respectively. Aboveground biomass was significantly higher 3 and 11 years after fire than in the unburned control treatment, but did not differ from the control one year after fire (Figure 11).

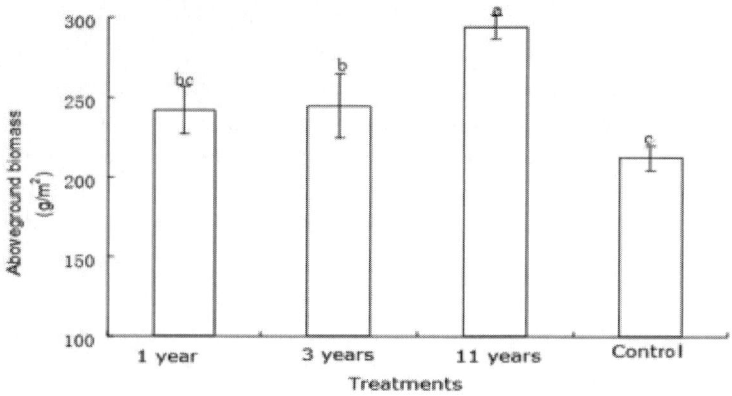

Figure 11. Differences in aboveground biomass (B) in different years after burning, and in the control treatment. The values are given as mean ± SE; the mean was calculated from 15 quadrats in each treatment.

Table 1. Species composition and summed dominance ratio (SDR) of the grassland community 1, 3, and 11 years after burning, and in the unburned control treatment

Plant family	Plant species	Control	Years after burn		
			1	3	11
Poaceae	*Stipa bungeana*	72.01	59.61	65.31	85.9
	Stipa grandis	68.55	64.08	70.45	57.7
	Leymus secalinus	34.65	45.02	35.47	24.7
	Poa sphondylodes	23.63	20.36	12.35	18.6
	Agropyron cristatum	23.25	37.77	33.64	27.6
	Poa subfastigiata Trin.	13.79	18.06	27.9	20.3
	Pennisetum centrasiaticum.	10.37		10.67	36.1
	Anthoxanthum nitens	34.32	14.62	23.73	-
	Elymus kamoji	-	28.75	-	45.4
Cyperaceae	*Carex rigesaens*	51.25	59.26	50.76	48.9
Leguminosae	*Melissitus ruthenicus*	42.93	25	43.83	29.6
Others	*Astragalus polycladus*	12.65	10.17	18.3	20.2
	Astragalus scaberrimus	5.09	11.54	10.87	-
	Oxytropis ochrantha	18.27	10.75	-	-
	Artemisia sacrorum	36.45	56.01	54.61	30.1
	Chrysanthemum lavandulifolium	23.92	40.62	33.72	21.6
	Leontopodium leontopodioides	21.58	32.48	15.42	18.1
	Heteropappus altaicus	41.16	31.72	30.17	14.9
	Saussurea japonica	17.72	28.21	5.49	26.7
	Thymus mongolicus	42.43	17.18	28.08	17.2
	Galium verum	24.93	17.2	22.62	35.2
	Potentilla bifurca	21.1	23.99	19.4	21.5
	Viola dissecta	23.95	19.97	12.57	19.1
	Potentilla acaulis	29.34	4.41	20.71	12.7
	Allium ramosum	14.88	16.97	14.02	11.1
	Polygala tenuifolia	18.73	7.3	10.96	20
	Stellera chamaejasme	8.55	8.99	9.83	9.32
	Scutellaria scordifolia var.villosissima	-	27.71	11.18	23.4
	Artemisia gansuensis	20.1	24.21	14.27	-
	Artemisia capillaris	12.9	9.16	20.68	-
	Adenophora ningxianica	-	14.92	17.68	22.4

Plant family	Plant species	Control	Years after burn		
			1	3	11
	Viola philippica	-	12.23	5.43	13.8
Others	*Dracocephalum heterophyllum*	9.75	11.18		11.9
	Androsace mariae	26.17		5.18	11.1
	Ixeridium sonchifolium	13.78		4.49	
	Ixeris chinense	9.37			7.28
	Olgaea tangutica			4.61	20.6
	Androsace erecta			12.92	7.67
	Bupleurum chinense		15.17	8.63	
	Salsola collina		5.76		17.6
	Neotorularia humilis		7.09		7.2
	Gentiana dahurica		41.00		
	Herminium monorchis				17.1
	Potentilla multifida				14.6
	Thesium refractum				9.83
	Gentiana squarrosa	7.41			
	Plantago depressa		7.08		
	Thalictrum petaloideum				5.96
Species richness		33	37	35	38

Table 2. Plant community diversity, evenness and richness indices for the following treatments: 1, 3, and 11 years after burning, and the unburned control. Different letters represent significant differences at $\alpha = 0.05$ levels. The values are given as mean ± SE; the mean was calculated from 15 quadrats in each treatment

Treatment	Shannon–Wiener index (*H*)	Evenness index (*E*)	Richness index (*R*)
1 year	1.97 ± 0.183^{a}	1.83 ± 0.20^{b}	14.78 ± 1.55^{a}
3 years	1.65 ± 0.214^{b}	2.94 ± 0.30^{a}	12.22 ± 1.32^{c}
11 years	1.74 ± 0.18^{b}	2.85 ± 0.25^{a}	13.67 ± 1.56^{ab}
Control	1.82 ± 0.18^{ab}	2.84 ± 0.31^{a}	12.44 ± 1.17^{bc}

Fire is used to control plant species composition, limit the cover of woody vegetation, and improve production of desirable forage grasses (Kitchen et al., 2009; Mouillot et al., 2002). In our study, fire increased species diversity and production compared with unburned grassland and these finding are consistent with studies in tall-grass prairies (Towne and Kemp 2003; Zhao et al., 2013). The reason could be that fire removed accumulated plant litter from the soil surface, benefiting plant growth by creating more favorable light and nutrient conditions (e.g., Knapp and Seastedt 1986). However, while fire can be beneficial for grasslands, burning management is now prohibited by the Chinese government and not used as a management practice anymore. Indeed, the frequency of natural fire has been increasing in China due to global warming and human activity, and these uncontrolled fires had significant negative impacts on the national stockbreeding economy (Liu et al., 2010), which lead to its abandonment.

Impacts of Grassland Management Practices on the Genetic Diversity of *Stipa Bungeana*

Genetic diversity within and among populations. One hundred plants from five different grassland management practices were sampled from Yunwu Mountain (Table 3). A total of 242 bands ranging from 100 to 2000bp were detected using 10 SRAP primer combinations in each of the five populations (one population per treatment) (Table 4). The number of bands per primer combination ranged from 19 to 29, with an average of 24. A total of 186 of the bands (76.86%) were polymorphic and the percentage of polymorphic bands produced by each primer ranged from 72.7% to 80.9%.

Table 3. Management practices and experimental treatments used for the study of genetic diversity in *Stipa bungeana* populations

Management practices	Population code	Year experiment started
Grazing exclusion	P1	1982
Grazing (2 sheep ha^{-1})	P2	1985
Mowing (once every two year)	P3	1982
Fire	P4	2000
Fertilization	P5	2010

Table 4. Primer combination and their polymorphism and diversity analysis

Primer combination	*t*	*p*	*P* %	Primer sequence
Me1/Em2	29	22	75.86	F: TGA GTC CAA ACC GGATA R: GAC TGC GTA CGA ATT TGC
Me4/Em1	21	17	80.95	F: TGA GTC CAA ACC GGACC R: GAC TGC GTA CGA ATT AAT
Me5/Em4	27	21	77.78	F: TGA GTC CAA ACC GGAAG R: GAC TGC GTA CGA ATT TGA
Primer combination	t	p	P %	Primer sequence
Me5/Em11	24	18	75.00	F: TGA GTC CAA ACC GGAAG R: GAC TGC GTA CGA ATT CTA
Me8/Em19	28	22	78.57	F: TGA GTC CAA ACC GGACT R: GAC TGC GTA CGA ATT TCA
Me11/Em19	22	17	77.27	F: TGA GTC CAA ACC GGAAC R: GAC TGC GTA CGA ATT TCA
Me12/Em8	19	14	73.68	F: TGA GTC CAA ACC GGAGA R: GAC TGC GTA CGA ATT CAC
Me13/Em12	26	20	76.92	F: TGA GTC CAA ACC GGAAG R: GAC TGC GTA CGA ATT CTC
Me13/Em14	22	16	72.73	F: TGA GTC CAA ACC GGAAG R: GAC TGC GTA CGA ATT CTT
ME6-X/ Em7	24	19	79.17	F: TGA GTC CTT TCC GGTAA R: GAC TGC GTA CGA ATT CAA
Total	242	186	76.86	
Average	24.2	18.6	76.86	

Abbreviations: *t* = number of total loci, *p* = number of polymorphic loci, P % = the percentage of polymorphic loci.

Table 5. Genetic diversity parameters of *Stipa bungeana* populations under different grassland management practices

Population code	*na*	*ne*	*h*	*H*
Grazing exclusion	1.1605	1.0611	0.0349	0.0311
Grazing	1.2890	1.1258	0.0723	0.1201
Mowing	1.3432	1.1702	0.0987	0.1454
Fire	1.2432	1.1023	0.0612	0.0946
Fertilization	1.2190	1.1145	0.0689	0.1065
Mean	1.2510	1.1148	0.0672	0.0995

Abbreviations: *na* = observed number of alleles; *ne* = effective number of alleles; *h* = Nei's gene diversity; *H* = Shannon's information index.

Within each population, the observed number of alleles (na) varied from 1.16 (P1) to 1.34 (P3) with an average of 1.25. The effective number of alleles (ne) ranged from 1.06 (P1) to 1.17 (P3) with a mean of 1.11. Nei's genetic diversity (h) varied from 0.03 (P1) to 0.10 (P2) with a mean value of 0.07. Shannon's information index (H) ranged from 0.03 (P1) to 0.15 (P2), with a mean value of 0.10. The results showed that mowed grassland (P3) had the highest level of genetic diversity, whereas the lowest genetic diversity occurred when grazing was excluded (P1) (Table 5).

Nybom (2004) found that the mean Nei's genetic diversity within plant populations was *h* = 0.22 (n = 20) for widespread species, with a mean of *h* = 0.25 (n = 37) for long-lived perennial species. In comparison with Nybom (2004), we found that Nei's genetic diversity of *S. bungeana* populations in our study was lower than average. The fragmented environment caused by human and natural disturbances may block gene flow among different populations, contributing to lower genetic diversity (Liu *et al.*, 2009). Moreover, over-grazing and mowing may further influence the genetic diversity among populations (Jing et al., 2013; Wu et al., 2010; Matlaga et al., 2004; Kleijn et al., 2002). Wu et al. (2010) indicated that the genetic diversity of *S. grandis* was lower under a mowing regime, because *S. grandis* was usually mowed before the seeds turned mature, which would have limited sexual recruitment while stimulating clonal reproduction. However, we found that the genetic diversity of *S. bungeana* was relatively high under grazing and mowing management practices compared with other management practices (Table 5). In grasslands where grazing was excluded, litter may have accumulated because of the high aboveground standing biomass (Cheng et al., 2012), inhibiting seed germination. Therefore *S. bungeana* probably grew by producing new ramets from genetically similar plants (Cheng et al., 2012; Jing et al., 2013). This could explain why grazing exclusion induced the lowest genetic diversity. In the grazing and mowing treatments, the timing of the disturbance potentially impacted the genetic diversity. In the grazed grassland, during mid–June, some mature seeds of different *S. bungeana* individuals can be spread by sheep ingestion and defecation (Mouissie et al., 2005). In the mowed grassland, litter accumulation is significantly reduced (data not shown) allowing seeds of different individual plants to reach the soil and germinate. In addition, mowing after the seeds turn mature may enhance seedling recruitment, and increases light availability at the ground level (Jensen and Gutekunst 2003; Kotowski et al., 2001). Together, these factors may result in an increase in genetic diversity. In the burned and fertilized treatments, the habitat characteristics (such as vegetation structure, height, standing biomass, absence of seed dispersion) were similar to the grazing exclusion grassland leading also to low genetic diversity.

Genetic Differentiation and Gene Flow Among Populations

We found that 31.3% of the total genetic variation existed among populations and 68.7% within populations (Table 6). The genetic differentiation coefficient is an index used to assess the genetic structure of a plant population. Compared to the genetic differentiation among populations, the genetic differentiation within populations of *S. bungeana* was higher. Previous studies indicated that environmental factors and geographical distance could determine the genetic structure, leading to higher genetic differentiation among populations (Wu et al., 2010; Peleg et al., 2008). For example, Zhao et al. (2006) found that the genetic diversity of S. *grandis* was influenced by the temperature and precipitation. The genetic structure was determined by variations of temperature and precipitation, implying that natural selection in terms of climatic variations had some influence on the pattern of genetic structure of plant populations (Wu et al., 2010). In the present study, the five differently managed grasslands were at the same site, adjacent to each other. Therefore, the opportunity for gene flow was high, causing low genetic differentiation among populations. In conclusion, based on the effects of grassland management practices on genetic diversity, we suggest that grazing and mowing regimes could be applied for the conservation or restoration grassland diversity.

Table 6. AMOVA of genetic variances within and among populations of *Stipa bungeana*

Source of variation	d.f.	Sum of squares	Variance component	Percentage variation	P value
Among groups	4	472.850	16.425	31.31	<0.001
Within groups	99	1658.686	17.831	68.69	<0.001
Total	103	2131.536	34.256	100	

* The variance components and P-values were tested statistically by nonparametric randomization tests using 1000 permutations.

SYNTHESIS

China has ~400 million hectares of various grassland types, accounting for ~41.7% of the country's land area (Ren et al., 2008). Recently, grassland degradation has become a serious problem in China. The grassland degradation was defined as a decrease in plant species diversity, plant height, vegetation cover and plant productivity (Han et al., 2008). The biodiversity of grassland refers to the number, abundance and composition of species, populations, and communities. Since recently, climate change is also affecting the biodiversity of grassland ecosystems (Jing et al., 2013; Bai et al., 2004; Mouillont et al., 2002). In order to restore the biodiversity of grasslands, a wide range of conservation techniques have been proposed and applied. As such, grazing exclusion, rotational grazing, mowing and fire disturbance play an important role in maintaining biodiversity in native grasslands (Jing et al., 2013; Liu et al., 2012; Wallis et al., 2007; Middleton et al., 2006; Duncan 2005; Billeter et al., 2003; Dolek et al., 2002; Mitlacher et al., 2002). Many studies have suggested that conservation techniques could influence plant species composition and production, but few studies have focused on the long-term effects. Therefore, in this study, we compiled a 30-year dataset on effects of grazing exclusion, grazing and mowing, and determined which management technique (and at frequency and intensity) best restores and maintains the biodiversity of semiarid grassland ecosystems.

Based on the results of the long-term experiments in northern China highlighted in this chapter, we suggest that grazing management and mowing should be recommended to restore and conserve biodiversity of semiarid grasslands regions. Indeed, while grazing exclusion could accelerate the restoration of degraded grasslands during the initial 15 years, species diversity and aboveground biomass decreased during later stages of succession. Rotational grazing and mowing were clearly more effective to maintain higher biodiversity of grasslands, with light grazing (2 sheep ha^{-1}) maintaining the highest species diversity and aboveground biomass. Moving once every two years resulted also in higher grassland species diversity and aboveground biomass than more frequent or no-mowing management and low-frequency mowing should be used as complement of light grazing or in replacement when grazing management is not possible. While fire management may stimulate biodiversity in grassland ecosystems, this practice is no longer possible in China due to global change and increasing frequency of uncontrolled natural fires. Besides an increase of species richness and productivity, grazing and mowing induced also the higher genetic diversity of *S. bungeana* and we highly recommend using both practices for the conservation of grassland genetic diversity in semiarid area in China.

ACKNOWLEDGMENTS

We thank Weijun Li, Xiuyun Ji, and the management bureau of Yunwushan Grassland Natural Reserve for their help with the field experiments.

REFERENCES

Akiyama, T. and Kawamura, K. (2007) Grassland degradation in China: methods of monitoring, management and restoration. *Grassand Science*, 53, 1–17.

Bai, Y.F., Han, X.G., Wu, J.G., Chen, Z.Z. and Li, L.H. (2004) Ecosystem stability and compensatory effects in the Inner Mongolia grassland. *Nature*, 431, 181–184.

Billeter, R.D., Hooftman, A.P. and Diemer, M. (2003) Differen tial and reversible responses of common fen meadow species to abandonment. *Applied Vegetation Sci*ence, 6, 3–12.

Bond, W.J., Woodward, F.I. and Midgley, G.F. (2005) The global distribution of ecosystems in a world without fire. *New Phytologist*, 165, 525–537.

Brockway, D.G., Gatewood, R.G. and Paris, R.B. (2002) Restoring fire as an ecological process in shortgrass prairie ecosystems: initial effects of prescribed burning during the dormant and growing seasons. *Journal Environmental Management*, 65,135–52.

Chapin, F.S., Zavaleta, E.S., Eviner, V.T., Naylor, R.L., Vitousek, P.M., Reynolds, H.L., Hooper, D.U., Lavorel, S., Sala, O.E., Hobbie, S.E., Mack, M.C. and Diaz, S. (2000) Consequences of changing biodiversity. *Nature*, 405, 234-242.

Chen, J.Q., Brosofske, K.D., Noormets, A., Crow, T.R., Bresee, M.K., Le Moine, J.M., Euskirchen, E.S., Mather, S.V. and Zheng, D. (2004) A working framework for quantifying carbon sequestration in disturbed land mosaics. *Environmental Management*, 33, 210-221.

Chen, X.W. and Li, L. (2004) Tree diversity change in remaining primary mixed broadleaved Korean pine forest under climate change and human activities. *Biological Conservation*, 13, 563-577.

Cheng, J., Cheng, J.M., Hu, T.M., Shao, H.B. and Zhang, J.M. (2012) Dynamic changes of *Stipa bungeana* steppe species diversity as better indicators for soil quality and sustainable utilization mode in Yunwu Mountain Nature Reserve, Ningxia, China. *Clean Soil Air Water*, 40, 127-133.

Collins, S. and Calabrese, L.B. (2012) Effects of fire, grazing and topographic variation on vegetation structure in tallgrass prairie. *Journal Vegetation Science*, 23, 563-575.

Connell, J. H. 1978. Diversity in tropical rainforests and coral reefs. *Science*, 199:1302-1310.

Cooper, T. McCann, Ballard, E. (2005) The effects of livestock grazing and recreation on Irish machair grassland vegetation. *Plant Ecology*, 181, 255-267.

Crawley, M.J., Johnston, A.E., Silvertown, J., Dodd, M., de Mazancourt, C., Heard, M.S., Henman, D.F. and Edwards, G.R. (2005) Determinants of species richness in the park grass experiment. *American Naturalist*, 165, 179-192.

DiTomaso, J.M., Brooks, M.L., Allen, E.B., Minnich, R., Rice, P.M. and Kyser, G.B. (2006) Control of invasive weeds with prescribed burning. *Weed Technology*, 20, 535-548.

Dolek, M. and Geyer, A. (2002) Conserving biodiversity on calcareous grasslands in the Franconian Jura by grazing: a comprehensive approach. *Biological Conservation*, 104, 351-360.

Dumont, B., Farruggia, A., Garel, J.P., Bachelard, P., Boitier, E. and Frain, M. (2009) How does grazing intensity influence the diversity of plants and insects in a species-rich upland grassland on basalt soils? *Grass Forage Science*, 64, 92-105.

Duncan, A. (2005) Farm Animals and Biodiversity. *Animal Sc*ience, 81, 187-188.

Excoffier, L., Smouse, P.E. and Quattro, J.M. (1992) Analysis of molecular variance inferred from metric distances among DNA haplotypes, application to human mitochondrial DNA restriction data. *Genetics*, 131, 479-491.

Georgia, H.D.S., Thomas, W.G., John, C.B. and Graham, A.T. (2004) Use of remote sensing techniques to determine the effects of grazing on vegetation cover and dune elevation at Assateague Island National Seashore: Impact of Horses," *Environmental Management*, 34, 642-649.

Gillen, R.L. and Sims, P.L. (2004) Stocking rate, precipitation, and herbage production on sand sagebrush-grassland. *Journal of Range Management*, 57, 148-152.

Golodets, C., Kigel, J. and Sternberg, M. (2010) Recovery of plant species composition and ecosystem function after cessation of grazing. *Plant and Soil*, 329, 365-378.

Gough, L., Osenberg, C.W., Gross, K.L. and Collins, S.L. (2000) Fertilization effects on species density and primary productivity in herbaceous plant communities. *Oikos*, 89, 428-439.

Guo, Q. F. (2005) Ecosystem maturity and performance. *Nature*, 435, E6.

Hamza, M.A. and Anderson, W.K. (2005) Soil compaction in cropping systems—a review of the nature, causes and possible solutions. *Soil Tillage Research*, 82, 121-145.

Han, G.D., Hao, X.Y., Zhao, M.L., Wang, M.J., Ellert, B.H., Willms, W. and Wang, M.J. (2008) Effect of grazing intensity on carbon and nitrogen in soil and vegetation in a meadow steppe in Inner Mongolia. *Agriculture Ecosystems and Environment*, 125, 21-32.

Han, J.G., Zhang, Y.J., Wang, C.J., Bai, W.M., Wang, Y.R., Han, G.D. and Li, L.H. (2008) Rangeland degradation and restoration management in China. *The Rangeland Journal*, 30, 233-239.

Hassan, R., Hamasha and Hensen, I. (2009) Seed germination of four Jordanian Stipa spp: differences in temperature regimes and seed provenances. *Plant Species Biology*, 24, 127-132.

He, N.P., Yu, Q., Wu, L., Wang, Y.S. and Han X.G. (2008) Carbon and nitrogen store and storage potential as affected by land-use in a Leymus chinensis grassland of northern China. *Soil Biology and Biochem*istry, 40, 2952-2959.

He, N.P. Zhang, Y.H. Yu, Q. Chen, Q.S. Pan, Q.M. Zhang, G.M. Han, X.G. (2011) Grazing intensity impacts soil carbon and nitrogen storage of continental steppe. *Ecosphere*, 2, 8. doi:10.1890/ES10-00017.1

Hickman, K.R., Hartnett, D.C., Cochran, R.C. and Owensby, C.E. (2004) Grazing management effects on plant species diversity in tallgrass prairie. *Journal of Range Management*, 57, 58-65.

Higgins, S.I., Bond, W.J., February, E.C., Bronn, A., Euston-Brown, D.I.W., Enslin, B., Govender, N., Rademan, L., O'Regan, S., Potgieter, A.L.F., Scheiter, S., Sowry, R., Trollope, L. and Trollope, W.S.W. (2007) Effects of four decades of fire manipulation on woody vegetation structure in savanna. *Ecology*, 88, 1119-1125.

Hobbs, R.J. and Huenneke, L.F. (1992) Disturbance, diversity, and invasion : implications for conservation. *Conservational Biology*, 6, 324-337.

Holdo, R.M. (2013) Effects of fire history and N and P fertilization on seedling biomass, Specific Leaf Area, and root:shoot ratios in a South African savannah. *South African Journal of Botanay*, 86, 5-8.

Holdo, R.M., Holt, R.D. and Fryxell, J.M. (2009) Grazers, browsers, and fire influence the extent and spatial pattern of tree cover in the Serengeti. *Ecological Applications*, 19, 95-109.

Hooper, D.U., III Chapin, F.S., Ewel, J.J., Hector, E.A., Inchausti, P., Lavorel, S., Lawton, J.H., Lodge, D.M., Loreau, M. and Naeem, S. (2005) Effects of biodiversity on ecosystem functioning: a consensus of current knowledge. *Ecological Monographs*, 75, 3-35.

Hummer, K.E. and Sugar, D. 1998. Pear gene bank information on world-wide web. Acta Horticulturae, 475, 117–121.

Jensen, K. and Gutekunst, K. 2003. Effects of litter on establishment of grassland plant species: the role of seed size and successional status. *Basic and Applied Ecology*, 4, 579-587.

Jing, Z.B., Cheng, J.M. and Chen, A. (2013) Assessment of vegetative ecological characteristics and the succession process during three decades of grazing exclusion in a continental steppe grassland. *Ecological Engineering*, 57,162-169.

Jing, Z.B., Cheng, J.M., Guo, C.H. and Wang, X.P. (2013) Seed traits, nutrient elements and assessment of genetic diversity for almond (Amygdalus spp.) endangered to China as revealed using SRAP markers. *Biochemical Systematics and Ecology*, 49, 51–57.

Jing, Z.B., Yu, J. and Cheng, J.M. (2013) Genetic diversity of a dominant species Stipa bungeana and its conservation strategy in the Loess Plateau of China. *Biochemical Systematics and Ecology,* 47, 126-131.

Kitchen, D., Blair, J.M. and Callaham Jr, M.A. (2009) Annual fire and mowing alter biomass, depth distribution, and C and N content of roots and soil in tallgrass prairie. *Plant and Soil*, 323, 235-247.

Kleijn, D. and Steinger, T. (2002) Contrasting effects of grazing and hay cutting on the spatial and genetic structure of Veratrum album, an unpalatable, long-lived, clonal plant species. *Journal of Ecology*, 90, 360-370.

Klein, J.A., Harte, J. and Zhao, X.Q. (2005) Dynamic and complex microclimate responses to warming and grazing manipulations. *Global Change Biology*, 11, 1440-1451.

Knapp, A.K. and Seastedt, T.R. (1986) Detritus accumulation limits productivity of tallgrass prairie. *Bioscience*, 36, 662-668.

Kolliker, R., Stadelmann, F.J., Reidy, B. and Nosberger, J. (1998). Fertilization and defoliation frequency affect genetic diversity of *Festuca partensis* Huds. In permanent grasslands. *Molecular Ecology*, 7, 1557-1567.

Kolos, A. and Banaszuk, P. (2013) Mowing as a tool for wet meadows restoration: Effect of long-term management on species diversity and composition of sedge-dominated wetland. *Ecological Engineering*, 55, 23-28.

Kotowski, W., van Andel, J., van Diggelen, R. and Hogendorf, J. (2001) Responses of fen plant species to groundwater level and light intensity. *Plant Ecology*, 155, 147-156.

Lamb, E.G. (2008) Direct and indirect control of grassland community structure by litter, resources and biomass. *Ecology*, 89, 216-225.

Lande, R. (1988) Genetics and demography in biological conservation. *Science*, 241, 1455-1460.

Lavoie, M., Starr, G., Mack, M.C., Martin, T.A. and Gholz, H.L. (2010) Effects of prescribed fire on understory vegetation, carbon pools, and soil nutrients in a longleaf pine-slash pine forest in Florida. *Natural Areas Journal*, 30, 82-94.

Lemen, C.A. and Clause, M.K. (1984) The effects of mowing on the rodent community of a native tall grass prairie in eastern Nebraska. *Natural Resource*, 5-10.

Li, G. Quiros, C.F. (2001) Sequence-related amplified polymorphism (SRAP), a newmarker system based on a simple PCR reaction: its application tomapping and gene tagging in Brassica. *Theoretical and Applied Genetics,* 103, 455–461.

Liira, J., Issak, M., Jogar, U., Mandoja, M. and Zobel, M. (2009) Restoration management of a floodplain meadow and its cost-effectiveness—the results of a 6-year experiment. *Annales Botanici Fennici,* 46, 397-408.

Lin, Y., Hong, M., Han, G.D., Zhao, M.L., Bai, Y.F. and Chang, S.X. (2010) Grazing intensity affected spatial patterns of vegetation and soil fertility in a desert steppe. *Agriculture Ecosystems and Environment,* 138, 282-292.

Liu, N., Zhang, Y.J., Chang, S.J., Kan, H.M. and Lin, L.J. (2012) Impact of grazing on soil carbon and microbial biomass in typical steppe and desert steppe of Inner Mongolia. *PloS One*, 7,e36434.

Liu, P., Huang, J.H., Sun, O.J. and Han, X.G. (2010) Litter decomposition and nutrient release as affected by soil nitrogen availability and litter quality in a semiarid grassland ecosystem. *Oecologia*, 162, 771-780.

Liu, W.S., Dong, M., Song, Z.P. and Wei, W. (2009) Genetic diversity pattern of *Stipa purpurea* populations in the hinterland of Qinghai–Tibet Plateau. *Annals of Applied Biology,* 154, 57-65.

Liu, W.X., Xu, W.H., Han, Y., Wang, C.H. and Wan, S.Q. (2007) Responses of microbial biomass and respiration of soil to topography, burning, and nitrogen fertilization in a temperate steppe. *Biology and* Fertility of *Soils,* 44, 259-268.

Liu, X.P., Zhang, J.Q., Cai, W.Y. and Tong, Z.J. (2010) Information diffusion-based spatio-temporal risk analysis of grassland fire disaster in northern China. *Knowledge-Based Systems*, 23, 53-60.

Maestre, F.T., Quere, J.L., Gotelli, N.J., Escudero, A., Ochoa, V., Delgado-Baquerizo, M., Garcia-Gomez, M., Bowker, M.A., Soliveres, S., Escolar, C., Garcia-Palacios, P., Berdugo, M., Valencia, E., Gozalo, B., Gallardo, A., Aguilera, L., Arredondo, T., Blones, J., Boeken, B., Bran, D., Conceicao, A.A., Cabrera, O., Chaieb, M., Derak, M., Eldridge, D.J., Espinosa, C.I., Florentino, A., Gaitan, J., Gatica, M.G., Ghiloufi, W., Gomez-Gonzalez, S., Gutierrez, J.R., Hernandez, R.M., Huang, X.W., Huber-Sannwald, E., Jankju, M., Miriti, M., Monerris, J., Mau, R.L., Morici, E., Naseri, K., Ospina, A., Polo, V., Prina, A., Pucheta, E., Ramirez-Collantes, D.A., Romao, R., Tighe, M., Torres-Diaz, C., Val, J., Veiga, J.P., Wang, D.L. and Zaady, E. (2012) Plant species richness and ecosystem multifunctionality in global drylands. *Nature*, 335, 214-218.

Malo, A.R. and Nicholson, S.E. (1990) A study of rainfall and vegetation dynamics in the African Sahel using normalized difference vegetation index. *Journal of Arid Enviroment*, 19, 1-13.

Masunga, G.S., Moe, S.R. and Pelekekae, B. (2013) Fire and Grazing Change Herbaceous Species Composition and Reduce Beta Diversity in the Kalahari Sand System. *Ecosystems*, 16, 252-268.

Matlaga, D. and Karoly, K. (2004) Long-term grazing effects on genetic variation in Idaho fescue. *Journal of Range Management*, 57, 275-279.

Mayer, R., Kaufmann, R., Vorhauser, K. and Erschbamer, B. (2009) Effects of grazing exclusion on species composition in high-altitude grasslands of the Central Alps. *Basic and Applied Ecology*, 10, 447-455.

Metera, E., Sakowski, T., Sloniewski, K. and Romanowicz, B. (2010) Grazing as a tool to maintain biodiversity of grassland-a review. *Animal Science Papers and Reports*, 28, 315-334.

Middleton, B.A., Holsten, B. and van Diggelen, R. (2006) Biodiversity management of fens and fen meadows by grazing, cutting and burning. *Applied Vegetation Science*, 9, 307-316.

Mitlacher, K., Poschlod, P., Rosén, E. and Bakker, J.P. (2002) Restoration of wooded meadows: a comparative analysis along a chronosequence on Öland (Sweden). *Applied Vegetation Science*, 5, 63-73.

Moe, S.R. and Wegge, P. (1997) The effects of cutting and burning on grass quality and axis deer (Axis axis) use of grassland in lowland Nepal. *Journal of Tropical Ecology*, 13, 279-92.

Mouillont, F., Rambal, S. and Joffre, R. (2002) Simulating climate change impacts on fire frequency and vegetation dynamics in a Mediterranean-type ecosystem. *Globle Change Biology*, 8, 423-437.

Mouissie, A.M., Veen, C.E.J.V.D., Veen, G.F. and Diggelen, R.V. (2005) Ecological correlates of seed survival after ingestion by Fallow Deer. *Functional Ecology*, 19, 284-290.

Nei, M. (1973) Analysis of gene diversity in subdivided populations. *Proceedings of the National Academy of Sciences,* 70, 3321-3323.

Nybom, H. (2004) Comparison of different nuclear DNA markers for estimating intraspecific genetic diversity in plants. *Molecular Ecology*, 13, 1143-1155.

Peleg, Z., Saranga, Y., Krugman, T., Abbo, S., Nevo, E. and Fahima, T. (2008) Allelic diversity associated with aridity gradient in wild emmer wheat populations. *Plant Cell Environment*, 31, 39-49.

Pierik, M., Ruijven, J.V., Bezemer, T.M., Geerts, R.H.E.M. and Berendse, F. (2011) Recovery of plant species richness during long-term fertilization of a species-rich grassland. *Ecology*, 92, 1393-1398.

Poptcheva, K., Schwartze, P., Vogel, A., Kleinebecker, T. and Holzel, N. (2009) Changes in wet meadow vegetation after 20 years of different management in a field experiment (North-West Germany). *Agriculture Ecosystems and Environment,* 134, 108-114.

Qiu, L. Wei, X. Zhang, X. Cheng, J. (2013) Ecosystem Carbon and Nitrogen Accumulation after Grazing Exclusion in Semiarid Grassland. *PLoS ONE,* 8:e55433. doi:10.1371/journal.pone.0055433

Reisch, C. and Scheitler, S. (2009) Disturbance by mowing affects clonal diversity: the genetic structure of Ranunculus ficaria (Ranunculuaceae) in meadows and forests. *Plant Ecology*, 201, 699-707.

Ren, J.Z., Hu, Z.Z., Zhao, J., Zhang, D.G., Hou, F.J., Lin, H.L. and Mu, X.D. (2008) A grassland classification system and its application in China. *The Rangeland Journal*, 30, 199-209.

Sarmiento, G., Pinillos, M., da Silva M.P. and Acevedo, D. (2004) Effects of soil water regime and grazing on vegetation diversity and production in a hyperseasonal savanna in the Apure Llanos, Venezuela. *Journal of Tropical Ecology*, 20, 209-220.

Schrama, M.J.J., Cordlandwehr, V., Visser, E.J.W., Elzenga, T.M., de Vries, Y. and Bakker, J.P. (2013) Grassland cutting regimes affect soil properties, and consequently vegetation composition and belowground plant traits. *Plant and Soil*, 366, 401-413.

Sivakumar, M.V.K. (2007) Interactions between climate and desertification. *Agricultural and Forest Meteorology,* 142, 143-155.

Slimani, H., Aidoud. A. and Roze, F. (2010) 30 Years of protection and monitoring of a steppic rangeland undergoing desertification, *Journal of Arid Environment*, 74, 685–691.

Spooner, P. and Briggs, S.V. (2008) Woodlands on farms in southern New South Wales: A longer-term assessment of vegetation changes after fencing. *Ecological Management and Restoration,* 9, 33-41.

Steffens, M., Kolbl, A., Totsche, K.U. and Kogel Knabner I. (2008) Grazing effects on soil chemical and physical properties in a semiarid steppe of Inner Mongolia (P.R. China). *Geoderma*, 143, 63-72.

Stolle, F., Chomitz, K.M., Lambin, E.F. and Tomich, T.P. (2003) Land use and vegetation fires in Jambi Province, Sumatra, Indonesia. *Forest Ecology and Management,* 179, 277-292.

Su, Y.Z., Li, Y.L., Cui, J.Y. and Zhao. W.Z. (2005) Influences of continuous grazing and livestock exclusion on soil properties in a degraded sandy grassland, Inner Mongolia, northern China. *Catena*, 59, 267-278.

Thomas, H.A., Gregory, P.A. and Mark, E.M. (2003) Changes in vegetation structure after long-term grazing in Pinyon-Juniper. *Ecosystems*, 6, 368-383.

Tilman, D., Wedin, D. and Knops, J. (1996) Productivity and sustainbility influenced by biodiversity in grassland ecosystems. *Nature*, 379, 718-720.

Towne, E.G. Kemp, K.E. (2003) Vegetation dynamics from annually burning tallgrass prairie in different seasons. *Journal of Range Management*, 56,185-192.

Tsutsumi, M., Shiyomi, M., Wang, Y., Tsuiki M. and Yu X. (2003) Species diversity of grassland vegetation under three different grazing intensities in the Heilongjiang Steppe of China. *Grassland Science*, 48, 510-516.

Turner, C.L., Seastedt, T.R. and Dyer, M.I. (1993) Maximization of aboveground grassland production: the role of defoliation frequency, intensity, and history. *Ecological Application*, 3, 175–186

Van Dyke, F., Van kley, S.E., Page, C.E. and Van beek, J.G. (2004) Restoration efforts for plant and bird communities in tallgrass prairies using prescribed burning and mowing. *Restoration. Ecology*, 12, 575-585.

Wallis de vries, M.F., Parkinson, A.E., Dulphy, J.P., Sayer, M. and Diana, E. (2007) Effects of livestock breed and grazing intensity on biodiversity and production in grazing systems. 4. Effects on animal diversity. *Grass and Forage Science*, 62, 185-197.

Wan, S., Hui, D. and Luo, Y. (2001) Fire effects on nitrogen pools and dynamics in terrestrial ecosystems: a meta-analysis. *Ecological Application*, 11, 1349-1365.

Wan, S., Luo, Y. and Wallace, L.L. (2002) Changes inmicroclimate induced by experimental warming and clipping in tallgrass prairie. *Global Change Biology*, 8, 754-768.

Wang, G.L., Liu, G.B. and Xu, M.X. (2009) Above- and belowground dynamics of plant community succession following abandonment of farmland on the Loess Plateau, China. *Plant and Soil*, 316, 227-239.

Wang, X.H., Yu, J.B., Zhou, D., Dong, H.F., Li, Y.Z., Lin, Q.X., Guan, B. and Wang, Y.L. (2012) Vegetative ecological characteristics of restored reed (*Phragmites australis*) wetlands in the Yellow River Delta, China. *Environmental Management*, 49, 325-333.

Wilkins, S., Keith, D. and Adam, P. (2003) Measuring success: evaluating the restoration of a grassy eucalypt woodland on the Cumberland Plain, Sydney, Australia. *Restoration Ecology*, 11, 489-503.

Wu, G.L., Du, G.Z., Liu, Z.H. and Thirgood, S. (2009). Effect of grazing exclusion and grazing on a Kobresia-dominated meadow in the Qinghai–Tibetan Plateau. *Plant and Soil*, 319, 115-126.

Wu, G.L., Liu Z.H., Zhang, L., Chen, J.M. and Hu, T.M. (2010) Long-term fencing improved soil properties and soil organic carbon storage in an alpine swamp meadow of western China. *Plant and Soil*, 332, 331-337.

Wu, J.B., Gao, Y.B. and Bao, X.Y. (2010) Genetic diversity of Stipa grandis P.Smirn populations across the species' range in the Inner Mongolia Plateau of China. *Biochemical Systematics and Ecology,* 1-7.

Yamamoto, Y., Kirita, H., Ohga, N. and Saito, Y. (1995) Extended summed dominace ratio, E-SDR, for comparison of grassland vegetation. *Japanese Society of Grassland Science,* 41, 37-41.

Yeh, F.C., Yang, R., Boyle, T.J., Ye, Z. and Xiyan, M. (2000) POPGENE 32, Microsoft Windows-based Freeware for Population Genetic Analysis, Version1.32 [CP/CD]. University of Alberta, Edmonton, Canada.

Yuan, J.Y., Ouyang, Z.Y., Zheng, H. and Xu, W.H. (2012) Effects of different grassland restoration approaches on soil properties in the southeastern Horqin sandy land, northern China. *Appled Soil Ecology*, 61, 34-39.

Zavaleta, E.S., Pasari, J.R., Hulvey, K.B. and Tilman, G.D. (2010) Sustaining multiple ecosystem functions in grassland communities requires higher biodiversity. *Proceedings of the National Academy of Sciences,* 107, 1443-1446.

Zhang, C., Xue, S., Liu, G.B. and Song, Z.L. (2011) A comparison of soil qualities of different revegetation types in the Loess Plateau, China. *Plant and Soil*, 347, 163-178.

Zhang, J.T. and Dong, Y.R. (2010) Factors affecting species diversity of plant communities and the restoration process in the loess area of China. *Ecological Eng*ineering, 36, 345-350.

Zhang, J.T., Ru, W.M. and Li, B. (2006) Relationships between vegetation and climate the Loess Plateau in China. *Folia Geobotanica*, 41, 151-163.

Zhao, H.L., Li, S.G., Zhang, T.H., Ohkuro, T. and Zhou, R.L. (2004) Sheep gain and species diversity in sandy grassland, Inner Mongolia. *Journal of Range Management*, 57, 187-190.

Zhao, L.P., Wu, G.L. and Shi, Z.H. (2013) Post-fire species recruitment in a semiarid perennial steppe on the Loess Plateau. *Australian Journal of Botany,* 61, 29-35.

Zhao, N.X., Gao, Y.B., Wang, J.L., Ren, A.Z. and Xu, H. (2006) RAPD diversity of Stipa grandis populations and its association with some ecological factors. *Acta Ecologica Sinica,* 26, 1312–1319.

Zhou, H.K., Tang, Y.H., Zhao, X.Q. and Zhou, L. (2006) Long-term grazing alters species composition and biomass of a shrub meadow on the Qinhai-Tibet plateau. *Pakistan Journal of Botany,* 38, 1055-1069.

In: Grassland Biodiversity and Conservation … ISBN: 978-1-61761-619-8
Editors: Pierre Mariotte and Paul Kardol

Chapter 6

GRAZING AND SPATIAL HETEROGENEITY: IMPLICATIONS FOR GRASSLAND STRUCTURE AND FUNCTION

Juliette M. G. Bloor *** and Julien Pottier***
Grassland Ecosystem Research Unit, INRA, Clermont-Ferrand, France

ABSTRACT

Grazed grasslands are dynamic, complex ecosystems characterised by patchy resource inputs and vegetation structure under constant modification by the management of domestic herbivores. Over the last decade there has been increasing recognition that plant and soil spatial heterogeneity has the potential for significant effects on biotic interactions, diversity and ecosystem properties. Nevertheless we know little about the overall trends of response to grazing-induced heterogeneity in grasslands systems. After an overview of how grazing animals promote spatial heterogeneity in grasslands, we synthesize literature to answer the question: does fine-scale grazing-induced spatial heterogeneity significantly affect field-scale grassland structure and function? We examine how spatial patterns of grazing components (defoliation, trampling and animal returns) mediate changes in plant community composition and field-scale diversity via spatial coexistence mechanisms. We then consider how these grazing components modify ecosystem function, reviewing the impacts of heterogeneity in defoliation and soil nutrients on above- and below-ground ecosystem processes. Overall, grazing-induced spatial heterogeneity appears to enhance plant and arthropod diversity in grasslands. The relatively few studies published on effects of spatial heterogeneity and grazed grassland function provide clear patterns of nitrogen losses in relation to urine inputs, but highlight the need for work on other aspects of grassland function. The importance of plant-soil feedbacks and effects of grazing-induced heterogeneity on plant functional traits, and ultimately on ecosystem properties, are discussed. We highlight gaps in current understanding and suggest that the study of spatial heterogeneity in managed grasslands is critical for accurate prediction of field-scale processes and their cascading effects on ecosystem service provision.

* Corresponding author: Juliette Bloor; email: juliette.bloor@clermont.inra.fr.

Keywords: Spatial heterogeneity, grazing, plant-soil feedbacks, above- and below-ground processes, ecosystem services

INTRODUCTION

Grazed grassland ecosystems provide key environmental services worldwide, as well as important economic and societal services via grazing-based livestock production systems (Gibson 2008). On a global scale, grasslands grazed by livestock represent 3.4 billion hectares and contribute to the livelihoods of over 800 million people (McGilloway 2005). By 2020, this agricultural sub-sector is expected to produce 30% of the value of the global agricultural output (Delgado 2005). In Europe, grassland ecosystems cover 80 million ha (22% of the EU-25 land area, EEA 2005), and sustain 100 million cows, 90 million sheep and 30 million horses (Eurostat 2010). Indeed, grazing is the cheapest source of feed for livestock, costs being higher for conserved herbage (hay, silage). Nevertheless, grazed grassland systems in Europe face threats from changing management practices and increasingly-common drought events associated with climate change (Burel & Garnier 2008; Bindi & Olesen 2010). Consequently grazed grasslands face the challenge of meeting multifunctional objectives and sustainable agricultural production against a background of changing environmental conditions and socio-economic uncertainty.

In Europe, the majority of grazed grasslands are multi-species, permanent grassland systems, situated in areas not used or not suitable for crop cultivation. Despite relatively low productivity, such semi-natural grasslands store 5-22g $C.m^{-2}.yr^{-1}$ on average in the soil (Soussana et al., 2010). At the same time, grasslands and associated grazing animals generate significant greenhouse gas emissions (CO_2, N_2O and CH_4 fluxes from livestock production estimated at 18% on a global-scale), and animal wastes can be a ground-water pollution issue (Steinfeld et al., 2006). Although there is a clear societal demand for accurate accounting of the greenhouse gas balance and development of mitigation options, C and N fluxes within grazed grasslands are difficult to predict due to the high spatial heterogeneity induced by grazing animals. Uneven patterns of animal returns and herbivory amplify intrinsic heterogeneity in vegetation structure linked to abiotic factors such as microtopography, hydrology and pedology (Adler et al., 2001). Spatial heterogeneity in grazed grasslands also has implications for biodiversity conservation via changes in small-scale habitat diversity (Mouissie et al., 2008; Helden et al., 2010). This raises a number of questions with regards to the application of grazing management. Should the enhancement of within-field heterogeneity be a management priority? Does spatial arrangement promote ecosystem function and/or biodiversity? Can spatial heterogeneity reduce system vulnerability to stochastic disturbances (climate perturbations, pest outbreaks)? Understanding how fine-scale pattern-process relationships are connected to broader-scale patterns of ecosystem structure and function represents a significant challenge in ecology (Lovett et al., 2005).

This chapter describes the consequences of grazing and its constituent components (defoliation, animal returns, trampling) for spatial heterogeneity in grasslands. After clarifying definitions and methods for studying spatial heterogeneity in grassland studies, we outline how grazing animals promote spatial heterogeneity in grasslands. We then review available literature on spatial heterogeneity in grazed grasslands to assess when and how grazing-induced spatial heterogeneity is important for ecosystem structure and function. We discuss research challenges and, where possible, suggest opportunities for progress.

Conceptual and Methodological Approaches to Spatial Heterogeneity

Spatial heterogeneity is a ubiquitous feature of all natural systems. It is both the signature of past ecological processes at different biological and geological scales, and the arena in which future processes will be realized. Historically, ecologists have tended to minimise the importance of spatial heterogeneity by averaging over various spatial scales; it is simpler to relate small-scale experiments to a homogeneous system than a heterogeneous system and homogeneous systems are easier to study (Strayer et al., 2003). However, there is increasing recognition that spatial heterogeneity in vegetation and soil has the potential for significant effects on biotic interactions, plant community dynamics and ecosystem functioning (Tilman 1988; Collins & Wein 1998; Hutchings et al., 2003; Lin et al., 2010).

Before we can analyze the causes and consequences of spatial heterogeneity in grazed grasslands, a few key concepts regarding the description of spatial heterogeneity require some introduction. As a first approximation, the degree of spatial heterogeneity in a given variable of interest can be measured with any statistical estimate of data variability, such as dispersion indices and measures of species turnover between sampling units (Whittaker 1960; Dale 1999). Measurements or studies of spatial heterogeneity using such non-spatial metrics can be considered as spatially implicit (Adler et al., 2001). In cases where a significant proportion of the observed variation is driven by spatial factors, this results in predictable arrangements in space. Indeed, spatial heterogeneity can take different distinctive forms or patterns, depending on the spatial dependence of variables from one location to another (Fortin & Dale 2005). In the case of single variables, spatial patterns can be classified as 'gradients', 'aggregated/patchy' or 'random' (Figure 1). When multiple variables are considered simultaneously, spatial patterns can be classified as aggregated, segregated or random.

Spatial data are of two types: (i) point pattern data where geographical coordinates are recorded for discrete entities (e.g. position of plant individuals in space), and (ii) regionalized variables where quantitative, semi-quantitative or qualitative variables are continuously distributed in space (e.g. soil nitrogen availability). Point patterns can be analysed with techniques such as second-order statistics and joint-count statistics (Fortin & Dale 2005). Regionalized variables can be described by various measures of spatial dependence including semivariograms, autocorrelograms, contagion indices and fractal analysis (Wiens 1989; Robertson & Gross 1994; Fortin & Dale 2005). Fractal analyses can also provide information on the scale-dependency of spatial patterns.

The problem of scale applies to the spatial pattern of organisms, where dispersal limitation may be a critical issue for population and species persistence (Hanski 1998; Engler et al., 2009), as well as to abiotic factors, biotic interactions and ecosystem process rates. Indeed, defining the relevant scale at which species distributions or ecological processes should be considered is a fundamental step to produce general predictions about spatial heterogeneity (Levin 1992; Chave 2013). For instance, the size of individual plants determines the scale of their competitive neighbourhood through depletion of local resources in a volume that depends on the shape and area of their canopy and rhizosphere (Ramseier & Weiner 2006). Moreover, neighbouring competition promotes the coexistence of species at larger scales (Tilman 1994). Biotic drivers also interact with abiotic factors to shape scale-dependent changes in spatial pattern of a given response variable. In work on restored grasslands, distribution of plant functional groups was found to be patchy at the fine-scale (a few meters) but to display a gradient pattern at the large scale (hundreds of meters) due to underlying gradients in soil texture (Pottier *et al.,* 2009). Understanding whether results vary as a function of scale or show consistent patterns in scaling effects is an important part of spatial heterogeneity assessments.

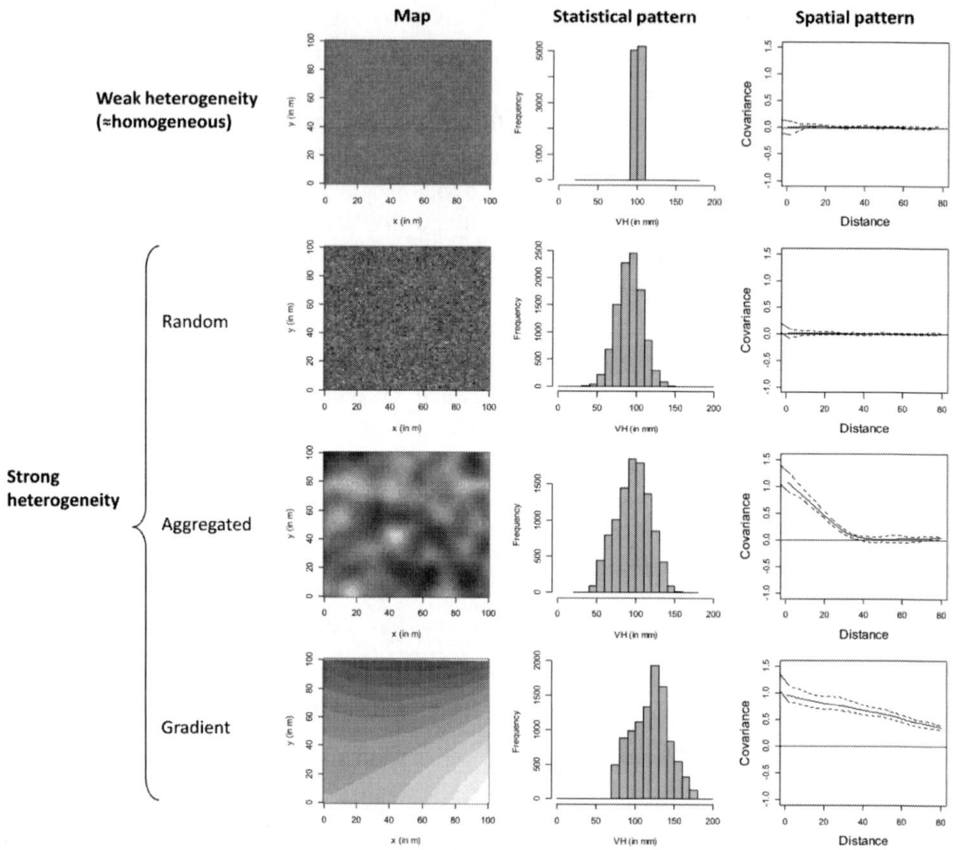

Figure 1. Examples of simulated spatial patterns of vegetation height (VH) within a hypothetical paddock. The statistical distribution of vegetation height allows homogeneous and heterogeneous cases to be distinguished. The spatial autocorrelation pattern, represented here with a covariogram, distinguishes between random, aggregated and gradient patterns.

In their overview of spatial methods in ecology, Fortin & Dale (2005) emphasize that meaningful spatial analysis requires: i) explicitly-defined variables or parameters of interest, ii) appropriate choices of scale and number of sampling units in order to allow relevant inference of the ecological phenomena, and iii) adapted spatial sampling schemes to capture spatial patterns. Detecting spatial patterns of regionalized variables is a function of the extent (area), resolution (size of individual observation units or grain) and spatial lag (distance between sampling locations) of the study (Figure 2). In the case of temperate grasslands grazed by large herbivores, we can consider that an extent covering a contiguous area where the stocking rate is constant (a paddock) is a useful standard. The grain size will vary with the grassland attribute under study and question. Given that it is generally impossible to exhaustively sample the targeted study area as a grid, the distance between sampling locations (spatial lag), should be carefully defined. Together with the sampling effort, this determines the statistical distribution of distances between sampling units along with the power and resolution of subsequent spatial exploration methods (Dungan et al., 2002). Design of an optimal sampling scheme to assess spatial heterogeneity reflects a trade-off between sampling locations that are too close together (redundancy of information) and sampling locations that are too far apart, with extraneous variation arising from processes at other spatial scales (Figure 3).

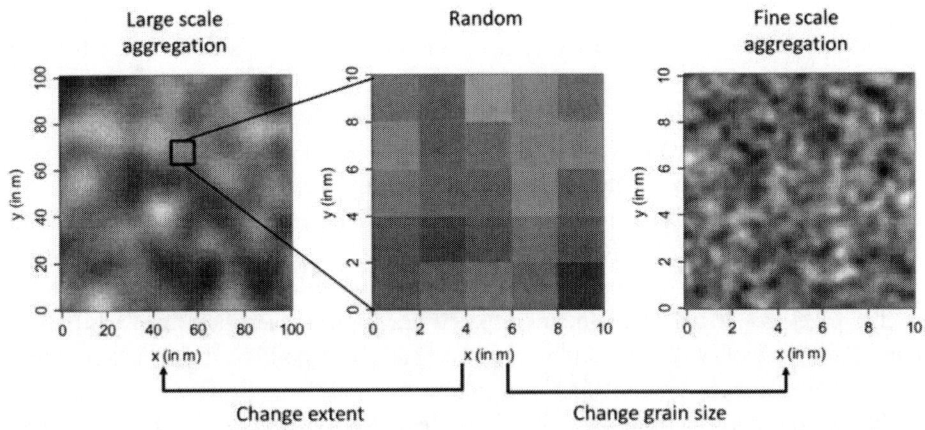

Figure 2. Pattern detection, scale and the observer. The perception of spatial pattern is highly sensitive to changing grain and extent of the sampling design.

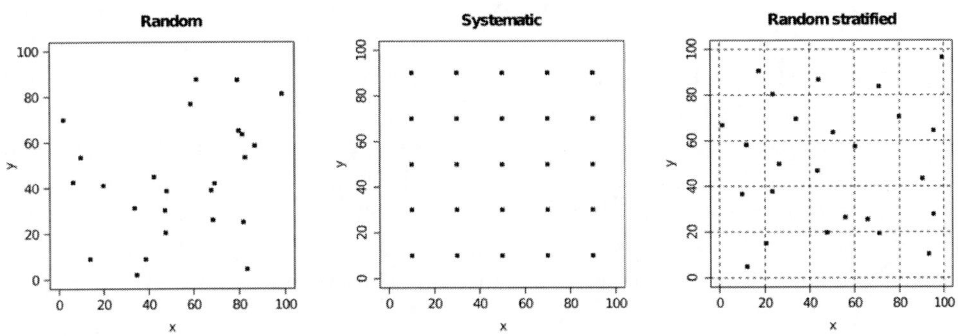

Figure 3. Common sampling designs for detecting spatial heterogeneity. Random sampling offers a larger variety of distances between sampling locations, and therefore a finer grain of spatial analyses compared with more regular spacing of sampling units, but may over- or under-sample some zones of the sampling domain. Stratified sampling schemes or nested grids may provide greater power for detecting particular types of spatial pattern (Fontin & Dale 2005).

Grazing as a Driver of Spatial Heterogeneity in Grassland Ecosystems

Grazed grasslands are dynamic, complex ecosystems under constant modification by the management of domestic herbivores (Parsons et al., 2000). These large mammals have direct effects on vegetation and soil via three main mechanisms: defoliation or removal of plant tissue, return of excreta (urine, dung) and trampling (Snaydon 1981). Moreover, herbivores have indirect effects on ecosystem structure and function due to plant-soil feedbacks, which modify above- and belowground properties over a variety of temporal and spatial scales (Bardgett & Wardle 2003).

Spatial heterogeneity in vegetation structure (height, cover, phenological state) is an inevitable consequence of animal grazing, as herbivores both create and respond to variation in vegetation state. Herbivore foraging leads to patches of short and tall stands of vegetation (grazed and ungrazed stands, respectively) as soon as vegetation biomass exceeds herbivore intake (Adler et al., 2001). Patchiness of vegetation structure is further exacerbated by selective animal foraging and social behavior (Parsons & Dumont 2003). Grazing typically maintains plants in a vegetative, nitrogen-rich condition, and positive feedbacks between grazing and plant quality promote the continued use of previously grazed patches. These grazing-induced changes in vegetation height may have indirect effects on microclimate and local soil conditions by altering plant growth rates, quality/quantity of above- and belowground litter inputs, root exudation and biotic interactions (Semmartin & Oesterheld 2001; Schrama et al., 2013). If maintained, patch grazing may result in local shifts in plant community composition with tall perennial grasses at a competitive disadvantage compared with shorter, more grazing-tolerant species (Adler et al., 2001).

In addition to plant tissue removal, grazing affects soil properties and vegetation patterns via animal returns. Herbivores remove high C:N material from large areas of grassland, much of it being returned to small areas as low C:N waste products in high concentrations (Afzal & Adams 1992). These animal returns have direct effects on soil properties, stimulating microbial activity and nutrient transformations, with consequences for neighbouring plant growth (Haynes & Williams 1993; Patra et al., 2005; Schrama et al., 2013). Furthermore, animal dung may have significant effects on local vegetation dynamics by creating favourable microhabitats for seed germination (Malo & Suárez 1995; Dai 2000). As with herbivore foraging, animal social behavior and spatial variation in animal activities (grazing, resting, attraction to water points) promotes an uneven spatial distribution in excretal patches across the grassland (Haynes & Williams 1993; White et al., 2001; Kohler et al., 2006a; Auerswald et al., 2009). This spatial heterogeneity of animal returns may be particularly pronounced at low stocking rates, in large fields or in continually grazed systems compared with rotational grazing (Dubeux 2005).

Animal returns may also interact with animal foraging behaviour to reinforce patterns of within-field heterogeneity. For example, large herbivores tend to avoid grazing in the immediate vicinity of dung pats, resulting in the development of taller vegetation close to dung: these tall patches may persist even after dung disappearance due to subsequent shifts in plant quality (and, hence palatability) (MacDiarmid & Watkin 1972; Gillet et al., 2010). This phenomenon is particularly marked in grasslands grazed by horses, where patterns of spatially-distinct, tall-grass latrines and areas of short grass are driven by both optimal foraging and avoidance of gastrointestinal parasites present in fecal patches (Fleurance et al., 2007).

Non-random animal movements and habitat-use underlie spatial heterogeneity in the third main component of grazing, i.e. trampling. Trampling by large herbivores exerts mechanical pressure and physical hoof damage on both the soil and vegetation, creating significant disturbance where animals pass repeatedly (tracks, watering points, resting points) and where stocking rates are high (Sheath & Carlson 1998; Drewry 2006). Presence of animals and trampling effects have been linked to changes in soil physical properties such as bulk density, porosity and water infiltration, as well as to reductions in grassland productivity (Singleton et al., 2000; Menneer & Ledgard 2005; Kurz et al., 2006). Nevertheless, trampling may also accelerate local nutrient transformations and vegetation dynamics by promoting the incorporation of organic matter into the soil and creating gaps in the vegetation which facilitate plant colonisation (Danell et al., 2006; Kohler et al., 2006b). Consequently the spatial heterogeneity in grazed grasslands encompasses both a mosaic of grazed/ungrazed patches of differing sward height, and areas affected by animal returns and trampling.

Consequences of Spatial Heterogeneity in Grazed Grassland Ecosystems

Although general patterns of grassland response to grazing have been the subject of comprehensive reviews and meta-analyses (Milchunas et al., 1988; Olff & Ritchie 1998; Danell et al., 2006; Bakker et al., 2006; Díaz et al., 2007a; McSherry & Ritchie 2013), consequences of spatial heterogeneity in grazed grasslands have not been similarly reviewed. We redress this by asking the question: does fine-scale grazing-induced spatial heterogeneity significantly affect field-scale grassland structure and function? As a basis for discussion, we surveyed the literature over the last 15 years to find studies that explicitly examined spatial heterogeneity in grasslands grazed by large herbivores. In particular, we sought to examine how spatial heterogeneity driven by different components of grazing (defoliation, animal returns or trampling) impacts plant diversity and biogeochemical cycling in temperate grasslands. A search was conducted in ISI® Web of Science with the timespan 1997 to 2013 using the keywords “grazing” and “grassland” and “spatial heterogeneity” or “spatial pattern” or “patch” or “beta diversity” or “spatial turnover”. This resulted in 629 references. All abstracts were looked through to eliminate irrelevant articles. Articles focusing on trees or shrubs were excluded, as were landscape-level studies where grasslands were considered as habitat patches in a large-scale mosaic. This evaluation resulted in a set of 144 references. We acknowledge that this is not an exhaustive list of all articles addressing heterogeneity in grazed grasslands, but our search terms allowed the gathering of the most relevant articles.

Short-listed studies from our literature search could be attributed to one of four main research themes (Figure 4). The majority of studies (44%) addressed species distributions, patterns of abundance or biotic interactions in relation to heterogeneity. One quarter of studies concerned interactions between vegetation heterogeneity and herbivore foraging behaviour. The remaining studies considered various aspects of heterogeneity and ecosystem function (production, C/N cycling), including patterns of spatial variation, and to a lesser extent, impacts of heterogeneity on ecosystem function.

For the purposes of this review, we do not present impacts of heterogeneous vegetation on vertebrate behavior but focus instead on plant community responses and ecosystem processes to identify overall trends and gaps in our understanding. To ensure a balanced view, we placed results from our literature search into the wider context of previously published literature wherever possible. Such appraisal is presented below.

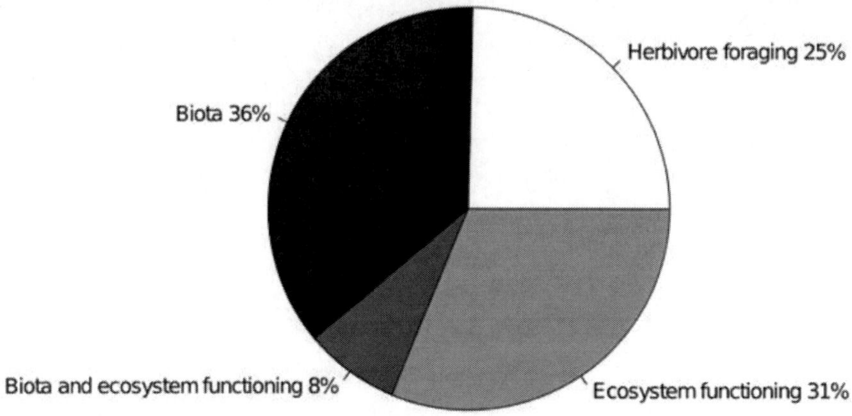

Figure 4. Results from an ISI® Web of Science search targeting spatial heterogeneity in grazed grasslands, with articles classified according to research topic (see text for full methodological details).

Impacts of Spatial Heterogeneity on Grassland Biota

Spatial heterogeneity is the cornerstone of many theoretical models of species coexistence (Chesson 2000; Barot & Gignoux 2004). It seems therefore reasonable to suppose that grazing-induced spatial heterogeneity could have significant implications for the diversity of grazed grasslands at the local-scale (alpha diversity within sampling units), within-field variation (beta diversity, between sampling units) or at the field-scale (considered here as gamma diversity, summing alpha and beta diversity). Heterogeneity-driven impacts on community structure and patterns of diversity may result from a number of different mechanisms including spatial niche partitioning and storage effects, patch dynamics or competition-colonisation trade-offs (Amarasekare 2003; Leibold et al., 2004). Here we describe the main ways in which spatial heterogeneity within grazed fields can impact populations or communities of organisms, and assess the experimental/empirical evidence in support of these ideas from the literature over the last 15 years (ISI® search described above). We focus primarily on plant communities, since the results of our literature search concerning spatial heterogeneity and biota showed a clear bias towards vascular plant species (62 articles of 75).

Assuming that species show ecological differences (e.g. ability to acquire and use resources, tolerance to defoliation, regeneration strategy), spatial heterogeneity may promote spatial niche partitioning and reduce interspecific competition for resources. Indeed, moderate grazing has long been assumed to benefit diversity by creating a mosaic of short and tall patches with contrasting growth forms and competitive interactions (Milchunas et al., 1988). Local changes in light and soil conditions due to trampling, defoliation and animal returns underlie patterns of differential plant growth and survival in space, promoting a shifting competitive hierarchy between co-existing plants (Grime 1979; Tilman 1985; Díaz et al.,

2004; Wright et al., 2004). In our literature search, we found 31 original articles concerning spatial variation of species responses in grazed grassland systems (Figure 5). Two articles specifically addressed the role of animal returns on species abundance and community composition using spatially-explicit approaches (Steinauer & Collins 2001, Gillet et al., 2010), demonstrating species-level changes both at the patch-level and in the immediate surroundings. Urine-induced changes in local-scale species abundance interacted with herbivore preferences for N-rich, palatable species in urine patches (Steinauer & Collins 2001). We did not find any articles examining species-level responses to spatial heterogeneity in defoliation directly, but several studies documented patch-level changes in community composition associated with species palatability, suggesting either local-scale, positive plant-plant interactions due to 'antiherbivore' refuge effects (Bisigato 2005; Estapé et al., 2013) or negative plant-plant interactions due to proximity to palatable species (Palmer et al., 2003).

In addition to direct effects on plant growth and survival, herbivore-mediated disturbances resulting from trampling or large dung pats have the potential to modify vegetation dynamics and influence species abundance via spatial heterogeneity in plant regeneration success (Bisigato & Bertiller, 2004). Impacts on plant regeneration may be further exacerbated by uneven patterns of defoliation, limiting seed production locally through removal of reproductive structures and modifying vegetative growth (Klimešová et al., 2008). Dung pats (and bare soil areas) may be recolonised by clonal ingress of surrounding plants or by seeds, either present in the soil or in the animal returns (Haynes & Williams 1993).

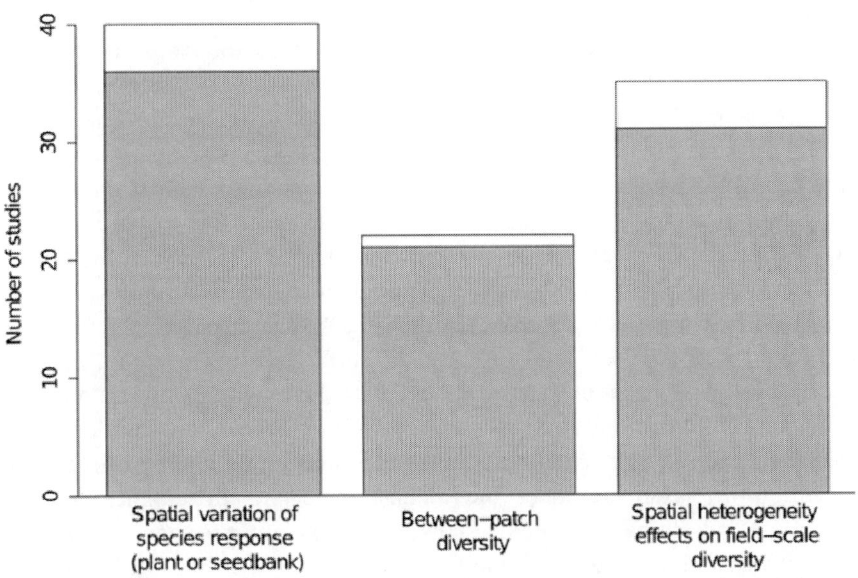

Figure 5. Results from an ISI® Web of Science search targeting effects of spatial heterogeneity on plant species and diversity in grazed grasslands, with articles classified according to topic. Shaded bars represent studies where grazing is present as either a gradient of intensity or binary factor (grazed/ungrazed); open bars represent studies that explicitly consider grazing components (i.e. defoliation, animal return or trampling).

Grazers are known to disperse seeds through exozoochory and endozoochory, although impacts of herbivore ingestion on subsequent seed germination may be mixed (Russi et al., 1992; Cosyns et al., 2005). Compared to the relatively large number of articles published on spatial variation in species responses to grazing in terms of abundance over the last 15 years, we found few studies reporting on the spatial heterogeneity of the seed bank or clonal reproduction under grazed conditions. These studies confirm that the regeneration potential of species shows patch-level variation in response to defoliation (Benot et al., 2011) and animal returns (Meissner & Facelli 1999; Malo et al., 2000; Hérault & Hiernaux 2004; Pazos & Bertiller 2007). However, we found only one example of work in grazed grasslands which assessed the realised regeneration success, showing greater spatial variation of success through seedling counts in grazed compared to ungrazed conditions (Bisigato & Bertiller 2004).

Patch-level variation in species abundance due to the ecological processes described above should translate into increased beta diversity (within-field variation of species composition) where multiple patch states are present. Our literature search revealed 13 articles, which measured beta diversity in grazed grasslands (Table 1). In general, grazing was associated with an increase in beta diversity, particularly in mesic conditions. Nevertheless, decreases or no significant changes in beta diversity were also observed in unproductive and/or over grazed habitats (Table 1). Indeed, results from tallgrass prairies suggest that grazing effects may show a humpback relationship with increasing levels of productivity (Frank 2005). Lack of consistent results between studies may reflect differences in stocking rate, herbivore selectivity and available forage which modify the degree of grazing-induced spatial heterogeneity (Dumont et al., 2011), as well as local site factors which may confound patterns of grazing-induced heterogeneity (Adler et al., 2001; Augustine 2003).

Between-patch variation associated with beta diversity and species turnover is integral to many theoretical mechanisms of species coexistence ('between-patch' mechanisms *sensu* Wilson 2011). In theory, such mechanisms promote field-scale diversity in grazed systems. When patches differ from each other in terms of their environmental characteristics on a sufficiently long time-scale, field-scale diversity is enhanced by spatial niche partitioning (reviewed by Barot & Gignoux 2004, Leibold et al., 2004).

An alternative mechanism is patch dynamics where patch attributes do not vary but instead species dynamics are governed by between-patch dispersal and local extinctions (Pickett & White 1985; Wilson 1994). Repeated disturbances by grazers may create a mosaic of patches, some with ruderal species, others with highly-competitive dominant species. Unlike the competition-colonisation model (see below), species are considered to have an equal chance to reach all patches. In this context, the maximum number of coexisting species is predicted to be sustained under intermediate levels of disturbance (Connell 1978). However, given that species commonly show a tradeoff between their competitive abilities and their dispersal abilities (Turnbull et al., 1999; Coomes & Grubb 2003), theoretical models have highlighted the positive consequences of this tradeoff for the maintenance of high species diversity (Competition-colonisation model, Levins & Culver 1971; Tilman 1994; Calcagno et al., 2006). Competitively-inferior species can coexist with competitively-superior species provided that the latter has lower colonisation capacity. Here, grazing-induced disturbance may be of great importance for the creation of open 'safe' sites for the competitively-inferior species. Finally, spatial heterogeneity can interact with fluctuating environmental conditions over time (Storage Effect Hypothesis, Chesson & Warner 1981).

Table 1. Summary features of studies addressing effect of grazing on the beta component of diversity within fields (ISI® Web of Science search for the timespan 1997-2013)

Habitat type	Grazing treatment	Herbivore	Co-factor	Grain[1]	Sampling design	Extent[2]	Grazing effect on beta-diversity	Reference
Arid	Grazed / ungrazed	sheep		$0.25\ m^2$	ten random quadrats	$50\text{-}200 m^2$	-	Meissner & Facelli (1999)
Costal dune	Grazed / ungrazed	catlle/pony/horse	soil acidity	$0.0625 m^2$	ten random quadrats	$64\ m^2$	- on low pH ; + on high pH	Kohyani et al. (2008)
Mediterranean	Ungrazed / moderate / heavy	cattle		$0.0625 m^2$	five random quadrats	$100\ m^2$	+	Golodets et al. (2011)
Savannah	Grazed(intensive) / ungrazed	many wildlife species	fire	$1 m^2$	four stratified random quadrats	$144\ m^2$	-	Masunga et al. (2012)
Semi-arid	Grazed(intensive) / ungrazed	cattle and sheep		$4\ m^2$	24-40 random quadrats	0.6 ha	0	Zhang (1998)
Subalpine	Stocking rate gradient	cattle, sheep and horse		point intercept	20 cm between points along a transect	250 m	+	Komac et al. (2011)
Subalpine	Grazed / ungrazed	sheep	grain size	$0.0025\text{-}1 m^2$	8-40 random quadrats	$100\ m^2$	-	Deléglise et al. (2011)
Temperate	Grazing intensity	sheep and deer		point intercept	40 evenly spaced points along a W shape transect	$100\ m^2$	+	DeGabriel et al. (2011)
Temperate	Grazed / ungrazed	bison	topographic position	$10 m^2$	five evenly spaced quadrats along a transect	50 m	+	Bakker et al. (2003)
Temperate	Grazed / mown	cattle	fertilisation and abiotic environment	$16.2 m^2$	three circular plots separated by 10 m	20 m	+	Klimek et al. (2008)
Temperate	Light / moderate / heavy	sheep		$0.001\ m^2$	Sierpinsky triangular, lag varied between 0.15 to 30m	$390\ m^2$	-	Dorrough et al. (2007)
Temperate	Grazed / ungrazed	bison	aboveground production	$1\ m^2$	five random quadrats	100 to $700\ m^2$	- at low and high productivity level; + at intermediate productivity level	Frank (2005)
Temperate	Grazed / ungrazed	cattle and horse		$0.0625 m^2$	25 random quadrats	1-2 ha	+	Marion et al. (2011)

[1] grain refers to the size of sampling units; [2] extent refers to the field area under study

Indeed, patchy defoliation and animal returns offer local and occasional opportunities for weak competitors to increase their growth rates and avoid competitive exclusion, provided that their decline is sufficiently slow under unfavorable conditions.

It is notable that all the studies documenting an increase in beta diversity in relation to grazing obtained from our literature search, also showed an increase in field-scale diversity. Patterns of grazing-induced changes in beta diversity were mirrored by those of gamma diversity in all cases. Of the 13 studies documenting beta diversity in grazed systems, one explicitly referred to the possible underlying coexistence mechanism (spatial niche partitioning, Bakker et al., 2003). In addition, a number of studies provided information on either spatial heterogeneity in species distributions and field-scale diversity (with no explicit measurement of beta diversity), or support for one of the 'between-patch' coexistence mechanisms listed above. The majority of these studies provide support for spatial niche partitioning, demonstrating positive relationships between spatial heterogeneity of resources and field diversity (Schlapfer et al., 1998; Augustine 2003; Bakker et al., 2003; Mikola et al., 2009; Yoshihara et al., 2010; Zhang et al., 2013). Results from four studies are consistent with patch dynamics, showing patch-stage transitions in successional phase (Bisigato & Bertiller 1997; Cipriotti & Aguiar 2005; Mcivor et al., 2005; Koerner & Collins 2013). Only one study examined the competition-colonisation hypothesis and found it was important in explaining community structure and vegetation dynamics for savannah (Porensky et al., 2013). Within the confines of our literature search, we found no studies addressing the storage effect resulting from grazing, although experimental methods for quantifying this mechanism have previously been proposed (Chesson & Warner 1981; Sears & Chesson 2007).

Given that field-scale diversity is driven by both alpha and beta diversity, the consequence of increased beta diversity on gamma diversity may be masked by decreases alpha diversity. This phenomenon is suggested by studies reporting high spatial heterogeneity of species but decreased or no change in field-scale diversity in response to grazing (Wang et al., 2002; Alados et al., 2007; Zhou et al., 2008). Alados and coauthors (2007) explain the discrepancy between beta and gamma diversity in their study by the fact that very few species can successfully colonise empty spaces created by herbivores. Further work is required to determine the relative importance of 'between-patch' mechanisms of species coexistence (spatial niche partitioning, patch dynamics, competition/colonization trade-off, Storage Effect) in grazed systems. Moreover, systematic partitioning of the alpha and beta components of field-scale diversity will provide valuable insights into the underlying within- and between-patch coexistence mechanisms induced by grazing.

In our literature search, the majority of studies examining biodiversity responses to grazing focused on plant species diversity. Nevertheless, changes in plant community structure may alter the soil microbial community by modifying the quantity and quality of resources that plants produce (Wardle et al., 2004), or by increasing the spatial variability of soil resources (Ettema & Wardle 2002). Although increasing plant diversity has been shown to have a positive effect on microbial diversity in many experimental studies, evidence from natural systems is mixed (Millard & Singh 2010). Animal returns have also been shown to modify the soil microbial community, and it is suggested that the high nutrient content and/or modified pH associated with animal waste selects for particular microorganisms (Mahmood & Prosser 2006). Our literature search counted only two articles dealing with microbes in the context of spatial heterogeneity (Northup et al., 1999; Sørensen et al., 2009). These studies indicate that greater microbial biomass is associated with tussocks compared with bare patches created by high grazing pressure (Northup et al., 1999), and that different grazing components (trampling and animal returns) interact on bacterial and fungal microbes (Sørensen et al., 2009).

Spatial heterogeneity of the structural characteristics of vegetation and community composition may also serve species of higher trophic levels. In our literature search we counted eight studies that considered arthropods. The overwhelming majority of these studies (seven out of eight) focused on aboveground arthropods (e.g. orthoptera, arachnids and beetles), with only one study considering belowground collembolans (Sørensen et al., 2009). Heterogeneous vegetation implies heterogeneous habitats for a multitude of arthropods, although species-specific responses to grazing make generalisations difficult (Joern & Laws 2013). We found several studies where the abundance of different groups of arthropods increased in tall vegetation patches, linked to protection from predators, increased food availability and favorable microclimates (Dittrich & Helden 2012; Spalinger et al., 2012). Presence of strong structural heterogeneity is often associated with higher arthropod diversity (Dennis et al., 1998; Reid & Hochuli 2007, Verdú et al., 2007, Dittrich & Helden 2012, Zhu et al., 2012), vegetation composition playing a minor role (Reid & Hochuli 2007, Zhu et al., 2012, but see van Klink et al., 2013). Grazing-induced spatial heterogeneity linked with soil nutrients and trampling has also been shown to increase soil nematode abundance (Sørensen et al., 2009) and modify abundance of some earthworm groups (Mikola et al., 2009, Sørensen et al., 2009). However, as with plants, overall effects of grazing on abundance and beta diversity of soil nematodes may vary depending on the degree of within-field, grazing-induced spatial heterogeneity (Veen et al., 2010).

Impacts of Spatial Heterogeneity on Grassland Function

Functional understanding of spatial pattern in heterogeneous systems requires study of the pools and fluxes of nutrients or materials between interacting patches with different attributes and spatial arrangement (Lovett et al., 2005). One approach to quantifying impacts of grazing-induced spatial heterogeneity on plant and soil processes involves the direct manipulation of spatial variability and/or pattern in grazing components (animal returns, defoliation, trampling), characterising patch attributes, between-patch transfers and whole-system performance. An alternative, more indirect approach relies on manipulative or correlative studies linking spatial heterogeneity in soil nutrients, botanical composition and sward structure (i.e. grazing-induced changes in ecosystem properties) with variation in ecosystem processes. Clarifying the relationship between spatial heterogeneity and ecosystem processes is a key step to determine when spatial heterogeneity is important, with implications for sampling strategies and upscaling procedures. To what extent do small-scale spatially-explicit patch interactions mediate larger-scale ecosystem function? Are average values of process rates sufficient to describe C/N fluxes across grazed fields?

Over the last two decades, there has been considerable interest in plant responses to soil nutrient heterogeneity at different scales of biological organisation (individuals, populations, communities; see reviews by Robinson 1994; Hutchings et al., 2003; Hodge 2006; García-Palacios et al., 2012). Using a variety of experimental techniques at the micro- and mesocosm level, these studies have demonstrated that soil heterogeneity can promote both morphological and physiological plant responses, modifying plant resource use, population yield and community-level productivity. More recently, soil nitrogen (N) heterogeneity has also been shown to moderate diversity-productivity relationships (Tylianakis et al., 2008), and to interact with global change drivers on above- and belowground plant biomass (García-Palacios et al., 2012). To date, however, work on soil nutrient heterogeneity has generally focused on homogeneous- and heterogeneous-treatment comparisons rather than spatially-explicit responses, patch transfers or effects of spatial pattern *per se* (but see Day et al., 2003; Monzeglio & Stoll 2005 as some examples). Furthermore, there are relatively few studies that explicitly examine plant and soil responses to spatial variability and pattern in animal returns, defoliation or trampling.

In the literature search described above, we found 17 original articles concerning impacts of animal returns on ecosystem function in the context of grazing and spatial heterogeneity (Table 2). The majority of articles assessed impacts of animal excreta on nitrogen cycling, with particular emphasis on gaseous or leaching losses of N. This is consistent with growing concerns about trace greenhouse gases and global warming, as well as agricultural contributions to ground water pollution (Steinfeld et al., 2006).

Table 2. Summary features of studies addressing impacts of animal returns on ecosystem function in the context of grazing and spatial heterogeneity (ISI® Web of Science search for the timespan 1997-2013)

Study	Study type	Animal returns	Herbivore model	Spatial Heterogeneity	Response variable
Wang et al. (2013)	F	U,D	Sheep	Implicit	CO_2 & CH_4 fluxes
Li et al. (2012)	M	U	Cattle	Explicit	N leaching
Hartmann et al. (2012)	F	U	Cattle	Implicit	N_2O fluxes, plant & soil C/N stocks
White-Leech et al. (2013)	F	U,D	Cattle	Explicit	Shoot biomass, crude protein content
Schnyder et al. (2009)	F	U,D	Cattle	Explicit	Soil N & P stocks
Saarijarvi & Virkajarvi (2009)	F	U,D	Cattle	Explicit	Plant & soil N stocks
Orwin et al. (2009)	F	U	Cattle	Explicit	Plant biomass, plant N & soil C/N stocks
Mikola et al. (2009)	F	U+D mix	Cattle	Implicit	Plant biomass, N & P; soil inorganic N
Wang (2008)	M	U	Cattle	Explicit	N leaching
Jewel et al. (2007)	M	D	Cattle	Explicit	Soil P stocks
Di & Cameron (2007)	L	U	Cattle	Implicit	NO_3-N leaching loss
Holst et al. (2007)	F	U, D	Sheep	Implicit	N_2O fluxes
Hutchings et al. (2007)	M	U,D	Cattle	Explicit	N_2O fluxes, NO_3-N leaching
Chardon et al. (2007)	L	D	Cattle	Implicit	Soil P stocks, P leaching
Anger et al. (2003)	F	U	Cattle	Implicit	N_2O fluxes
Semmartin & Oesterheld (2001)	F	U (urea-N)	n/a	Implicit	Plant biomass
Silva et al. (1999)	L	U	Cattle	Implicit	NO_3-N leaching loss

F, Field; M, Modelling; L, Lysimeter; U, Urine; D, Dung.

Impacts of cattle excreta on nutrient stocks or transfers were most frequently assessed (14 articles), with a bias towards impacts of urine (11 cases including urine versus six with dung). Nine of the studies in Table 2 were found to be spatially implicit, comparing parameters measured in or under patches of dung and urine with control soil unaffected by animal returns. The remaining eight articles were spatially explicit, either measuring response parameters across space (e.g. White-Leech et al., 2013), recording plant protein content in biomass at varying distances from the edge of dung patches) or accounting for spatial configuration of excreta patches and response areas in simulation studies (e.g. Li et al., 2012).

Taken together, the publications on impacts of cattle urine provide a clear illustration of the importance of grazing-induced spatial heterogeneity on grassland ecosystem function. Urine patches are known to represent a high input of N, readily hydrolyzed to plant-available forms; these patches are associated with increased plant N yield as well as increased N_2O emissions and N leaching losses (Anger et al., 2003; Saarijärvi & Virkajärvi 2009). Lateral diffusion of urine-N increases plant growth and/or N yield beyond the area of the urine patch, with significant effects detected up to 30 cm away from the patch edge (Saarijärvi & Virkajärvi 2009; White-Leech et al., 2013). In contrast, N leaching losses decrease with increasing distance away from the centre of the urine patch (Wang 2008). However, urine patch size modifies soil N retention, with larger urine patches promoting greater plant N uptake and biomass production (Orwin et al., 2009). Modelling studies confirm that urine patches promote paddock-scale N losses via N_2O emissions or N leaching, and show that calculation of N leaching using evenly-distributed or average urine patches (in terms of volume and N concentration) significantly underestimate grassland N losses (Hutchings et al., 2007; Li et al., 2012).

Whilst spatial heterogeneity in animal returns clearly has an important influence on plant and soil processes, results from experimental studies suggest that defoliation is the main driver of grazing effects on plant biomass and N stocks in cattle-grazed systems (Mikola et al., 2009). Consequently spatial heterogeneity in defoliation may also have a significant impact on ecosystem processes, particularly if patch-level variation in sward height results in patch-level convergence in other plant traits via environmental filtering (Götzenberger et al., 2012). Nevertheless, our literature search found few studies on the topic of spatial heterogeneity in defoliation and grassland function. Previous work has shown that defoliation increases the light penetration to the base of the plant canopy in artificially-created patches, with greater increases in larger patches (Semmartin & Oesterheld 2001).

It seems reasonable to expect that defoliation-induced changes in microclimate may affect temperature- or moisture-dependent process rates at the local scale. Comparisons of short and tall vegetation patches with different levels of grazing pressure have revealed differences in litter decomposition rates (Vaieretti et al., 2013) and soil water infiltration (McIntyre & Tongway 2005) between patch-types but no differences in soil inorganic N or N mineralisation rates (Rossignol et al., 2006; Vaieretti et al., 2010). However, none of these spatially-implicit studies controlled for species composition, and therefore the observed patch effects may be confounded by indirect effects of plant diversity rather than direct effects of defoliation.

Biodiversity is widely considered to play a key role for ecosystem functions such as productivity, stability and resistance to invasions (Loreau & Hector 2001; Hooper et al., 2005). In the previous section, we highlighted the potential for spatial heterogeneity to modify patterns of both patch-scale and field-scale diversity in grazed grasslands. Such grazing-induced changes in floristic composition may have cascading effects on ecosystem processes via biodiversity-ecosystem functioning relationships and complementarity effects (Hooper et al., 2005; Loreau 2010). For example, vegetation patches of different floristic composition have been shown to partly coincide with spatial variation in soil N mineralisation (Rossignol et al., 2006) and N_2O fluxes (Klumpp et al., 2010) in grazed grasslands. In addition, communities which are more diverse in species or functional groups are predicted to have greater stability against environmental perturbations, driven by redundancy of functionally-similar species or compensation by better adapted species which buffer ecosystem processes under changing conditions (Hooper et al., 2005; Bloor & Bardgett 2012). To our knowledge, the significance of spatial patterns in plant diversity (i.e. within-field beta diversity) for grazed grassland ecosystem function remains to be investigated. Given that changes in plant traits and species abundance may lead to significant changes in ecosystem function before species loss occurs (Hillebrand et al., 2008), understanding the consequences of altered functional diversity within plant communities is also critical for the assessment of ecosystem processes in heterogeneous, grazed grasslands.

Functional diversity is defined as the value, range and relative abundance of plant traits present in a given community (Díaz et al., 2007b). Numerous authors have advocated the use of a trait-based framework for a mechanistic understanding of biodiversity-ecosystem function relationships, reducing the complexity of multi-species communities to a few key traits (Lavorel & Garnier 2002; Díaz et al., 2007b). This approach rests on the measurement of morphological, physiological or phenological plant features to scale up from individual plants to ecosystem function, targeting plant traits which vary in response to environmental conditions (response traits) and/or affect ecosystem processes (effect traits) (Violle et al., 2007). For example, leaf dry matter content (LDMC), an index of leaf tissue density related to plant nutrient retention and growth, provides valuable information on plant resource-use strategies at the community level (Lavorel & Garnier 2002). Work on leaf functional traits has provided convincing evidence for a spectrum of resource-use strategies among plants; this resource-use spectrum opposes exploitative species (high specific leaf area and leaf N content, low LDMC) with conservative species characterised by the opposite leaf traits (Diaz et al., 2004; Wright et al., 2004). Recent work also suggests that variation in plant traits and resource-use strategy are strongly linked to microbial traits and belowground processes (García-Palacios et al., 2012; Grigulis et al., 2013). Indeed, the cascading effects of covariation in plant and microbial traits on ecosystem function may underlie trade-offs between ecosystem services such as biomass production and nutrient retention (Grigulis et al., 2013).

We suggest that spatial heterogeneity could also be included in conceptual models linking plant traits and ecosystem functioning, particularly for cases where grazing-induced spatial heterogeneity is associated with discrete patch matrices such as excreta patches or grazing lawns (Figure 6). Here, we choose to follow the trait-based approach of Lavorel and Garnier (2002), which articulates environmental response and ecosystem effects through species traits. In this model, the plant traits expressed at a given time in space reflect the outcome of both abiotic and biotic filters which constrain local species persistence from a larger, regional pool of species (Lavorel & Garnier 2002).

In grazed systems, large herbivores are expected to constrain within-patch trait distributions with cascading effects for plant and soil structure and function (Grigulis et al., 2013). Moreover, grazing activities may promote spatial pattern in between-patch variation of plant traits (Figure 6), with implications for coexistence of different plant strategies at the field-scale, and the effects of plant diversity on ecosystem function. By assessing community responses to key components of spatial heterogeneity, and their concomitant effects on ecosystem processes, it may be possible to disentangle the relative importance of spatial pattern and patch attributes on ecosystem function by using simple plant-based indicators.

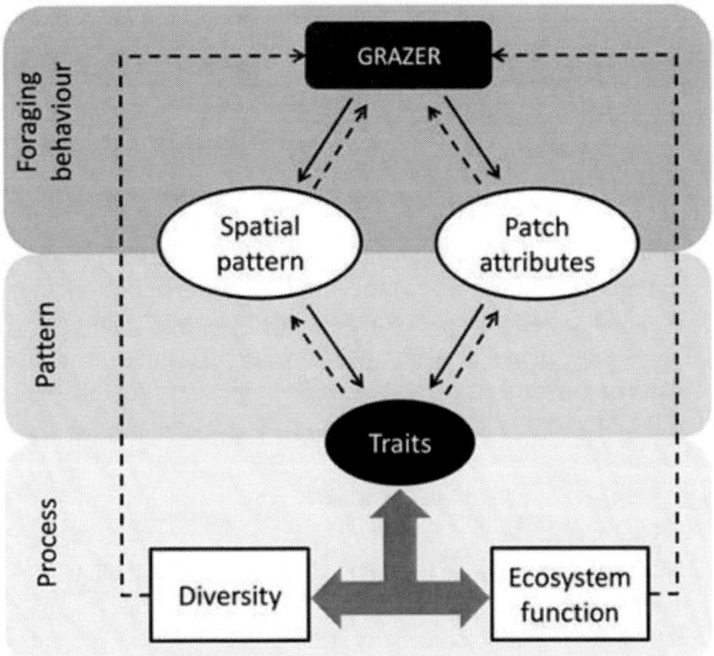

Figure 6. Conceptual model including spatial heterogeneity into a trait-based framework linking plant responses and ecosystem function. Grazing components (defoliation, animal returns, trampling) are expected to constrain within-patch trait distributions with cascading effects for plant and soil structure and function (*sensu* Lavorel & Garnier 2002). Moreover, grazing may promote spatial pattern in variation of plant traits. Both patch attributes and spatial pattern influence trait distribution at the field-scale, which may have consequences for ecosystem function and diversity, as well as herbivore grazing behaviour. Direct effects are given by solid lines and feedback effects are given by dashed lines.

Static versus Dynamic Patterns of Spatial Heterogeneity

Up until this point we have emphasized the potential consequences of spatial heterogeneity expressed as a static or stable spatial pattern, with occasional references to spatial dynamics of plant species. Both empirical and theoretical studies indicate that selective grazing by large herbivores can promote the creation of relatively stable patches within the pasture (Adler et al., 2001; Rossignol et al., 2011; Dumont et al., 2012), although the degree of seasonal and interannual stability of grazed/ungrazed patches may vary depending on animal type, stocking rate and site productivity (Dumont et al., 2012). Clearly, these static spatial patterns are most likely to be associated with concurrent shifts in within-field structure and function. However, grazed grasslands are inherently heterogeneous in both space and time, not least due to unpredictable animal excretion.

In theory, spatial heterogeneity may interact with temporal complexity and the duration or stability of patch types, modifying the strength and direction of biotic interactions as well as flows and process rates (Strayer et al., 2003). For example, temporal variation in resource availability is considered to have profound consequences for population dynamics, community organisation and ecosystem processes. This is of particular relevance for patchy resource inputs such as animal returns which represent a pulse of resources; variation in the magnitude, duration and frequency of occurrence of inputs at a given location may be equally, if not more important than spatial pattern *per se* (Yang et al., 2008; Venail et al., 2011). Limited experimental evidence from tropical savannas also suggests that reciprocal interactions between large grazers and vegetation patterns may modify pattern-process relationships over time (Augustine 2003). As yet, the importance of dynamic spatial patterns for temperate grassland function remains to be determined, and most likely requires the development of new modelling and analytical tools. Spatially-explicit, mechanistic models have already been developed for a certain number of demographic parameters and nutrient transfers (metacommunity and meta-ecosystem models, Loreau et al., 2003; Leibold & Norberg 2004). These ideas, often applied to theoretical systems, could provide the basis for spatially-explicit grassland models. In parallel, new experiments manipulating the spatial organisation of plants, soil properties and grazer components (defoliation, trampling and animal returns) will allow the calibration, validation, and hypothesis testing associated with such model development.

CONCLUSION AND FUTURE RESEARCH DIRECTIONS

The consideration of spatial heterogeneity in ecological studies presents conceptual and methodological challenges across multiple spatial scales (patch, field, landscape). This is of particular relevance to grazed grasslands, where management practices enhance within-field spatial variation in biotic and abiotic factors. Although we have a good understanding of what causes spatial heterogeneity in grazed grasslands, there is a clear need for improved understanding of what aspects of spatial heterogeneity are most important and when spatial heterogeneity matters. Not only will greater knowledge of the functional role of spatial heterogeneity in grazed grasslands provide valuable information for a system of economic and environmental interest, but it will also provide a test of general ecological theory that is applicable to other terrestrial ecosystems.

Despite strong theoretical grounds for examining spatial heterogeneity in grazed grasslands, we found relatively few studies explicitly addressing impacts of spatial heterogeneity driven by defoliation, trampling or animal returns in the literature. The studies concerning spatial heterogeneity and grazed grassland structure or function showed a clear bias towards taxonomic approaches to diversity and N cycling respectively. In general, descriptive studies were more frequent than hypothesis-driven experimental studies, and spatially-explicit approaches were under-represented. Our review indicates that we are still a long way from being able to provide simple management recommendations regarding the enhancement of grassland multi-functionality via within-field heterogeneity. Limited evidence suggests that heterogeneity-induced increases in biodiversity may occur at the expense of field-scale nitrogen recycling efficiency, but this remains to be confirmed for different management practices (e.g. animal type, stocking rates) and for other nutrients.

In view of our findings, we suggest three areas in which advances could help overcome current knowledge gaps: (1) Extend studies of taxonomic diversity to include functional trait and phylogenetic approaches, including soil fauna as well as aboveground plants and invertebrates; (2) Incorporate animal-plant-soil components into experiments to determine the hierarchy of spatial heterogeneity effects and to identify possible feedbacks between interacting factors; (3) Account for the temporal dynamics in grazed systems and potential interactions with spatial heterogeneity. Given that different methods and scales of study may have significant impacts on the results obtained, we suggest that the definition of a standard methodological framework may facilitate generalisations across different grazing systems (see Dumont et al., 2012 for one example of methods for multi-site and multi-scale comparisons). Moreover, understanding the ecological implications of spatial heterogeneity at the ecosystem-scale most likely requires an integrated modelling framework taking into account the spatial distribution of animal-plant-soil interactions. Use of spatial models may help identify domains where relationships do or do not scale. Ultimately, these gains in knowledge should lead to new data and tools relevant to biodiversity, sustainable agriculture and the management of semi-natural grassland ecosystems.

REFERENCES

Adler, P.B., Raff, D. & Lauenroth, W.K. (2001) The effect of grazing on the spatial heterogeneity of vegetation. *Oecologia*, 128, 465–479.

Afzal, M. & Adams, W.A. (1992) Heterogeneity of soil mineral nitrogen in pasture grazed by cattle. *Soil Science of America Journal*, 56, 1160–1166.

Alados, C.L., Aich, A. El, Komac, B., Pueyo, Y. & García-Gonzalez, R. (2007) Self-organized spatial patterns of vegetation in alpine grasslands. *Ecological Modelling*, 201, 233–242.

Amarasekare, P. (2003) Competitive coexistence in spatially structured environments: a synthesis. *Ecology Letters*, 6, 1109–1122.

Anger, M., Hoffmann, C. & Kühbauch, W. (2003) Nitrous oxide emissions from artificial urine patches applied to different N-fertilized swards and estimated annual N_2O emissions for differently fertilized pastures in an upland location in Germany. *Soil Use and Management*, 19, 104–111.

Auerswald, K., Mayer, F. & Schnyder, H. (2009) Coupling of spatial and temporal pattern of cattle excreta patches on a low intensity pasture. *Nutrient Cycling in Agroecosystems*, 88, 275–288.

Augustine, D. (2003) Spatial heterogeneity in the herbaceous layer of a semi-arid savanna ecosystem. *Plant Ecology*, 167, 319–332.

Bakker, C., Blair, J. & Knapp, A. (2003) Does resource availability, resource heterogeneity or species turnover mediate changes in plant species richness in grazed grasslands? *Oecologia*, 137, 385–391.

Bakker, E.S., Ritchie, M.E., Olff, H., Milchunas, D.G. & Knops, J.M.H. (2006) Herbivore impact on grassland plant diversity depends on habitat productivity and herbivore size. *Ecology letters*, 9, 780–8.

Bardgett, R.D. & Wardle, D.A. (2003) Herbivore-mediated linkages between aboveground and belowground communities. *Ecology*, 84, 2258–2268.

Barot, S. & Gignoux, J. (2004) Mechanisms promoting plant coexistence: can all the proposed processes be reconciled? *Oikos*, 106, 185–192.

Benot, M.-L., Bonis, A., Rossignol, N. & Mony, C. (2011) Spatial patterns in defoliation and the expression of clonal traits in grazed meadows. *Botany*, 89, 43–54.

Bindi, M. & Olesen, J.E. (2010) The responses of agriculture in Europe to climate change. *Regional Environmental Change*, 11, 151–158.

Bisigato, A.J. & Bertiller, M.B. (1997) Grazing effects on patchy dryland vegetation in northern Patagonia. *Journal of Arid Environments*, 36, 639–653.

Bisigato, A.J. & Bertiller, M. (2004) Temporal and micro-spatial patterning of seedling establishment. Consequences for patch dynamics in the southern Monte, Argentina. *Plant Ecology*, 174, 235–246.

Bisigato, A.J., Bertiller, M., Ares, J. & Pazos, G. (2005) Effect of grazing on plant patterns in arid ecosystems of Patagonian Monte. *Ecography*, 28, 561–572.

Bloor, J.M.G. & Bardgett, R.D. (2012) Stability of above-ground and below-ground processes to extreme drought in model grassland ecosystems: Interactions with plant species diversity and soil nitrogen availability. *Perspectives in Plant Ecology, Evolution and Systematics*, 14, 193–204.

Bolker, B.M., Pacala, S.W. & Neuhauser, C. (2003) Spatial dynamics in model plant communities: what do we really know? *The American naturalist*, 162, 135–48.

Burel, F. & Garnier, E. (2008) Les effets de l'agriculture sur la biodiversité. *Agriculture et biodiversité: valoriser les synergies* (ed X. Le Roux), Rapport d'expertise ESCo.

Calcagno, V., Mouquet, N., Jarne, P. & David, P. (2006) Coexistence in a metacommunity: the competition-colonization trade-off is not dead. *Ecology Letters*, 9, 897–907.

Chave, J. (2013) The problem of pattern and scale in ecology: what have we learned in 20 years? *Ecology letters*, 16 Suppl 1, 4–16.

Chesson, P.L. (2000) Mechanisms of maintenance of species diversity. *Annual Review of Ecology and Systematics*, 31, 343–366.

Chesson, P.L. & Warner, R.R. (1981) Environmental variability promotes coexistence in lottery competitive systems. *The American Naturalist*, 117, 923–943.

Cipriotti, P.A. & Aguiar, M.R. (2005) Effects of grazing on patch structure in a semi-arid two-phase vegetation mosaic. *Journal of Vegetation Science*, 16, 57–66.

Collins, B. & Wein, G. (1998) Soil resource heterogeneity effects on early succession. *Oikos*, 82, 238–245.

Connell, J.H. (1978) Diversity in Tropical Rain Forests and Coral Reefs. *Science*, 199, 1302–1310.

Coomes, D.A. & Grubb, P.J. (2003) Colonization, tolerance, competition and seed-size variation within functional groups. *Trends in Ecology & Evolution*, 18, 283–291.

Cosyns, E., Delporte, A., Lens, L. & Hoffmann, M. (2005) Germination success of temperate grassland species after passage through ungulate and rabbit guts. *Journal of Ecology*, 93, 353–361.

Dai, X. (2000) Impact of cattle dung deposition on the distribution pattern of plant species in an alvar limestone grassland. *Journal of Vegetation Science*, 11, 715–724.

Dale, M.R.T. (1999) *Spatial Pattern Analysis in Plant Ecology*. Cambridge University Press.

Danell, K., Bergström, R., Duncan, P. & Pastor, J. (2006) *Large Herbivore Ecology, Ecosystem Dynamics and Conservation*. Cambridge University Press.

Day, K.J., Hutchings, M.J. & John, E.A. (2003) The effects of spatial pattern of nutrient supply on yield, structure and mortality in plant populations. *Journal of Ecology*, 91, 541–553.

DeGabriel, J.L., Albon, S.D., Fielding, D. a., Riach, D.J., Westaway, S. & Irvine, R.J. (2011) The presence of sheep leads to increases in plant diversity and reductions in the impact of deer on heather. *Journal of Applied Ecology*, 48, 1269–1277.

Delgado, C. (2005) Rising demand for meat and milk in developing countries: implications for grasslands-based livestock production. *Grasslands: a global resource. Proceedings of the twentieth International Grassland Congress, Dublin, Ireland,* (ed D.A. McGilloway), pp. 29–38. Wageningen Academic Publishers, Wageningen.

Dennis, P., Young, M. & J, G. (1998) Distribution and abundance of small insects and arachnids in relation to structural heterogeneity of grazed, indigenous grasslands. *Ecological Entomology*, 23, 253–264.

Díaz, S., Hodgson, J.G., Thompson, K., Cabido, M., Cornelissen, J.H.C., Jalili, A., et al. (2004) The plant traits that drive ecosystems: Evidence from three continents. *Journal of Vegetation Science*, 15, 295.

Díaz, S., Lavorel, S., de Bello, F., Quétier, F., Grigulis, K. & Robson, T.M. (2007a) Incorporating plant functional diversity effects in ecosystem service assessments. *Proceedings of the National Academy of Sciences of the United States of America*, 104, 20684–9.

Díaz, S., Lavorel, S., McIntyre, S., Falczuk, V., Casanoves, F., Milchunas, D.G., Skarpe, C., Rusch, G.M., Sternberg, M., Noy-Meir, I., Landsberg, J., Zhang, W., Clark, H. & Campbell, B.D. (2007b) Plant trait responses to grazing–a global synthesis. *Global Change Biology*, 12, 1–29.

Dittrich, A.D.K. & Helden, A.J. (2012) Experimental sward islets: the effect of dung and fertilisation on Hemiptera and Araneae. *Insect Conservation and Diversity*, 5, 46–56.

Drewry, J.J. (2006) Natural recovery of soil physical properties from treading damage of pastoral soils in New Zealand and Australia: A review. *Agriculture, Ecosystems & Environment*, 114, 159–169.

Dubeux, J.C.. (2005) *Management Strategies to Improve Nutrient Cycling in Grazed Pensacola Bahiagrass Pastures*. University of Florida, Gainsville.

Dumont, B., Carrère, P., Ginane, C., Farruggia, A., Lanore, L., Tardif, A., Decuq, F., Darsonville, O. & Louault, F. (2011) Plant–herbivore interactions affect the initial direction of community changes in an ecosystem manipulation experiment. *Basic and Applied Ecology*, 12, 187–194.

Dumont, B., Rossignol, N., Loucougaray, G., Carrère, P., Chadoeuf, J., Fleurance, G., Bonis, A., Farruggia, A., Gaucherand, S., Ginane, C., Louault, F., Marion, B., Mesléard, F. & Yavercovski, N. (2012) When does grazing generate stable vegetation patterns in temperate pastures? *Agriculture, Ecosystems & Environment*, 153, 50–56.

Dungan, J.L., Perry, J.N., Dale, M.R.T., Legendre, P., Fortin, M., Jakomulska, A., Miriti, M., Rosenberg, M.S. & Fortin, S. (2002) A balanced view of scale in spatial statistical analysis. *Ecography*, 25, 626–640.

Engler, R., Randin, C.F., Vittoz, P., Czáka, T., Beniston, M., Zimmermann, N.E. & Guisan, A. (2009) Predicting future distributions of mountain plants under climate change: does dispersal capacity matter? *Ecography*, 32, 34–45.

Estapé, S., Semmartin, M. & Paruelo, J. (2013) Refuge effect of an unpalatable forb on community structure and grass morphology in a temperate grassland. *Plant Ecology*, 214, 363–372.

Ettema, C.H. & Wardle, D.A. (2002) Spatial soil ecology. *Trends in Ecology and Evolution*, 17, 177–183.

Eurostat. (2010) *Agricultural Statistics Main Results 2008–09.*

Fleurance, G., Duncan, P., Fritz, H., Cabaret, J., Cortet, J. & Gordon, I.J. (2007) Selection of feeding sites by horses at pasture: Testing the anti-parasite theory. *Applied Animal Behaviour Science*, 108, 288–301.

Fortin, M.-J. & Dale, M.R.T. (2005) *Spatial Analysis A Guide for Ecologists.* Cambridge University Press.

Frank, D.A. (2005) The interactive effects of grazing ungulates and aboveground production on grassland diversity. *Oecologia*, 143, 629–34.

García-Palacios, P., Maestre, F.T., Bardgett, R.D. & de Kroon, H. (2012) Plant responses to soil heterogeneity and global environmental change. *Journal of Ecology*, 100, 1303–1314.

Gibson, D.J. (2008) *Grasses and Grassland Ecology.* Oxford University Press.

Gillet, F., Kohler, F., Vandenberghe, C. & Buttler, A. (2010) Effect of dung deposition on small-scale patch structure and seasonal vegetation dynamics in mountain pastures. *Agriculture, Ecosystems & Environment*, 135, 34–41.

Götzenberger, L., de Bello, F., Bråthen, K.A., Davison, J., Dubuis, A., Guisan, A., et al. (2012) Ecological assembly rules in plant communities-approaches, patterns and prospects. *Biological reviews of the Cambridge Philosophical Society*, 87, 111–27.

Grigulis, K., Lavorel, S., Krainer, U., Legay, N., Baxendale, C., Dumont, M., et al. (2013) Relative contributions of plant traits and soil microbial properties to mountain grassland ecosystem services. *Journal of Ecology*, 101, 47–57.

Grime, J.P. (1979) *Plant Strategies and Vegetation Processes.* Wiley.

Haining, R. (1990) *Spatial Data Analysis in the Social and Environmental Sciences.* Cambridge University Press.

Hanski, I. (1998) Metapopulation dynamics. *Nature*, 396, 41–49.

Haynes, R.J. & Williams, P.H. (1993) Nutrient cycling and soil fertility in the grazed pasture ecosystem. *Advances in Agronomy*, 49, 119–200.

Helden, A.J., Anderson, A., Sheridan, H. & Purvis, G. (2010) The role of grassland sward islets in the distribution of arthropods in cattle pastures. *Insect Conservation and Diversity*, 3, 291–301.

Hérault, B. & Hiernaux, P. (2004) Soil seed bank and vegetation dynamics in Sahelian fallows; the impact of past cropping and current grazing treatments. *Journal of Tropical Ecology*, 20, 683–691.

Hillebrand, H., Bennett, D.M. & Cadotte, M.W. (2008) Consequences of dominance: a review of evenness effects on local and regional ecosystem processes. *Ecology*, 89, 1510–1520.

Hodge, A. (2006) Plastic plants and patchy soils. *Journal of Experimental Botany*, 57, 401–11.

Hooper, D.U., Chapin, I.F.., Ewell, J.J., Hector, A., Inchausti, P., Lavorel, S. et al. (2005) Effects of biodiversity on ecosystem functioning: a consensus of current knowledge. *Ecological Monographs*, 75, 3–35.

Hutchings, M.J., John, E.A. & Wijesinghe, D.K. (2003) Toward understanding the consequences of soil heterogeneity for plant populations and communities. *Ecology*, 84, 2322–2334.

Hutchings, N.J., Olesen, J.E., Petersen, B.M. & Berntsen, J. (2007) Modelling spatial heterogeneity in grazed grassland and its effects on nitrogen cycling and greenhouse gas emissions. *Agriculture, Ecosystems & Environment*, 121, 153–163.

Joern, A. & Laws, A.N. (2013) Ecological mechanisms underlying arthropod species diversity in grasslands. *Annual Review of Entomology*, 58, 19–36.

Klimešová, J., Latzel, V., de Bello, F. & Groenendael, J.M. Van. (2008) Plant functional traits in studies of vegetation changes in response to grazing and mowing: towards a use of more specific traits. *Preslia*, 80, 245–253.

van Klink, R., Rickert, C., Vermeulen, R., Vorst, O., Wallis, DeVries, M.F. & Bakker, J.P. (2013) Grazed vegetation mosaics do not maximize arthropod diversity: Evidence from salt marshes. *Biological Conservation*, 164, 150–157.

Klumpp, K., Bloor, J.M.G., Ambus, P. & Soussana, J.-F. (2010) Effects of clover density on N_2O emissions and plant-soil N transfers in a fertilised upland pasture. *Plant and Soil*, 343, 97–107.

Koerner, S.E. & Collins, S.L. (2013) Small-scale patch structure in North American and South African grasslands responds differently to fire and grazing. *Landscape Ecology*, 28, 1293–1306.

Kohler, F., Gillet, F., Gobat, J.-M. & Buttler, A. (2006a) Effect of cattle activities on gap colonization in mountain pastures. *Folia Geobotanica*, 41, 289–304.

Kohler, F., Gillet, F., Reust, S., Wagner, H.H., Gadallah, F., Gobat, J.-M. & Buttler, A. (2006b) Spatial and Seasonal Patterns of Cattle Habitat use in a Mountain Wooded Pasture. *Landscape Ecology*, 21, 281–295.

Kurz, I., O'Reilly, C.D. & Tunney, H. (2006) Impact of cattle on soil physical properties and nutrient concentrations in overland flow from pasture in Ireland. *Agriculture, Ecosystems & Environment*, 113, 378–390.

Lavorel, S. & Garnier, E. (2002) Predicting changes in community composition and ecosystem functioning from plant traits: revisiting the Holy Grail. *Functional Ecology*, 16, 545–556.

Leibold, M. a., Holyoak, M., Mouquet, N., Amarasekare, P., Chase, J.M., Hoopes, M.F., Holt, R.D., Shurin, J.B., Law, R., Tilman, D., Loreau, M. & Gonzalez, A. (2004) The metacommunity concept: a framework for multi-scale community ecology. *Ecology Letters*, 7, 601–613.

Leibold, M. a. & Norberg, J. (2004) Biodiversity in metacommunities: Plankton as complex adaptive systems? *Limnology and Oceanography*, 49, 1278–1289.

Levin, S.A. (1992) The problem of pattern and scale in ecology. *Ecology*, 73, 1943–1967.

Levins, R. & Culver, D. (1971) Regional coexistence of species and competition between rare species. *Procceding of the National Academy of Sciences*, 68, 1246–1248.

Li, F.Y., Betteridge, K., Cichota, R., Hoogendoorn, C.J. & Jolly, B.H. (2012) Effects of nitrogen load variation in animal urination events on nitrogen leaching from grazed pasture. *Agriculture, Ecosystems & Environment*, 159, 81–89.

Lin, Y., Hong, M., Han, G., Zhao, M., Bai, Y. & Chang, S.X. (2010) Grazing intensity affected spatial patterns of vegetation and soil fertility in a desert steppe. *Agriculture, Ecosystems & Environment*, 138, 282–292.

Loreau, M. (2010) Linking biodiversity and ecosystems: towards a unifying ecological theory. *Philosophical transactions of the Royal Society of London. Series B, Biological sciences*, 365, 49–60.

Loreau, M. & Hector, A. (2001) Partitioning selection and complementarity in biodiversity experiments. *Nature*, 412, 72–6.

Loreau, M., Mouquet, N. & Holt, R.D. (2003) Meta-ecosystems: a theoretical framework for a spatial ecosystem ecology. *Ecology Letters*, 6, 673–679.

Lovett, G.M., Jones, C., Turner, M.G. & Weathers, K.C. (eds). (2005) *Ecosystem Function in Heterogeneous Landscapes*. Springer.

MacDiarmid, B.N. & Watkin, R.B. (1972) The cattle dung patch. 2. Effect of a dung patch on the chemical status of a soil and ammonia losses from the patch. *Grass and Forage Science*, 27, 43–48.

Mahmood, S. & Prosser, J.I. (2006) The influence of synthetic sheep urine on ammonia oxidizing bacterial communities in grassland soil. *FEMS microbiology ecology*, 56, 444–454.

Malo, J., Jimenez, B. & Suárez, F. (2000) Herbivore dunging and endozoochorous seed deposition in a Mediterranean dehesa. *Journal of Range Management*, 53, 322–328.

Malo, J.E. & Suárez, F. (1995) Establishment of pasture species on cattle dung: the role of endozoochorous seeds. *Journal of Vegetation Science*, 6, 169–174.

McGilloway, D.A. (ed). (2005) *Grassland: a Global Resource. Proceedings of the Twentieth International Grassland Congress, Dublin, Ireland*. Wageningen Academic Publishers.

McIntyre, S. & Tongway, D. (2005) Grassland structure in native pastures: links to soil surface condition. *Ecological Management and Restoration*, 6, 43–50.

Mcivor, J.G., McIntyre, S., Saeli, I. & Hodgkinson, J.J. (2005) Patch dynamics in grazed subtropical native pastures in south-east Queensland. *Austral Ecology*, 30, 445–464.

McSherry, M.E. & Ritchie, M.E. (2013) Effects of grazing on grassland soil carbon: a global review. *Global change biology*, 19, 1347–57.

Meissner, R.A. & Facelli, H.M. (1999) Effects of sheep exclusion on the soil seed bank and annual vegetation in chenopod shrublands of South Australia. *Journal of Arid Environments*, 42, 117–128.

Menneer, J.C. & Ledgard, S.F. (2005) The effects of treading by dairy cows during wet soil conditions on white clover productivity, growth and morphology in a white clover–perennial ryegrass pasture. *Grass and Forage*, 60, 46–58.

Mikola, J., Setälä, H., Virkajärvi, P., Saarijärvi, K., Ilmarinen, K., Voigt, W. & Vestberg, M. (2009) Defoliation and patchy nutrient return drive grazing effects on plant and soil properties in a dairy cow pasture. *Ecological Monographs*, 79, 221–244.

Milchunas, D.G., Sala, O.E. & Lauenroth, W.K. (1988) A generalized model of the effects of grazing by large herbivores on grassland community structure. *The American Naturalist*, 132, 87–106.

Millard, P. & Singh, B.K. (2010) Does grassland vegetation drive soil microbial diversity? *Nutrient Cycling in Agroecosystems*, 88, 147–158.

Monzeglio, U. & Stoll, P. (2005) Spatial patterns and species performances in experimental plant communities. *Oecologia*, 145, 619–28.

Mouissie, A.M., Apol, M.E.F., Heil, G.W. & van Diggelen, R. (2008) Creation and preservation of vegetation patterns by grazing. *Ecological Modelling*, 218, 60–72.

Northup, B.K., Brown, J.R. & Holt, J. A. (1999) Grazing impacts on the spatial distribution of soil microbial biomass around tussock grasses in a tropical grassland. *Applied Soil Ecology*, 13, 259–270.

Olff, H. & Ritchie, M. (1998) Effects of herbivores on grassland plant diversity. *Trends in Ecology & Evolution*, 13, 261–265.

Orwin, K.H., Bertram, J.E., Clough, T.J., Condron, L.M., Sherlock, R.R. & O'Callaghan, M. (2009) Short-term consequences of spatial heterogeneity in soil nitrogen concentrations caused by urine patches of different sizes. *Applied Soil Ecology*, 42, 271–278.

Palmer, S.C.F., Hester, A.J., Elston, D.A., Gordon, I.J. & Hartley, S.E. (2003) The perils of having tasty neighbors: grazing impacts of large herbivores at vegatation boundaries. *Ecology*, 84, 2877–2890.

Parsons, A.J., Carrère, P. & Schwinning, S. (2000) Dynamics of heterogeneity in a grazed sward. *Grassland ecophysiology and grazing ecology* (ed G. Lemaire), pp. 289–316. CABI.

Parsons, A.J. & Dumont, B. (2003) Spatial heterogeneity and grazing processes. *Animal research*, 52, 161–179.

Patra, A.K., Abbadie, L., Clays-Josserand, A., Degrange, V., Grayston, S.J., Loiseau, P., Louault, F., Mahmood, S., Nazaret, S., Philippot, L., Poly, F., Prosser, J.I., Richaume, A. & Le Roux, X. (2005) Effects of grazing on microbial functional groups involved in soil N dynamics. *Ecological Monographs*, 75, 65–80.

Pazos, G.E. & Bertiller, M.B. (2007) Spatial patterns of the germinable soil seed bank of coexisting perennial-grass species in grazed shrublands of the Patagonian Monte. *Plant Ecology*, 198, 111–120.

Pickett, S.T.A. & White, P.S. (1985) *The Ecology of Natural Disturbance and Patch Dynamics*. Academic Press.

Porensky, L.M., Wittman, S.E., Riginos, C. & Young, T.P. (2013) Herbivory and drought interact to enhance spatial patterning and diversity in a savanna understory. *Oecologia*, 173, 591–602.

Pottier, J., Bédécarrats, A. & Marrs, R.H. (2009) Analysing the spatial heterogeneity of emergent groups to assess ecological restoration. *Journal of Applied Ecology*, 1248–1257.

Ramseier, D. & Weiner, J. (2006) Competitive effect is a linear function of neighbour biomass in experimental populations of *Kochia scoparia*. *Journal of Ecology*, 94, 305–309.

Reid, A.M. & Hochuli, D.F. (2007) Grassland invertebrate assemblages in managed landscapes: Effect of host plant and microhabitat architecture. *Austral Ecology*, 32, 708 718.

Robertson, G.P. & Gross, K.L. (1994) Assessing the heterogeneity of belowground resources: quantifying pattern and scale. *Plant Exploitation of Environmental Heterogeneity* (eds M.M. Caldwell & R.W. Pearcy), pp. 237–253. Academic Press, New York.

Robinson, D. (1994) The responses of plants to non-uniform supplies of nutrients. *New Phytologist*, 127, 635–674.

Rossignol, N., Bonis, A. & Bouzillé, J.-B. (2006) Consequence of grazing pattern and vegetation structure on the spatial variations of net N mineralisation in a wet grassland. *Applied Soil Ecology*, 31, 62–72.

Rossignol, N., Chadoeuf, J., Carrère, P. & Dumont, B. (2011) A hierarchical model for analysing the stability of vegetation patterns created by grazing in temperate pastures. *Applied Vegetation Science*, 14, 189–199.

Russi, L., Cocks, P.S. & Roberts, E.H. (1992) The fate of legume seeds eaten by sheep from a Mediterranean grassland. *Journal of Applied Ecology*, 29, 772–778.

Saarijärvi, K. & Virkajärvi, P. (2009) Nitrogen dynamics of cattle dung and urine patches on intensively managed boreal pasture. *The Journal of Agricultural Science*, 147, 479–491.

Schlapfer, M., Zoller, H. & Korner, C. (1998) Influences of mowing and grazing on plant species composition in calcareous grassland. *Botanica Helvetica*, 108, 57–67.

Schrama, M., Veen, G.F., Bakker, E.S., Ruifrok, J.L., Bakker, J.P. & Olff, H. (2013) An integrated perspective to explain nitrogen mineralization in grazed ecosystems. *Perspectives in Plant Ecology, Evolution and Systematics*, 15, 32–44.

Sears, A.L.W. & Chesson, P. (2007) New methods for quantifying the spatial storage effect: an illustration with desert annuals. *Ecology*, 88, 2240–7.

Semmartin, M. & Oesterheld, M. (2001) Effects of grazing pattern and nitrogen availability on primary productivity. *Oecologia*, 126, 225–230.

Sheath, G.W. & Carlson, W.T. (1998) Impact of cattle treading on hill land: 1. Soil damage patterns and pasture status. *New Zealand Journal of Agricultural Research*, 41, 271–278.

Singleton, P.L., Boyes, M. & Addison, B. (2000) Effect of treading by dairy cattle on topsoil physical conditions for six contrasting soil types in Waikato and Northland, New Zealand, with implications for monitoring. *New Zealand Journal of Agricultural Research*, 43, 559–567.

Snaydon, R.W. (1981) The ecology of grazed pastures. *Grazing Animals* (ed F.H.W. Morley), pp. 13–31. Elsevier.

Sørensen, L.I., Mikola, J., Kytöviita, M.-M. & Olofsson, J. (2009) Trampling and spatial heterogeneity explain decomposer abundances in a sub-arctic grassland subjected to simulated reindeer grazing. *Ecosystems*, 12, 830–842.

Soussana, J.F., Tallec, T. & Blanfort, V. (2010) Mitigating the greenhouse gas balance of ruminant production systems through carbon sequestration in grasslands. *Animal*, 4, 334–50.

Spalinger, L.C., Haynes, A.G., Schütz, M. & Risch, A.C. (2012) Impact of wild ungulate grazing on Orthoptera abundance and diversity in subalpine grasslands. *Insect Conservation and Diversity*, 5, 444–452.

Steinauer, E.M. & Collins, S.L. (2001) Feedback loops in ecological hierarchies following urine deposition in tallgrass prairie. *Ecology*, 82, 1319–1329.

Steinfeld, H., Gerber, P., Wassenaar, T., Castel, V., Rosales, M. & de Haan, C. (2006) *Livestock's Long Shadow. Environmental Issues and Options*. FAO, Rome.

Strayer, D.L., Ewing, H.A. & Bigelow, S. (2003) What kind of spatial and temporal details are required in models of heterogeneous systems? *Oikos*, 102, 654–662.

Tilman, D. (1985) The resource-ratio hypothesis of plant succession. *American Naturalist*, 125, 827–852.

Tilman, D. (1988) *Dynamics and Structure of Plant Communities*. Princeton University Press, Princeton, NL.

Tilman, D. (1994) Competition and biodiversity in spatially structured habitats. *Ecology*, 75, 2–16.

Tilman, D. & Kareiva, P.M. (1997) *Spatial Ecology: The Role of Space in Population Dynamics and Interspecific Interactions*. Princeton University Press.

Turnbull, L., Rees, M. & Crawley, M. (1999) Seed mass and the competition/colonization trade-off: a sowing experiment. Journal of Ecology, 87, 899–912.

Tylianakis, J.M., Didham, R.K., Bascompte, J. & Wardle, D.A. (2008) Global change and species interactions in terrestrial ecosystems. *Ecology Letters*, 11, 1351–1363.

Vaieretti, M.V., Cingolani, A.M., Harguindeguy, N.P., Gurvich, D.E. & Cabido, M. (2010) Does decomposition of standard materials differ among grassland patches maintained by livestock? *Austral Ecology*, 35, 935–943.

Vaieretti, M.V., Cingolani, A.M., Pérez Harguindeguy, N. & Cabido, M. (2013) Effects of differential grazing on decomposition rate and nitrogen availability in a productive mountain grassland. *Plant and Soil*, 371, 675–691.

Veen, G.F., Olff, H., Duyts, H. & van der Putten, W.H. (2010) Vertebrate herbivores influence soil nematodes by modifying plant communities. *Ecology*, 91, 828–35.

Venail, P.A., Kaltz, O., Olivieri, I., Pommier, T. & Mouquet, N. (2011) Diversification in temporally heterogeneous environments: effect of the grain in experimental bacterial populations. *Journal of Evolutionary Biology*, 24, 2485–95.

Verdú, J.R., Moreno, C.E., Sánchez-Rojas, G., Numa, C., Galante, E. & Halffter, G. (2007) Grazing promotes dung beetle diversity in the xeric landscape of a Mexican Biosphere Reserve. *Biological Conservation*, 140, 308–317.

Violle, C., Navas, M.-L., Vile, D., Kazakou, E., Fortunel, C., Hummel, I. & Garnier, E. (2007) Let the concept of trait be functional! *Oikos*, 116, 882–892.

Wang, F. (2008) Modeling the spatial distribution of nitrogen leaching from dairy farmland. *Vadose Zone Journal*, 7, 439.

Wang, Y., Shiyomi, M., Tsuiki, M., Tsutsumi, M., Yu, X. & Yi, R. (2002) Spatial heterogeneity of vegetation under different grazing intensities in the Northwest Heilongjiang Steppe of China. *Agriculture, Ecosystems & Environment*, 90, 217–229.

Wardle, D.A., Bardgett, R.D., Klironomos, J.N., Setälä, H., van der Putten, W.H. & Wall, D.H. (2004) Ecological linkages between aboveground and belowground biota. *Science*, 304, 1629–33.

White, S.L., Sheffield, R.E., Washburn, S.P., King, L.D. & Green, J.T. (2001) Spatial and time distribution of dairy cattle excreta in an intensive pasture system. *Journal of Environmental Quality*, 30, 2180–2187.

White-Leech, R., Liu, K., Sollenberger, L.E., Woodard, K.R. & Interrante, S.M. (2013) Excreta deposition on grassland patches. II. Spatial pattern and duration of forage responses. *Crop Science*, 53, 696.

Whittaker, R.H. (1960) Vegetation of the Siskiyou Mountains, Oregon and California. *Ecological Monographs*, 30, 279–338.

Wiens, J.A. (1989) Spatial scaling in ecology. *Functional Ecology*, 3, 385–397.

Wilson, J.B. (1994) The'intermediate disturbance hypothesis' of species coexistence is based on patch dynamics. *New Zealand Journal of Ecology*, 18, 176–181.

Wilson, J.B. (2011) The twelve theories of co-existence in plant communities: the doubtful, the important and the unexplored. *Journal of Vegetation Science*, 22, 184–195.

Wright, I.J., Reich, P.B., Westoby, M., Ackerly, D.D., Baruch, Z., Bongers, F., et al. (2004) The worldwide leaf economics spectrum. *Nature*, 428, 821–7.

Yang, L.H., Bastow, J.L., Spence, K.O. & Wright, A.N. (2008) What can we learn from resource pulses? *Ecology*, 89, 621–34.

Yoshihara, Y., Okuro, T., Buuveibaatar, B., Undarmaa, J. & Takeuchi, K. (2010) Complementary effects of disturbance by livestock and marmots on the spatial heterogeneity of vegetation and soil in a Mongolian steppe ecosystem. *Agriculture, Ecosystems & Environment*, 135, 155–159.

Zhang, H., Gilbert, B., Wang, W., Liu, J. & Zhou, S. (2013) Grazer exclusion alters plant spatial organization at multiple scales, increasing diversity. *Ecology and Evolution*, 3, 3604–3612.

Zhou, Z., Sun, O.J., Luo, Z., Jin, H., Chen, Q. & Han, X. (2008) Variation in small-scale spatial heterogeneity of soil properties and vegetation with different land use in semiarid grassland ecosystem. *Plant and Soil*, 310, 103–112.

Zhu, H., Wang, D., Wang, L., Bai, Y., Fang, J. & Liu, J. (2012) The effects of large herbivore grazing on meadow steppe plant and insect diversity. *Journal of Applied Ecology*, 49, 1075–1083.

In: Grassland Biodiversity and Conservation … ISBN: 978-1-61761-619-8
Editors: Pierre Mariotte and Paul Kardol

Chapter 7

EFFECTS OF LAND-USE AND GLOBAL CHANGE ON GROWTH PERFORMANCE AND INTERACTIONS BETWEEN A DOMINANT AND A RARE SEMI-DRY GRASSLAND SPECIES

Monika Partzsch*
Institute of Biology/Geobotany and Botanical Garden,
Martin-Luther-University Halle-Wittenberg, Germany

ABSTRACT

Major environmental changes, and an associated decline in biodiversity, are taking place as the result of changes in traditional land-use. Climate changes, such as warming and nitrogen deposition, are also leading to alterations in plant-plant interactions, which are key factors in the regulation of species composition, productivity, structure and dynamics in plant communities and ecosystems. The positive (facilitation) and negative (competition) interactions between co-existing species are manifold, and depend both on plant traits and on prevailing environmental conditions.

Two competition experiments were carried out in order to investigate the effects of environmental change on plant-plant interactions in semi-dry grasslands, with the study species being the expanding grass *Festuca rupicola* and the rare dicot *Dianthus carthusianorum*. The experimental approach followed a replacement design, in which plants were placed in pots with constant densities but with varying mixture compositions. In the first outdoor pot experiment (2 years), the effect of land-use was tested by simulating grazing/mowing by clipping. Land abandonment was simulated by leaving the plants unclipped. In a second growth chamber experiment (9 months), the effects of N deposition and warming were tested both under nutrient-poor (simulating natural habitat conditions) and nutrient-rich (simulating nutrient enrichment) conditions. One growth chamber simulated ambient temperature during the vegetation period (20°C by day / 10°C by night) whereas the other simulated predicted future conditions (22/15°C). The parameters of biomass, leaf number, number of flowers and panicle length were measured, and interactions calculated thereafter.

* Corresponding author: Monika Partzsch, monika.partzsch@botanik.uni-halle.de.

The results of the first experiment showed that the vegetative and generative parameters of *F. rupicola* significantly decreased in clipped treatments, and significantly increased with increasing numbers of the dicot present in each pot. The number of flowers of *D. carthusianorum* was reduced by a marginally significant amount when mixed with the grass. However, the biomass was not significantly affected by clipping or mixture composition. The plant-plant interactions suggested facilitation of *F. rupicola* and competitive to neutral effects on *D. carthusianorum*. In general, the abandonment of traditional land-use (grazing or mowing) leads to the expansion of *F. rupicola* and implies a decline in the co-existing rare dicot over time.

The results of the second experiment showed that plant performance was species-dependent. After nutrient addition, the biomass and the number of leaves of both species increased significantly (by a factor of ten), and the species interacted strongly. Warming strongly promoted the growth of *D. carthusianorum*, whereas the biomass and the number of leaves decreased significantly with decreasing density of the grass. *Festuca rupicola* responded in a different way to elevated temperature: plant biomass was not affected, but the number of leaves was significantly reduced. As such, the plant-plant interactions switched after warming and the grass was negatively affected and the dicot facilitated.

The results showed that, under warming, the dominance of the grass, induced by the abandonment of grazing, will be reduced and that the dicot will become more prominent. Therefore, our results also suggest that maintenance of grazing is important for the conservation of species-rich grassland communities in face of future climate change.

Keywords: Competition, *Dianthus carthusianorum,* facilitation, *Festuca rupicola*, land-use change, N deposition, plant-plant interactions, warming

INTRODUCTION

Major global environmental change is underway, the main processes of which are summarized by the collective term "global change". These include climate warming, N deposition and CO_2-enrichment. In addition, increasing landscape fragmentation and changes in traditional land-use practices play an important role. All these processes affect interactions between co-existing species (Tylianakis et al., 2008). Plant-plant interactions play a key role in the regulation of species composition, productivity and structure, and in the dynamics of plant communities and ecosystems (Fowler 1986; Callaway 1995; Whitford 2002; Brooker 2006). The interactions between the constructing species are manifold, and depend on both environmental conditions and specific plant traits (Holmgren et al., 1997; Callaway & Walker 1997; Callaway 2007). In the past, such interactions were usually identified as being negative in terms of competition for light, nutrients, water, space, and pollinators. More recent plant ecologists have increasingly also focused on positive interactions (Brooker et al., 2008), in which neighboring plants protect each other from the impact of competitors, herbivores and climatic extremes, and allocate additional resources through canopy leaching. Co-existing plants lead to nutrient enrichment, microbial and mycorrhizal networks (del-Val & Crawley 2005; Kuijper et al., 2005; Weigelt et al., 2007; Veblen 2008; Kardol et al., 2010). As such, competition is most common in benign ecosystems (Grime 1973; Huston 1979), whereas facilitation mostly occurs in stressful ecosystems (Bruno et al., 2003) such as savannas, deserts, dry grasslands, scrublands, and salt marshes (Belsky 1994; Pugnaire & Luque 2001; Callaway et al., 2002; Maestre et al., 2003; Maestre & Cortina 2004).

Franz (1952, 1953) described the basic biocoenological principle in which species richness in plant communities results from the long-term continuous development of habitats under consistent environmental conditions, leading to balanced and stable communities. In the case of dry grassland communities, a major environmental factor has been traditional land-use practices such as grazing by sheep and goats or mowing, which dates back to medieval times (Pott 1996) and has led to the development of semi-natural species-rich communities in the landscape of porphyry outcrops (central Germany) over a long period of time (Richter et al., 2003). However, following the political upheaval of 1990, traditional land-use practices have been in continual decline as a result of economic changes (Partzsch 2000, 2001). Consequently, the formerly species-rich dry grasslands have been transformed into species-poor communities dominated by grasses such as *Arrhenatherum elatius* (Dostálek & Frantík 2008), *Brachypodium pinnatum* (Willems 1987; Bobbink et al., 1988), *Bromus erectus* (Bornkamm 2006), *Festuca* ssp. (Partzsch 2000, 2001; Klimaschewski et al., 2006), *Poa angustifolia* (Partzsch 2000, Wesche et al., 2005) and *Stipa* ssp. (Enyedi et al., 2008). This leads to a strong decline in rare and vulnerable dicotyledonous species, which appears to be a global phenomenon (Balmer & Erhardt 2000; Poschlod et al., 2005; Römermann et al., 2005; Chýlová & Münzbergová 2008; Silva et al., 2008).

Beside land-use change, an ongoing process of global change characterized by various processes including warming and eutrophication is underway. During the past 100 years, the global average temperature has increased by approximately 0.6°C and, depending on the scenario, a rise of between 1.1 and 6.4°C is forecast for the year 2100 (Intergovernmental Panel on Climate Change [IPCC], 2007). A meta-analysis of 143 studies of several plant and animal species has shown that a significant impact from global warming is already discernible and may be disrupting connectedness among species. This may lead to a reformulation of communities reflecting species changes, extirpations and extinctions (Root et al., 2003; Tylianakis et al., 2008). In addition, agriculture, combustion of fossil fuels and other human activities have substantially altered the global cycle of nitrogen (N), leading to increasing availability and mobility of N over large global regions, with long-lasting environmental consequences (Vitousek et al., 1997). The critical N load of semi-dry grasslands ranges between 14 and 25 kg. $ha^{-1}.yr^{-1}$ (De Jong et al., 1998), whereas the annual deposition in the dry region of central Germany during the 1990's was between 46 and 51 $kg.ha^{-1}.yr^{-1}$ (Körschens & Mahn 1995). Such long-term N enrichment leads to competitive exclusion of characteristic grassland species by more nitrophilic plants, especially under formerly oligotrophic to mesotrophic habitat conditions (Willems & Bobbink 1990; Bobbink et al., 1998). As such, N enrichment has resulted in a change in species composition in semi-dry grasslands, an increase in yield and a promotion of grass species (Bornkamm 2006; Horswill et al., 2008; Stevens et al., 2010). Moreover, Dupré et al. (2010) have argued that the effects on plant-plant interactions caused by land-use change are intensified by global changes such as warming and nitrogen deposition.

A number of studies have demonstrated that plant-plant interactions are affected by climate conditions. It has been shown that interactions switch from facilitation to competition when a dry and hot year was followed by a wet and cool year (Greenlee & Callaway 1996). In subalpine and alpine communities, with increasing altitude and changing climate conditions, the frequency of facilitative interactions increased (Choler et al. 2001; Kikvidze et al., 2006).

Positive interactions may therefore intensify with increasing abiotic stress, even within a single growing season. Facilitative effects during dry years in saltmarsh ecosystems have been related to inter-annual variations in climate (Bertness & Ewanchuk 2002). Bruno et al. (2003) pointed out that positive interactions are widespread in natural communities and are a major driving factor of plant community dynamics and ecosystem processes. Increasing environmental severity leads to a shift from a dominance of competitive interactions to a dominance of facilitative interactions, and suggests that the global shift toward increasingly drier Mediterranean-type ecosystems will result in greater facilitation between species (Brooker 2006).

Nature conservation has mostly focused on the impact of global change on rare or endangered species (Maschinski et al., 2006), but a reduction in abundance of dominant and widespread species can have important consequences in terms of habitat quality and ecosystem functioning (Hovenden et al., 2008). As such, the focal species of this chapter are the widespread and expanding grass *Festuca rupicola* and the rare dicot *Dianthus carthusianorum.* Both semi-dry grassland species are native and co-exist in the steppe-like ecosystem of the dry region of central Germany, which is characterized by nutrient-poor soils and warm, dry climate conditions (Meusel 1940; Mahn 1965). As a result of shifts away from traditional land-use during the last 20 years, the abundance of *F. rupicola* has increased in these communities from ca. 25% to ca. 90%, and has turned the former species-rich communities into species-poor stands dominated by this grass species (Partzsch 2000, 2001; Wesche et al., 2005). By comparison, the abundance of the dicotyledonous species *D. carthusianorum* is less than 1% in the studied communities. Based on this observation, the question of how the two species interact, with and without traditional land-use by grazing or mowing, was addressed using a manipulative competition experiment. The experiment was conducted under field conditions using different mixture compositions of the two species (monocultures and mixtures). The hypothesis that N deposition and warming also lead to a change in plant-plant interactions was tested by a further competition experiment with both species in different mixture compositions (monocultures and mixtures) exposed to nutrient-poor and nutrient-rich environments at ambient and elevated temperatures in climate growth chambers. Both experiments should help in estimating whether there is a change in growth performance and biotic interactions (positive or negative) between the two species after land-use change and/or after nutrient enrichment and warming, and in evaluating the effect of these processes on the biodiversity of grassland communities.

Material and Methods

Study Species

Dianthus carthusianorum L. (Clusterhead; Caryophyllaceae) is a perennial, 15-45 cm high hemicryptophyte. *Festuca rupicola* Heuff. (Gramineae) is a dominant perennial, 15-80 cm high tussock grass (Figure 1). Both species are distributed in submeridional to southern temperate Europe (Jäger 2011) and mainly occur together in the communities of Festuco-Brometea. The abundance of *Dianthus carthusianorum* is currently declining, and this rare dicot is protected by law in Germany (Korneck et al., 1996). The grass is common and has spread extensively during the last 20 years. It is currently migrating into coexisting grassland communities (Klimaschewski et al., 2006). As a result of the change of traditional land-use practice, the abundance of *F. rupicola* has increased in the communities from ca. 25% to more than 90%, and has transformed the former species-rich communities into species-poor stands dominated by this species (Wesche et al., 2005). Across the different communities, the abundance of *D. carthusianorum* varied between + and 1 on the Braun-Blanquet scale in the vegetation surveys (Partzsch 2000).

Both species are very similar in germination biology: the seeds are not dormant and germinate quickly after dispersal (Partzsch 2010, 2011a). Mature seeds of both species were collected from semi-dry grassland in the surroundings of Halle (Saale) at the end of June 2007 and 2008.

Study Site

The landscape of porphyry outcrops is located in central Germany (51°31' - 51°35' N and 11°50' - 11°56' E), at an elevation of 75-135 m a.s.l. (Figure 2). With a mean annual precipitation of 473 mm and an annual mean temperature of 9.2°C, the region belongs to the driest areas of Germany (Döring 2004).

The geomorphology is defined by former glacial and periglacial activities, and by the erosive impact of the river Saale. The bedrock is acidic porphyry, which is widely covered by alkaline loess deposits. These are however eroded in several places, exposing isolated rock outcrops. Edaphic conditions are characterized by patchy mosaics of acidic Rankers and Syrosems, acidic Cambisols, neutral to slightly alkaline (Para-) Chernosems and occasional alkaline Rendzinas.

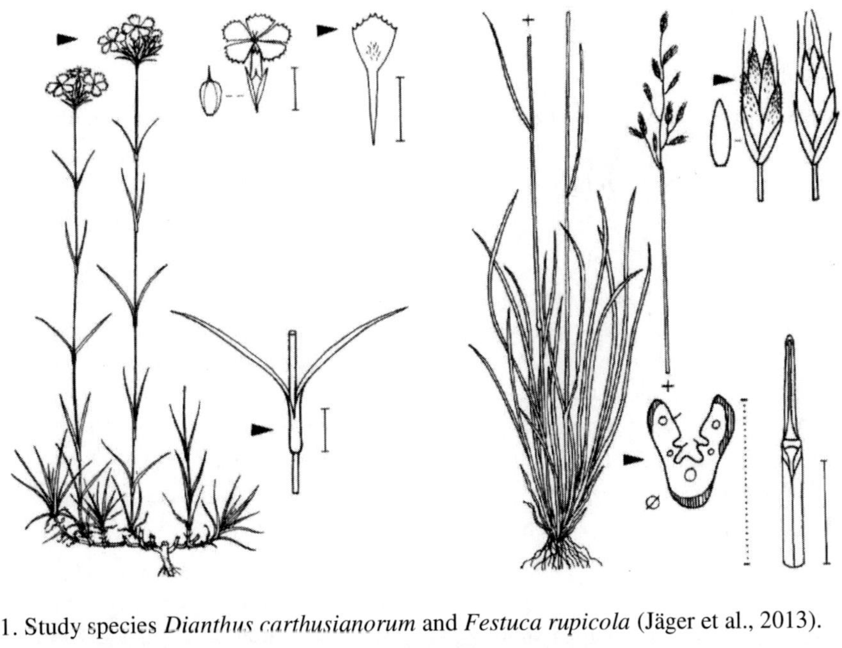

Figure 1. Study species *Dianthus carthusianorum* and *Festuca rupicola* (Jäger et al., 2013).

Figure 2. Landscape of the porphyry outcrops located in central Germany (Saxony-Anhalt; Photo: Partzsch 1993).

The edaphic and microclimatic conditions have resulted in a variety of small-scale vegetation mosaics, where communities belong mostly to the Festuco-Brometea, the Sedo-Scleranthetea and the Nardo-Callunetea (Partzsch 2000; Wesche et al., 2005) and are mostly endangered and registered on the Red List (Schubert et al., 2004).

The region was settled as early as 6500 BC. Agricultural use has intensified ever since, especially on the Chernosems. The fertile soils found over loess deposits have been cultivated with crops, whereas less productive sites on the porphyry outcrops have been used as grazing pasture for sheep. Such farming practices have been an important precondition for the development and existence of semi-natural dry grasslands and dwarf shrub heaths. However, following the political upheavals of 1990, traditional land-use practices have been in continual decline due to economic reasons (Partzsch 2000, 2001).

Experiment on the Effect of Land-use Change

In order to investigate the interactions between *D. carthusianorum* with *F. rupicola,* a competition experiment following a replacement design (De Wit 1960; Gibson et al., 1999) was conducted, where the proportion of species within mixtures varied, with a constant density maintained throughout. In August 2007, seeds of the two species were sown in plastic pots in a compost-sand mixture (2:3, pH value 7.0) and cultivated in the greenhouse. In September, the young plants were transferred to larger pots (9 plants each; diameter 22cm; height 19 cm), which were filled with the same compost-sand mixture and placed outdoors at the Botanical Garden in Halle (Saxony-Anhalt, central Germany). The experiment comprised four treatments with nine individuals per pot: [1] including monoculture of *D. carthusianorum* (hereafter called D9), [2] monoculture of *F. rupicola* (F9), [3] mixtures of the two species with two-thirds *D. carthusianorum* and one-third *F. rupicola* (D6F3) and [4] the same mixture with one-third *D. carthusianorum* and two-thirds *F. rupicola* (D3F6). Individuals were planted in three lines of three, with the less abundant species being placed in the middle. Distances between individual plants and lines within a pot were kept constant (5 cm). Each treatment was replicated ten times (n = 10), resulting in a total of 40 pots. Pots were located randomly; positions were changed every 2 to 3 weeks throughout the vegetation period, the plants were watered as required, and weeds were removed when necessary. At the end of June 2008, the plants were clipped in half of the pots (n = 5) within the four mixture compositions. The clipped pots simulated traditional land-use practice in the form of grazing or mowing, and the unclipped pots simulated abandonment. In July 2009, all pots were harvested and the aboveground biomass, number and length of flowering stems and number of flowers per individual (for *D. carthusianorum*) and length of panicles (for *F. rupicola*) measured.

Experiment on the Effect of N Deposition and Warming

In order to investigate the effect of plant-plant interactions between the two species under future temperature and nutrient conditions, a further competition experiment was conducted in climate-controlled growth chambers. The experiment comprised six mixture composition treatments with five individuals per pot: [1] monoculture of *D. carthusianorum* (D5), [2] monoculture of *F. rupicola* (F5), [3] mixture of *D. carthusianorum* (D) with 4 individuals and of *F. rupicola* (F) with one individual (i.e. D4F1), [4] mixture of D3F2, [5] mixture of D2F3 and [6] mixture of D1F4. At the same time, the mixtures and monocultures were tested under two nutrient levels: pots with nutrient-poor soil substrate (pH 5.9; N 80 mg/l, P_2O_5 80 mg/l, K_2O 80 mg/l) simulating current habitat conditions, and pots with nutrient-rich soil substrate using NPK granule fertilizer (Basacote Plus 9M, 15% N; produced by COMPO GmbH & KG, Münster, Germany) simulating N deposition enrichment. One gram of fertilizer was added to each pot in order to simulate an annual N deposition of ca. 50 $kg.ha^{-}1.yr^{-1}$. The experiments were carried out simultaneously in two growth chambers (type VB 1514, Vötsch Industrietechnik, Germany) in order to simulate summertime climate conditions in the dry region of central Germany. Each mixture composition was replicated eight times, resulting in a total of 192 pots (6 mixture composition treatments × 2 temperature treatments (ambient, elevated) × 2 nutrient treatments (low, high).

A light regime (illumination 185 $\mu mol.s^{-1}.m^{-2}$) of 10 hours light and 10 hours darkness was used, in which the light phase started and ended with a twilight period (illumination 80 $\mu mol.s^{-1}.m^{-2}$) of two hours. The temperature regime in the first chamber simulated the approximate ambient summertime conditions (20°C during the day and 10°C during the night), while the second chamber simulated the temperature conditions arising as the result of global warming (22/15°C), by increasing the temperature by approximately 3.5°C. During the twilight phases, the temperature was gradually increased or decreased. Relative humidity was kept at 65% during the day and 80% at night. The specific regimes were chosen because it is predicted that future nighttime radiant emission for central Germany will be much less than it is today (Easterling et al., 1997) and that nighttime temperatures will increase to a greater degree than daytime temperatures. The pots were rearranged and randomly placed every week to mitigate effects of location in the chambers (The chambers were brand-new with guaranteed identical conditions). The pots were watered as required and the whole experiment lasted nine month.

After nine months, aboveground biomass and number of leaves per individual as well as the number of flowering stems and flowers per individual of *D. carthusianorum* (which only flowered under nutrient rich conditions) were measured.

Data Analysis

Positive and negative interactions between the two plant species were calculated as the Relative Interaction Index (Armas et al., 2004):

$$RII = (P{+}N - P{-}N)/(P{+}N + P{-}N),$$

where P+N is the performance of the target plant in the presence of neighbours, and P-N is the performance of the target plant in the absence of neighbours. The values range from –1 to 1 and are symmetrical around zero, with positive values indicating facilitation and negative values indicating competition. The calculation of the RII is based on the aboveground biomass per individual.

All plant parameter data collected were checked for normality and log-transformed when necessary to meet the assumptions of analysis of variance (ANOVA). A parametric one-way ANOVA with HSD Tukey`s post hoc test ($p<0.05$) was used to compare the differences between the monocultures and mixtures. For the comparison of the performance of the plant parameters dependent on different environmental factors, multi-factorial analyses of variance were used. A two-way ANOVA was employed in testing the effect of clipped and unclipped treatments (with grazing and without) and the different mixture composition as fixed factors, and a three-way ANOVA for testing the effects of global change factors: nutrient levels (low, high), temperature (ambient, elevated) and mixture composition (monoculture, mixtures) as fixed factors. The fitness parameters of the number of flowering stems and flowers per individual of *D. carthusianorum* were tested using non-parametric Kruskal-Wallis tests, followed by pairwise *Mann-Whitney U-tests*, as the data was not normally distributed. Statistical analyses were performed using SPSS 19.0.

Results

Land-Use Change Effects on Plant Growth, Reproduction and Plant-Plant Interactions

The biomass of *D. carthusianorum* after two years of growth showed no significant differences between monocultures and mixtures (Figure 3, Table 1). In addition, the clipped and unclipped treatments did not significantly differ, and there was no interactive effect of clipping and mixture composition. By contrast, *F. rupicola* showed significant differences between the monoculture and mixtures, with a decrease in biomass with an increase in the number of *F. rupicola* individuals in each pot. There was also a significant difference between clipping treatments, with the clipped grass individuals producing less biomass than the unclipped.

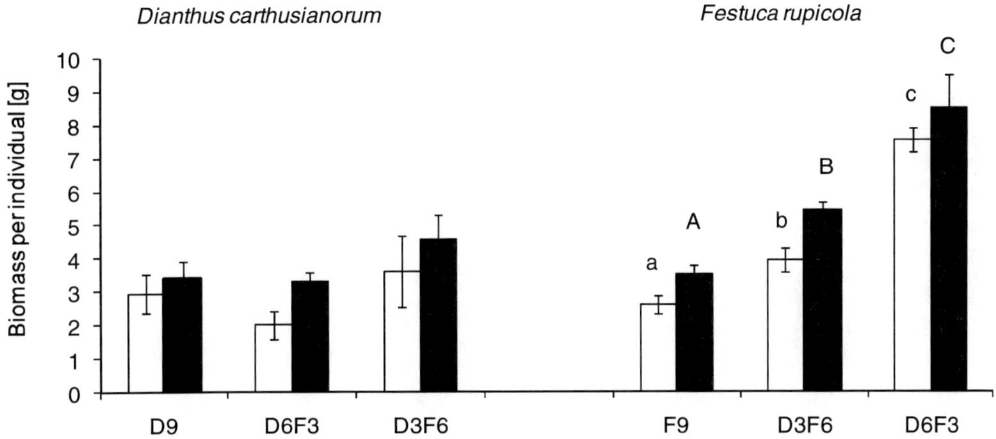

Figure 3. Biomass per individual [g] in monocultures and mixtures of *Dianthus carthusianorum* (D) and *Festuca rupicola* (F) with (□) and without (■) clipping (mean + SE; bars with different letters are significantly different: small letters: clipped; large letters: unclipped). Monocultures: D9 and F9; nine individuals of *D. carthusianorum* or *F. rupicola*, respectively. Mixtures: D6F3: six individuals of *D. carthusianorum* with three individuals of *F. rupicola*, D3F6: three individuals of *D. carthusianorum* with six individuals of *F. rupicola*.

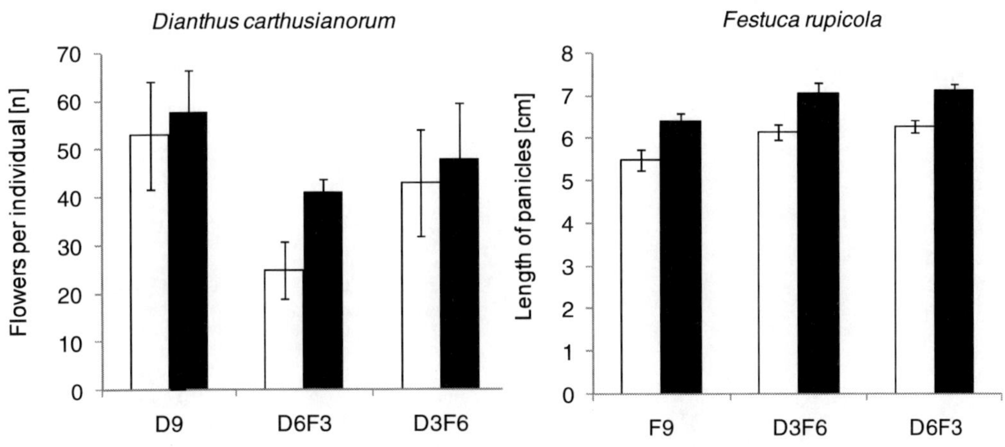

Figure 4. Number of flowers per individual [g] of *Dianthus carthusianorum* (D) and length of panicles [cm] of *Festuca rupicola* (F) in monocultures and mixtures with (□) and without (■) clipping (mean + SE; bars with different letters are significantly different; small letters: clipped; large letters: unclipped). Monocultures: D9 and F9; nine individuals of *D. carthusianorum* or *F. rupicola*, respectively. Mixtures: D6F3: six individuals of *D. carthusianorum* with three individuals of *F. rupicola*, D3F6: three individuals of *D. carthusianorum* with six individuals of *F. rupicola*.

Table 1. Two-way ANOVAs testing effects of clipping (with and without) and mixture composition (*D. carthusianorum*: D5, D6F, D3F6; *F. rupicola*: F9, D3F6, D6F3) and interaction effects on plant parameters in the second year of the competition experiment

	Dianthus carthusianorum		*Festuca rupicola*	
Source of variation	F	P	F	P
Biomass per individual				
Clipping (C)	2.989	0.097	8.028	**0.009**
Mixture composition (M)	2.456	0.107	54.083	**< 0.0001**
C x M	0.202	0.819	0.226	0.800
Fitness parameter				
Clipping (C)	1.343	0.258	31.158	**< 0.0001**
Mixture composition (M)	2.989	0.069	8.412	**0.002**
C x M	0.264	0.770	0.018	0.982
Relative Interaction Index (RII)				
Clipping (C)	1.705	0.210	1.059	0.319
Mixture composition (M)	1.775	0.201	47.307	**< 0.0001**
C x M	0.066	0.800	2.136	0.163

Table 2. Three-way ANOVAs testing effects of nutrient (low, high), temperature (ambient, elevated) and mixture composition (*D. carthusianorum*: D1F4, D2F3, D3F2, D4F1, D5; *F. rupicola*: D4F1, D3F2, D2F3, D1F4, F5) and interacting effects on biomass, number of leaves per individual and the RII of both grassland species

Source of variation	df	Biomass per individual [g]		Number of leaves per individual		df	Relative Interaction Index (RII)	
		F	P	F	P		F	P
Dianthus carthusianorum								
Nutrient (N)	1	543.337	**<0.0001**	612.566	**<0.0001**	1	5.268	**0.024**
Temperature (T)	1	100.510	**<0.0001**	140.092	**<0.0001**	1	24.389	**<0.0001**
Mixture composition (M)	4	0.414	0.800	3.304	**0.013**	3	0.219	0.883
N x T	1	61.784	**<0.0001**	112.795	**<0.0001**	1	82.987	**<0.0001**
N x M	4	1.474	0.213	1.810	0.130	3	1.810	0.149
T x M	4	2.824	**0.027**	1.622	0.172	3	1.655	0.181
N x T x M	4	12.040	**<0.0001**	2.414	0.052	3	9.586	**<0.0001**
Error	140					112		
Festuca rupicola								
Nutrient (N)	1	379.806	**<0.0001**	367.245	**<0.0001**	1	2.489	0.117
Temperature (T)	1	0.670	0.414	20.181	**<0.0001**	1	10.553	**0.002**
Mixture composition (M)	4	1.954	0.105	2.037	0.092	3	1.487	0.222
N x T	1	3.401	0.067	6.299	**0.013**	1	24.527	**<0.0001**
N x M	4	0.981	0.420	4.587	**0.002**	3	0.896	0.446
T x M	4	1.323	0.264	1.000	0.410	3	0.390	0.760
N x T x M	4	2.769	**0.030**	4.758	**0.001**	3	0.957	0.416
Error	140					112		

The number of flowers of *D. carthusianorum* was reduced to a marginally significant degree in mixtures with the grass, but was not affected by clipping (Figure 4, Table 2). By contrast, the panicle length of *F. rupicola* significantly decreased in the monoculture as compared to the mixtures, with clipped individuals producing significantly shorter panicles than unclipped individuals. However, no significant interactive effect resulted from clipping and mixture composition.

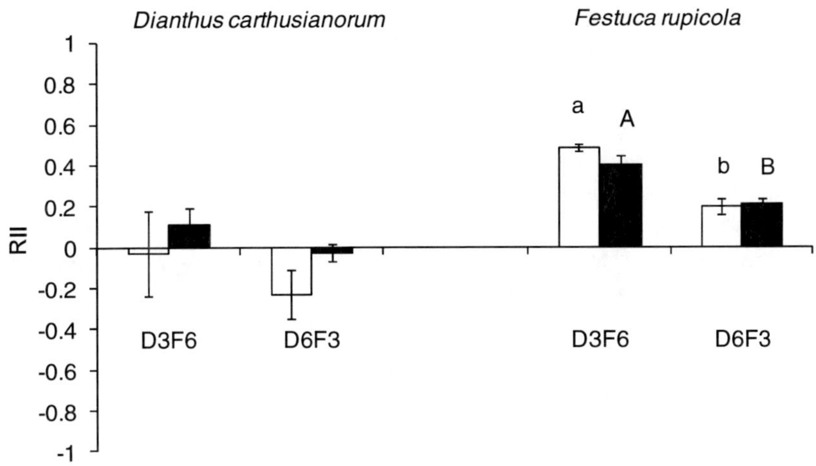

Figure 5. Relative Interaction Index (RII) of *Dianthus carthusianorum* (D) and *Festuca rupicola* (F) in mixtures of both grassland species with (□) and without (■) clipping in the second year (mean + SE; bars with different letters are significantly different; small letters: clipped; large letters: unclipped). Positive values indicate facilitation, negative values indicate competition. Mixtures: D3F6: three individuals of *D. carthusianorum* with six individuals of *F. rupicola*, D6F3: six individuals of *D. carthusianorum* with three individuals of *F. rupicola*.

The Relative Interaction Index (RII) of *D. carthusianorum* was not significantly affected by clipping, mixture composition or their interaction (Figure 5, Table 1). In mixtures with the grass, *D. carthusianorum* was inhibited or rarely affected, so that the plant-plant interactions were more or less neutral. By contrast, *F. rupicola* was facilitated by the presence of *D. carthusianorum,* with greater effects at lower densities of *D. carthusianorum* (D3F6). As such, there were significant effects from mixture composition, but not from clipping or interaction between clipping and mixture composition.

Effect of N Deposition and Warming on Plant Growth, Reproduction and Plant-Plant Interaction

Under nutrient-poor conditions, the biomass and number of leaves of *D. carthusianorum* did not significantly differ with mixture composition at ambient temperature (Figure 6a,b). In addition, at elevated temperature, no difference was observed in biomass, whereas the number of leaves did differ. After nutrient addition (Figure 6c,d), biomass and number of leaves increased approximately tenfold. At ambient temperature, the biomass per individual of *D. carthusianorum* increased significantly with increasing number of *D. carthusianorum* individuals per pot (Figure 6c), but the number of leaves did not vary (Figure 6d). However, elevated temperature caused a strong increase in biomass and leaf number in comparison to ambient conditions, and both parameters significantly decreased with increasing number of *D. carthusianorum* individuals per pot.

Under nutrient-poor conditions, *Festuca rupicola* showed no significant differences in biomass between the mixture composition at ambient and elevated temperatures (Figure 7a). The number of leaves varied significantly at ambient temperature, and was highest in the pots with a high number of *F. rupicola* individuals (Figure7b). Nutrient addition led to a tenfold increase in biomass and a threefold increase in leaf number (Figure 7c,d). At ambient temperature, biomass and number of leaves significantly decreased with increasing number of *F. rupicola* individuals per pot. At elevated temperature, both parameters generally decreased, but did not significantly vary with mixture composition.

The results of the three-way ANOVA showed that biomass and number of leaves of *D. carthusianorum* were significantly affected by the interactive effects of nutrient and temperature (Table 2). There were also significant interactive effects resulting from temperature x mixture composition and nutrient x temperature x mixture composition on the biomass of *D. carthusianorum*. Surprisingly, the biomass of *F. rupicola* was only significantly affected by the interactive effects of nutrient x temperature x mixture composition, and the single factor nutrient. Moreover, the number of leaves was significantly reduced by warming, and was significantly affected by all interactions that included nutrient as a factor.

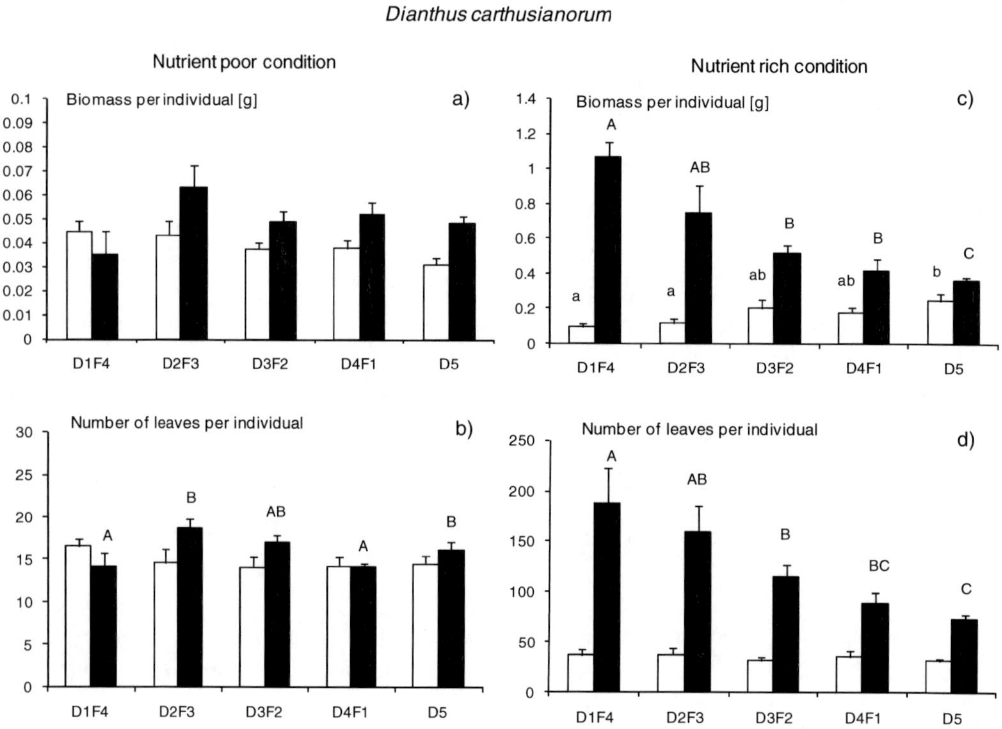

Figure 6. Performance of biomass and number of leaves per individual of *Dianthus carthusianorum* under nutrient-poor (left) and nutrient-rich (right) conditions at ambient (□) and elevated (■) temperatures (mean + SE; bars with different letters are significantly different; small letters: ambient temperature, large letters: elevated temperature). Mixtures D1F4, D2F3, D3F2 and D4F1: one, two, three or four individuals of *D. carthusianorum* (D) with four, three, two or one individuals of *F. rupicola* (F), monoculture D5: five individuals of *D. carthusianorum* (D); ordered by increasing density of *D. carthusianorum*.

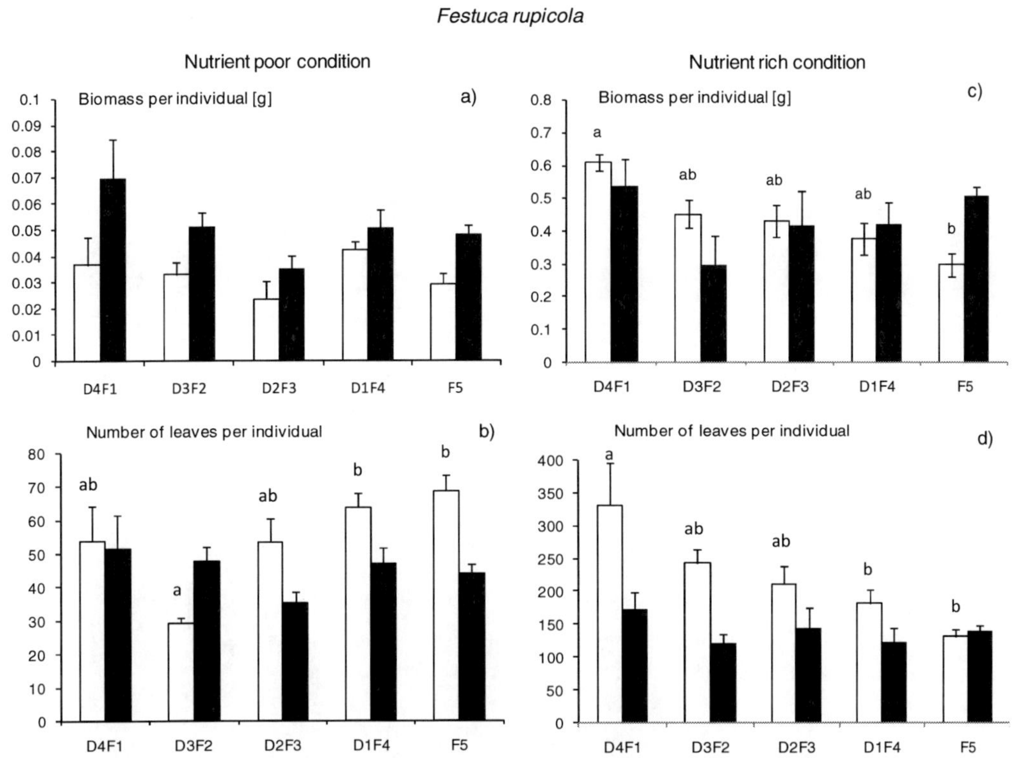

Figure 7. Performance of biomass und number of leaves per individual of *Festuca rupicola* under nutrient-poor (left) and nutrient-rich (right) conditions at ambient (□) and elevated (■) temperatures (mean + SE; bars with different letters are significantly different; small letters: ambient temperature, large letters: elevated temperature). Mixtures D4F1, D3F2, D2F3, D1F4: four, three, two or one individuals of *D. carthusianorum* (D) with one, two, three or four individuals of *F. rupicola* (F), monoculture F5: five individuals of *F. rupicola* (F); ordered by increasing density of *F. rupicola*.

The positive and negative plant-plant interactions between the two study species were generally low under nutrient-poor conditions (Figure 8, left). At ambient temperature, *D. carthusianorum* was slightly facilitated and *F. rupicola* inhibited, but without significant differences between mixture composition. Under elevated temperature, plant interactions were more or less neutral, and differed significantly only for *D. carthusianorum*. Under nutrient-rich conditions (Figure 8, right), *D. carthusianorum* was negatively affected by competition with increasing grass density, whereas *F. rupicola* was significantly facilitated with increasing number of *D. carthusianorum* individuals in each pot. At elevated temperature the plant-plant interaction reversed, with *D. carthusianorum* switching to facilitation. The facilitation effects rose with increasing number of *F. rupicola* individuals in each pot, and *F. rupicola* became inhibited.

The results of the three-way ANOVA showed that the RII of *D. carthusianorum* was significantly affected by interactive effects of nutrient x temperature and nutrient x temperature x mixture compositions, as well as the single factors nutrient and temperature (Table 2). Only the interactive effect of nutrient x temperature and the factor temperature significantly influenced the RII of *F. rupicola*.

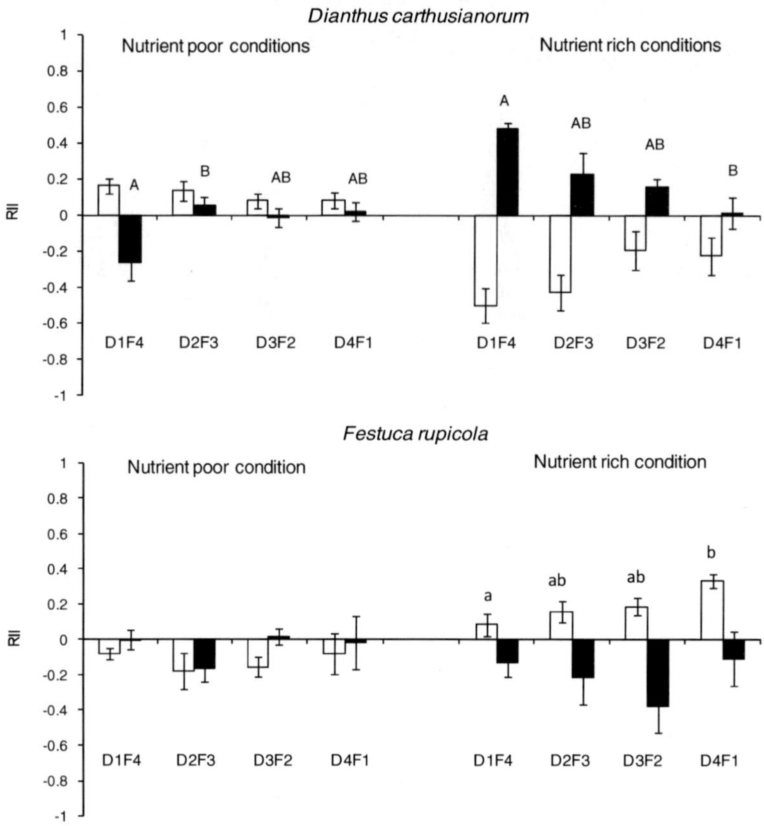

Figure 8. Relative Interaction Index (RII) of *Dianthus carthusianorum* (D) and *Festuca rupicola* (F) in the mixtures of both grassland species under nutrient-poor (left) and nutrient-rich (right) conditions at ambient (□) and elevated temperatures (■). Positive values indicate facilitation, negative values indicate competition (mean + SE; bars with different letters are significantly different; small letters: ambient temperature, large letters: elevated temperature). Mixtures D1F4, D2F3, D3F2, D4F1: one, two, three or four individuals of *D. carthusianorum* (D) with four, three, two or one individuals of *F. rupicola* (F); ordered by increasing density of *D. carthusianorum* and decreasing density of *F. rupicola*.

Dianthus carthusianorum started flowering under nutrient-rich conditions after the four month, but *F. rupicola* did not. At ambient temperature, *D. carthusianorum* produced flowering stems and flowers only in the mixtures D3F2, D4F1 and D5, where the number of *F. rupicola* individuals was reduced (Figure 9). At elevated temperature, *D. carthusianorum* flowered in all treatments except D3F2. The highest numbers of flowering stems and flowers were produced in the pots with the highest number of *F. rupicola* individuals (D1F4).

DISCUSSION

Land-Use Change Impacts on Species Conservation

The results of the two-year field experiment showed that the expanding grass *F. rupicola* was significantly reduced in biomass and reproduction (shorter panicles) by clipping, which simulates land-use practice in the form of grazing. This suggests that the competitive ability of *F. rupicola* increased significantly even after the first year of abandonment. However, both plant parameters were affected differently by the density of the respective species in the pots. The grass suffered to a much greater degree as a result of intraspecific competition in the monoculture than as a result of interspecific competition in the mixtures, since the biomass was significantly reduced with increasing density of *F. rupicola* in the pots. Also, the grass produced significantly more generative biomass in the mixtures with *D. carthusianorum* than in the monoculture. By contrast, *D. carthusianorum* produced fewer flowers in the mixtures, which showed that interspecific competition with the grass had a stronger negative effect on the generative growth of the dicot. Over time, a reduction in flower numbers can lead to seed limitation, following by a decline in biodiversity. This is in accordance with Aarssen & Keogh (2002) and Freckleton et al. (2009), who argued that generative growth in terms of fecundity is a more sensitive indicator of vigour than biomass production.

The findings of the first experiment agree with the results of del-Val & Crawley (2005), who found a competitive hierarchy in grasslands in which the monocotyledonous species, such as grasses, were much stronger competitors than dicots. Moreover, Carlsen et al. (2001) showed in a competition experiment in California that the rare endemic annual dicot *Amsinckia grandiflora* (Boraginaceae) was pushed back by competition from exotic annual grasses, but that it developed better in the presence of the native perennial grass *Poa secunda*. However, both exotic and native grasses caused a reduction in biomass and number of flowers in the dicot. In another experiment, the competition experienced by the low growing species *Alyssum montanum* (Brassicaceae) increased with increasing density of *F. rupicola* (Partzsch et al., 2011). On the other hand, the rare semi-dry grassland species *Campanula glomerata* was particularly facilitated when mixed with *F. rupicola*, whereas in the presence of *Poa angustifolia,* it was strongly inhibited (Partzsch & Bachmann 2011).

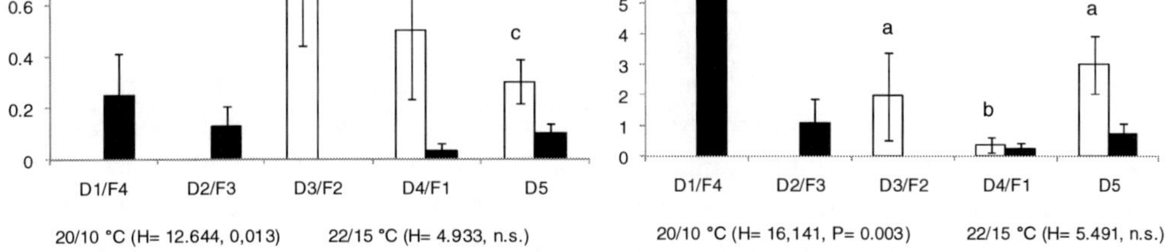

Figure 9. Number of flowering stems and flowers per individual of *Dianthus carthusianorum* under nutrient-rich conditions at ambient (□) and elevated (■) temperatures (mean + SE). The results of the Kruskal-Wallis test are shown (H- and P-values; bars with different letters are significantly different; small letters: ambient temperature). Mixtures D1F4, D2F3, D3F2 and D4F1: one, two, three or four individuals of *D. carthusianorum* (D) with four, three, two or one individuals of *F. rupicola* (F), monoculture D5: five individuals of *D. carthusianorum* (D); ordered by increasing density of *D. carthusianorum*.

This indicates that the competitive effect of grass species on dicotyledonous species depends on the differing growth forms and strategy types of the species. As such, the switch between facilitation and competition is influenced by strategy type as well as by environmental factors (Chen et al., 2009). The study species of this chapter, *D. carthusianorum,* belongs to the intermediate competition-stress-ruderal-strategy type (CSR-type) and possesses lower competitive ability than the tussock grass *F. rupicola*, which is a stress tolerant competitor (CSR-type) with greater competitive ability (Grime 1979; Frank & Klotz 1990). This is reflected in the Relative Interaction Index (RII), which showed that *F. rupicola* was facilitated in the mixture with *D. carthusianorum,* and that the dicot was more or less suppressed by the grass in the second year. Also, after two years, a trend was already visible in which clipping transforms the negative interaction into a positive interaction in the dicot. This suggests that long-term abandonment of the grasslands would lead to a stronger decline in *D. carthusianorum,* as already observed in the field (Partzsch 2000, 2001).

Several studies have demonstrated that land-use affects the development of particular plant species in different ways. On the one hand, grazing or mowing can cause either a reduction (Edwards 1985; Kluth et al., 2003; Hejcman et al., 2010), or an increase in biomass and number of flowers (Mueggler 1972; Georgiadis et al., 1989; Esmaeili et al., 2009). This indicates that plant species respond in different ways to land-use change, as a result of differences in plant traits and competitive ability. Mariotte et al. (2013) found that species loss after abandonment of pasturing was mainly the result of a higher reduction in the number of subordinate species as a consequence of the increasing proportion of dominant species. This is similar to the present results with the expansion of the dominant grass *F. rupicola* and the decline of the rare dicot *D. carthusianorum*. The abandonment of grazing in recent decades counteracts the long-term development of species-rich grasslands communities since the early Middle Ages, and leads to a progressive decline in biodiversity in vulnerable grassland communities (Collins et al., 1998; Partzsch 2001, 2011b; Pykälä 2003, Mariotte et al., 2013). A reintroduction of traditional land-use in the form of extensive grazing would promote the regrowth of rare species, such as *D. carthusianorum*, due to the ensuing reduction in competition from dominant species such as *F. rupicola*.

Global Change Impacts on Dominant and Rare Species

The two semi-dry grassland species *D. carthusianorum* and *F. rupicola* responded in different ways to the manipulated factors: N deposition, warming and mixture composition. While N deposition showed strong promoting effects on the growth of both species, the responses to warming and mixture composition, and the interaction effects of all three factors, were different. Under nutrient-poor conditions, biomass was not affected by the differing mixture compositions under current and predicted climate conditions. This shows that the growth was low and not affected neither by positive or negative interactions nor interspecific or intraspecific competition. The plant-plant interactions were nearly neutral. After the addition of NPK fertilizer, the higher nutrient uptake led to a significant (ca. tenfold) increase in the growth of both semi-dry grassland species, with the species reaching the adult stage much earlier than under natural, nutrient-poor habitat conditions. At the same time, the species started to interact. *Dianthus carthusianorum* increased in biomass with increasing density in the pots, showing that interspecific competition is higher than intraspecific competition. However, *F. rupicola* responded in the opposite way, which suggests that the grass is a stronger competitor than the dicot at ambient temperature. These findings suggest higher competitive abilities of *F. rupicola* and facilitative effects of *D. carthusianorum*.

Warming enhanced biomass and number of leaves of *D. carthusianorum* by a factor of ca. two to ten, depending on mixture composition, with a strongly decreasing effect with increasing density of *D. carthusianorum* in the pots. Strong interactive effects of nutrient level with temperature, and nutrient level with temperature and mixture composition suggest that, under the predicted conditions of global change through N deposition and warming, the rare dicot *D. carthusianorum* will be heavily promoted. By contrast, the biomass production of *F. rupicola* was not affected by warming. However, the number of leaves was significantly reduced (by around half) in the different mixtures as the result of better growth of the co-existing *D. carthusianorum*. These species-specific effects agree with the results of Zhang et al. (2008), who found, in a similar experiment, that biomass and height of *Festuca rubra* decreased, while *Trifolium pratense* remained unchanged with increasing temperature and nutrient addition. Results from arctic, alpine and mediterranean ecosystems also showed that particular species increased in growth height and biomass, but that other species where inhibited after warming (Kuodo & Suzuki 2003; Llorens et al., 2004). In accordance with the present results of *F. rupicola,* Hovenden et al. (2008) reported that seedlings of the temperate grass *Austrodanthonia caespitosa* descended from "warmed plants" were 20% smaller, which suggests that climate warming will most likely reduce population growth or distribution of this dominant grass. However, biomass loss by warming can be caused by temperature-sensitive respiration processes (Yoshida 1981; Amthor 2000), which reduce the amount of assimilates available for growth and yield (Monteith 1981). This will increase as the result of predicted higher nighttime temperatures in the study area (Easterling et al., 1997).

Species-specific responses to warming can result in shifts in competitive priority (Zhang et al., 2008), which is confirmed by the results of this chapter. The plant-plant interactions were significantly affected by warming and switched from competition (ambient temperature) to facilitation (elevated temperature) for *D. carthusianorum,* which increased with the number of grass individuals in the pot. However, *F. rupicola* reacted in an opposing manner. The general performance of the rare dicot *D. carthusianorum* under elevated temperature suggests that this species is more adapted to the predicted global change, especially warming, than the dominant grass *F. rupicola*, although both species are typical representatives of semi-dry grassland communities. Shifts from facilitation to competition and vice versa, as a result of shifts in temperature, were described by Greenlee & Callaway (1996), Choler et al. (2001), Bertness & Ewanchuk (2002) and Kikvidze et al. (2006). The promotion of the rare dicot *D. carthusianorum* by global warming suggests a positive effect on biodiversity in the future.

The different growth performance and interactions of the study species may be caused by different physiological and morphological plant traits (Zhang et al., 2008). As such, life-form, life history and strategy type (after Grime 1979) of co-existing species, as well as the character of stress factors, which are distinguished between resource (e.g. water, nutrient, light) and non-resource stress factors (e.g. temperature, wind, salinity, soil structure), determine the interplay between facilitation and competition (Grime 1977; Chen et al., 2009; Maestre et al., 2009). The interactive effects of nutrient levels and temperature showed that the combination of both stress factors reduced the competitive ability of the grass, and increased the competitive ability of the dicot. This agrees with the findings of Sala et al.

(2000), who showed that multiple environmental changes can lead to synergistic or antagonistic effects.

Under current climate conditions, the fitness of the dicot *D. carthusianorum* was reduced with increasing numbers of grass individuals in each pot. This suggests strong interspecific competition, which mainly affects fecundity and leads to a decrease in the number of flowers, fruits and seeds per fruit (Fréville & Silbertown 2005). After warming, the growth performance of *F. rupicola* was reduced and competition between both species declined: the dicot also flowered and fruited in the pots with higher numbers of grass individuals. However, the number of flowering stems and the number of flowers of *D. carthusianorum* were lower under warmed than under ambient conditions, which suggests that the vegetative and generative parameters react in different ways to global change. This agrees with Klanderud & Totland (2005), who reported that both nutrient availability and warming often show synergistic effects on plant performance, and that warming had only minor effects on vegetative growth but increased the height of flowering stems of *Thalictrum alpinum*. Findings of this experiment showed opposite mechanisms for *D. carthusianorum,* whose growth was strengthened by warming, whereas no effects were observed on reproductive outputs.

Conclusion and Implications

The results of this chapter show that the change of traditional land-use through the loss of grazing can lead to the expansion of the grass *F. rupicola,* and possibly cause the decline of co-existing dicots through increasing competitive abilities of grasses. By contrast, processes of global change, such as N deposition and warming, lead to a change in species composition of semi-dry grasslands in such a way, that the dominant grass *F. rupicola* can become inhibited and the rare dicot *D. carthusianorum* becomes facilitated. The difference between the investigated processes is that the abandonment of pasturing or mowing takes place suddenly, mostly due to economical reasons, whereas N deposition and warming proceed gradually over long periods of time, both affecting plant-plant interactions quickly or slowly, respectively. The elevated temperature of ca. 3.5°C used in this experiment will be reached in the dry region of central Germany within the next five to seven decades, on the basis of forecasts derived from the prognoses of the Intergovernmental Panel on Climate Change (IPCC, 2007). As such, the present negative effect on biodiversity caused by the abandonment of grazing cannot be compensated by the predicted positive effect of global change on biodiversity in the near future. However, a contemporary reintroduction of extensive management with sheep and goats can counteract the decline and loss of rare species, since grazing reduces the density of dominant species and trampling leads to spatial heterogeneity (gaps), which promote niche regeneration for rare species (Kalamees & Zobel 2002; Mariotte et al., 2013). This would further promote the conservation of species-rich semi-dry grasslands, in that the fortunes of rare species would be improved, since they may be more resistant to global change. However, plant-plant interactions between grasses and dicots need to be investigated further in order to provide a wider and more robust understanding of their dynamics under different environmental conditions.

ACKNOWLEDGMENT

I would like to thank Christine Voigt and the students Karin Schrieber, Jennifer Klatt, Silvana Tornack and Kathrin Pohle, who managed the experiments throughout and helped to collect data. In addition, I would like to thank Isabell Hensen, Ernst-Gerhard Mahn and Anselm Krumbiegel for their valuable comments on the manuscript, and Chris Meighan for checking my English.

REFERENCES

Aarssen, L. W. & Keogh, T. (2002) Conundrums of competitive ability in plants: what to measure? *Oikos* 96, 531-541.

Amthor, J. S. (2000) Direct effect of elevated CO_2 on nocturnal *in situ* leaf respiration in nine temperate deciduous tree species is small. *Tree Physiology* 20, 139-144.

Armas, C., Ordiales, R. & Pugnaire, F. I. (2004) Measuring plant interactions: a new comparative index. *Eco*logy 85, 2682-2686.

Balmer, O. & Erhardt, A. (2000) Consequences of succession on extensively grazed grassland for central European butterfly communities: rethinking conservation practices. *Conservation Bio*logy 14, 746-757.

Belsky, A. J. (1994) Influences of trees on savanna productivity: tests of shade, nutrients, and tree grass competition. *Ecology* 75, 922-932.

Bertness, M. D. & Ewanchuk, P. J. (2002) Latitudinal and climate-driven variation in the strength and nature of biological interactions in New England salt marshes. *Oecologia* 132, 392-401.

Bobbink, R., Hornung, M. & Roelofs, J. G. M. (1998) The effects of air-borne nitrogen pollutants on species diversity in natural and semi-natural European vegetation. *Plant Ecology* 86, 717-738.

Bornkamm, R. (2006) Fifty years vegetation development of a xerothermic calcareous grassland in Central Europe after heavy disturbance. *Flora* 201, 249-267.

Brooker, R. W. (2006) Plant-plant interactions and environmental change. *New Phytologist* 171, 271-284.

Brooker, R. W., Maestre, F. T., Callaway, R. M., Lortie, C. L., Cavieres, L. A., Kunstler, G., et al. (2008) Facilitation in plant communities: the past, the present, and the future. *Journal of Ecology* 96, 18-34.

Bruno, J. F., Stachowicz, J. J. & Bertness, M. D. (2003) Inclusion of facilitation into ecological theory. *Trends in Ecology & Evolution* 18, 119-125.

Callaway, R. M. (1995) Positive interactions among plants. *The Botanical Review* 61, 306-349.

Callaway, R. M. (2007) *Positive interactions and interdependence in plant communities*. Springer, Dordrecht.

Callaway, R. M. & Walker, L. R. (1997) Competition and facilitation: a synthetic approach to interactions in plant communities. *Ecology* 78, 1958-1965.

Callaway, R. M., Brooker, R. W., Choler, P., Kikvidze, Z., Lortie, C. J., Michalet, R., Paolini, L., Pugnaire, F. I., Newingham, B., Aschehoug, E. T., Armas, C. & Cook, B. J. (2002) Positive interactions among alpine plants increase with stress. *Nature* 417, 844-848.

Carlsen, T. M., Menke, J. W. & Pavilk, B. M. (2001) Reducing competitive suppression of a rare annual forb by restoring native California perennial grasslands. *Restoration Ecology* 8, 18-29.

Chen, S. Y., Xu, J., Maestre, F. T., Chu, C. J., Wang, G. & Xiao, S. (2009) Beyond dual-lattice models: Incorporating plant strategies when modeling the interplay between facilitation and competition along environmental severity gradients. *Journal of Theoretical Biology* 258, 266-273.

Choler, P., Michalet, R. & Callaway, R. M. (2001) Facilitation and competition on gradients in alpine plant communities. *Ecology* 82, 3295-3308.

Chýlová, T. & Münzbergová, Z. (2008) Past land-use co-determines the present distribution of dry grassland plant species. *Preslia* 80, 183-198.

Collins, S. L., Knapp, A. K., Briggs, J. M., Blair, J. M. & Steinauer, E. M. (1998) Modulation of diversity by grazing and mowing in native tall grass prairie. *Science* 280, 754-747.

De Jong, M., Fangmeier, A. & Jäger, H. J. (1998) Reaktionen von drei Süßgrasarten mit unterschiedlichen Nährstoffansprüchen auf steigende NH_3-Konzentrationen und NH_4^+-Gaben. – Ergebnisse aus zwei Vegetationsperioden. *Verhandlungen Gesellschaft für Ökologie* 28, 373-380.

De Wit, C. T. (1960) On competition. *Verslagen van Landbouwkundige Onderzoekingen* 66, 1-82.

del-Val, E. & Crawley, M. J. (2005) What limits herb biomass in grasslands: competition of herbivory? *Oecologia* 142, 202-211.

Döring, J. (2004) Zu den Klimaverhältnissen im östlichen Harzvorland. *Hercynia NF* 37, 137-154.

Dostálek, J. & Frantík, T. (2008) Dry grassland plant diversity conservation using low-intensity sheep and goat grazing management: case study in Prague (Czech Republic). *Biodiversity and Conservation* 17, 1439-1454.

Dupré, C., Stevens, C. J., Ranke, T., Bleeker, A., Peppler-Lisbach, C., Gowing, D. J. G., et al. (2010) Changes in species richness and composition in European acidic grasslands over the past 70 years: the contribution of cumulative atmospheric nitrogen deposition. *Global Change Biology* 16, 344-357.

Easterling, D. R., Horton, B., Jones, P. D., Peterson, T. C., Karl, T. R., Parker, D. E., et al. (1997) Maximum and minimum temperature trends for the globe. *Science* 277, 363-367.

Edwards, J. (1985) Effects of herbivory by moose on flower and fruit production of *Aralia nudicaulis*. *Journal of Ecology* 73, 861-868.

Enyedi, Z. M., Ruprecht, E. & Deak, M. (2008) Long-term effects of the abandonment of grazing on steppe-like grasslands. *Applied Vegetation Science* 11, 53-60.

Esmaeili, M. M., Bonis, A., Bouzillé, J.-B., Mony, C. & Benot, M.-L. (2009) Consequence of ramet defoliation on plant clonal propagation and biomass allocation: Example of five rhizomatous grassland species. *Flora* 204, 25-33.

Fowler, N. (1986) The role of competition in plant communities in arid and semiarid regions. *Annual Review of Ecology and Systematics* 17, 89-110.

Frank, D. & Klotz, S. (1990) Biologisch-ökologische Daten zur Flora der DDR. *Wissenschaftliche Beiträge Martin-Luther Universität, Halle-Wittenberg.* 1990/32 (P41).

Franz, H. (1952/53) Dauer und Wandel der Lebensgemeinschaften. *Verein Verbreitung naturwissenshafticher Kenntnisse* 93, 27-45.

Freckleton, R. P., Watkinson, A. R. & Rees, M. (2009) Measuring the importance of competition in plant communities. *Journal of Ecology* 97, 379-384.

Fréville, H. & Silvertown, J. (2005) Analysis of interspecific competition in perennial plants using Life Table Response Experiments. *Plant Ecology* 176, 69-78.

Georgiadis, N. J., Ruess, R. W., Mc Naughton, S. J. & Western, D. (1989) Ecological conditions that determine when grazing stimulates grass production. *Oecologia* 81, 316-322.

Gibson, D. J., Connolly, J., Hartnett, D. C. & Weidenhamer, J. D. (1999) Designs for greenhouse studies of interactions between plants. *Journal of Ecology* 87, 1-16.

Greenlee, J. T. & Callaway, R. M. (1996) Abiotic stress and the relative importance of interference and facilitation in montane bunchgrass communities in western Montana. *The American Naturalist* 148, 386-396.

Grime, J. P. (1973) Competitive exclusion in herbaceous vegetation. *Nature* 242, 344-347.

Grime, J. P. (1977) Evidence for the existence of three primary strategies in plants and its relevance to ecological and evolutionary theory. *The American Naturalist* 111, 1169-1194.

Grime, J. P. (1979) *Plant strategies and vegetation processes*. Chichester, New York, Brisbane, Toronto.

Hejcman, M., Češková, M., Schellberg, J. & Pätzold, S. (2010) The Rengen Grassland Experiment: Effect of Soil Chemical Properties on Biomass Production, Plant Species Composition and Species Richness. *Folia Geobotanica* 45, 125-142.

Holmgren, M., Scheffer, M. & Huston, M. A. (1997) The interplay of facilitation and competition in plant communities. *Ecology* 78, 1966-1975.

Horswill, P., O'Sullivan, O., Phoenix, G. K., Lee, J. A. & Leake, J. R. (2008) Base cation depletion, eutrophication and acidification of species-rich grasslands in response to long-term simulated nitrogen deposition. *Environmental Pollution* 155, 336-349.

Hovenden, M. J., Wills, A. E., Chaplin, R. E., Vander Schoor, J. K., Williams, A. L., Osanai, Y. & Newton, P. C. D. (2008) Warming and elevated CO_2 affect the relationship between seed mass, germinability and seedling growth in *Austrodanthonia caespitosa*, a dominant Australian grass. *Global Change Biology* 14, 1633-1641.

Huston, M. (1979) General hypothesis of species-diversity. *The American Naturalist* 113, 81-101.

IPCC (2007) *Intergovernmental Panel on Climate Change*. Fourth Assessment Report. Working Group I Report on "The Physical Science Basis".

Jäger, E. J. (ed.) (2011) *Exkursionsflora von Deutschland, begründet von W. Rothmaler. Gefäßpflanzen*: *Grundband*. Spektrum, Akad. Verlag, Heidelberg.

Jäger, E. J., Wesche, K., Ritz, C., Müller, F. & Welk, E. (2013) *Rothmaler - Exkursionsflora von Deutschland, Gefäßpflanzen: Atlasband.* Spektrum, Akad. Verlag, Heidelberg.

Kalamees, R. & Zobel, M. (2002) The role of the seed bank in gap regeneration in a calcareous grassland community. *Ecology* 83, 1017-1025.

Kikvidze, Z., Khetsuriani, L., Kikodze, D. & Callaway, R. M. (2006) Seasonal shifts in competition and facilitation in subalpine plant communities of the central Caucasus. *Journal of Vegetation Science* 17, 77-82.

Klanderud, K. & Totland, O. (2005) The relative importance of neighbours and abiotic environmental conditions for population dynamic parameters of two alpine plant species. *Journal of Ecology* 93, 493-501.

Klimaschewski, B., Evers, C. & Brandes, D. (2006) Untersuchungen zur Einwanderung von Festuco-Brometea- und Koelerio-Corynephoretea-Arten in Brachflächen. *Arbeiten aus dem Institut für Landschaftsökologie Münster* 15, 97-111.

Kluth, S., Kruess, A. & Tscharntke, T. (2003) Influence of mechanical cutting and pathogen application on the performance and nutrient storage of *Cirsium arvense*. *Journal of Applied Ecology* 40, 334-343.

Kardol, P., Cregger, M. A., Campany, C. E. & Classen, A. T. (2010) Soil ecosystem functioning under climate change: plant species and community effects. *Ecology* 91, 767-781.

Korneck, D., Schnittler, M. & Vollmer, I. (1996) Rote Liste der Farn- und Blütenpflanzen. *Schriftenreihe für Vegetationskunde* 28, 1-187.

Körschens, M. & Mahn, E.G. (1995) *Strategien zur Regeneration belasteter Agrarökosysteme des mitteldeutschen Schwarzerdegebietes*. B.G. Teubner Verlagsgesellschaft, Stuttgart, Leipzig.

Kuijper, D. P. J., Dubbeld, J. & Bakker, J. P. (2005) Competition between two grass species with and without grazing over a productivity gradient. *Plant Ecology* 179, 237-246.

Kuodo, G. & Suzuki, S. (2003) Warming effects on growth, production, and vegetation structure of alpine shrubs: a five-year experiment in northern Japan. *Oecologia* 135, 280-287.

Llorens, L., Penuelas, J., Estiarte, M. & Bruna, P. (2004) Contrasting growth changes in two dominant species of a Mediterranean shrubland submitted to experimental drought and warming. *Annals of Botany* 94, 843-853.

Mariotte, P., Buttler, A., Kohler., F., Gilgen, A. K. & Spiegelberger, T. (2013) How do subordinate and dominate species in semi-natural mountain grasslands relate to productivity and land-use change? *Basic and Applied Ecology* 14, 217-224.

Maestre, F. T. & Cortina, J. (2004) Do positive interactions increase with abiotic stress? A test from the semiarid steppe. *Proceedings of the Royal Society London, B* 271, 331-333.

Maestre, F. T., Bautista, S. & Cortina J. (2003). Positive, negative, and net effects in grass-shrub interactions in Mediterranean semiarid grasslands. *Ecology* 84, 3186-3197.

Maestre, F. T., Callaway, R. M., Valladares, F. & Lortie, C. J. (2009) Refining the stress-gradient hypothesis for competition and facilitation in plant communities. *Journal of Ecology* 97, 199-205.

Mahn, E. G. (1965) Vegetationsaufbau und Standortverhältnisse der kontinental beeinflußten Xerothermrasengesellschaften Mitteldeutschlands. *Abhandlungen der Sächsischen Akademie der Wissenschaften, Leipzig, Mathematisch-naturwissenschaftliche Klasse* 49(1), 1-138.

Maschinski, J., Baggs, J. E., Quintana-Ascencio, P. F. & Menges, E. S. (2006) Using population viability analysis to predict the effects of climate change on the extinction risk of an endangered limestone endemic shrub, Arizona cliffrose. *Conservation Biology* 20, 218-228.

Meusel, H. (1940) Die Grasheiden Mitteleuropas. Versuch einer vergleichend-pflanzengeographischen Gliederung. *Botanisches Archiv* 41, 337-519.

Monteith, J. L. (1981) Evaporation and surface temperature. *Quarterly Journal of the Royal Meteorological Society* 107, 1-27.

Mueggler, W. F. (1972) Influence of competition on the response of Bluebunch Wheatgrass to clipping. *Journal of Range Management* 25, 88-92.

Partzsch, M. (2000) Die Porphyrkuppenlandschaft des Unteren Saaletals – Strukturwandel ihrer Vegetation in den letzten vier Jahrzehnten. *Tuexenia* 20, 153-187.

Partzsch, M. (2001) Die Porphyrkuppenlandschaft des Unteren Saaletals – Vergleich der Vegetation in Abhängigkeit von Flächengröße und Genese der Porphyrkuppen. *Botanisches Jahrbuch für Systematik, Pflanzengeschichte und Pflanzengeographie* 123, 1-45.

Partzsch, M. (2010) Zur Keimungsbiologie von zehn ausgewählten xerothermen Grasarten. *Hercynia N.F.* 43, 299-317.

Partzsch, M. (2011a) Zur Keimungsbiologie acht ausgewählter Xerothermrasenarten - Teil 2: Caryophyllaceae. *Hercynia N.F.* 44, 127-144.

Partzsch, M. (2011b) Does land-use change affect the interaction between two dry grassland species? *Flora* 206, 550-558.

Partzsch, M. & Bachmann, U. (2011) Is *Campanula glomerata* threatened by competition from expanding grasses? – Results of a long-term experiment. *Plant Ecology* 212, 251-261.

Partzsch, M., Piesch, C. & Hensen, I. (2011) A comparative study of germination behaviour and plant performance of two dry grassland species. *Folia Geobotanica* 46, 35-48.

Poschlod, P., Bakker, J.P. & Kahmen, S. (2005) Changing land-use and its impact on biodiversity. *Basic and Applied Ecology* 6, 93-98.

Pott, R. (1996) Die Entwicklungsgeschichte und Verbreitung xerotherme Vegetationseinheiten in Mitteleuropa unter dem Einfluß des Menschen. *Tuexenia* 16, 337-369.

Pugnaire, F. L. & Luque, M. T. (2001) Changes in plant interactions along a gradient of environmental stress. *Oikos* 93, 42-49.

Pykälä, J. (2003) Effects of restoration with cattle grazing on plant species composition and richness of semi-natural grasslands. *Biodiversity and Conservation* 12, 221-2224.

Richter, B., Partzsch, M. & Hensen, I. (2003) Vegetation, Kultur- und Nutzungsgeschichte der xerothermen Hügellandschaft bei Mücheln/Wettin (Sachsen-Anhalt). *Hercynia N.F.* 36, 91-121.

Römermann, C., Dutoit, T., Poschlod, P. & Buisson, E. (2005) Influence of former cultivation on the unique Mediterranean steppe of France and consequences for conservation management. *Biological Conservation* 121, 21-33.

Root, T. L., Price, J. T., Hall, K. R., Schneider, S. H., Rosenzweig, C. & Pounds, J.A. (2003) Fingerprints of global warming on wild animals and plants. *Nature* 421, 57-59.

Sala, O. E., Chapin, F. S., Armesto, J. J., Berlow, E., Bloomfield, J., Dirzo, R., et al. (2000) Global Biodiversity Scenarios for the Year 2100. *Science* 10, 1770-1774.

Schubert, R., Frank, D., Herdam, H., Hilbig, W., Jage, H., Karste, G., et al. (2004) Rote Liste der Farn- und Blütenpflanzengesellschaften des Landes Sachsen-Anhalt. *Berichte des Landesamtes für Umweltschutz Sachsen-Anhalt* 39, 111-122.

Silva, J. P., Toland, J., Jones, W., Eldridge, J., Thorpe, E. & O`Hara, E. (2008) *LIFE and Europe´s grasslands: Restoring a forgotten habitat.* Luxembourg, European Communities.

SPSS (2011). *SPSS for Windows 19.0.* SPSS, Chicago.

Stevens, C. J., Thompson, K., Grime, J. P., Long, C. J. & Gowing, D. J. G. (2010) Contribution of acidification and eutrophication to declines in species richness of calcifuge grasslands along a gradient of atmospheric nitrogen deposition. *Functional Ecology* 24, 478-484.

Tylianakis, J. M., Didham, R. K., Bascompte, J. & Wardle, D. A. (2008) Global change and species interaction in terrestrial ecosystems. *Ecological Letters* 11, 1351-1363.

Veblen, K. E. (2008) Season- and herbivore-dependent competition and facilitation in a semiarid savanna. *Ecology* 89, 1532-1540.

Vitousek, P. M., Aber, J. D., Howarth, R. W., Likens, G. E., Matson, P. A., Schindler, D. W., et al. (1997) Human alteration of the global nitrogen cycle: sources and consequences. *Ecological Applications* 7, 737-750.

Weigelt, A., Schumacher, J., Walther, T., Bartelheimer, M., Steinlein, T. & Beyschlag, W. (2007) Identifying mechanisms of competition in multi-species communities. *Journal of Ecology* 95, 53-64.

Wesche, K., Partzsch, M., Krebes, S. & Hensen, I. (2005) Gradients in dry grassland and heath vegetation on rock outcrops in eastern Germany – an analysis of a large phytosociological data set. *Folia Geobotanica* 40, 341-356.

Whitford, G. (2002) *Ecology of desert systems.* Academic Press, London.

Willems, J. H. (1987) Increasing dominance of *Brachypodium pinnatum* (L.) Beauv. in chalk grasslands: a threat to species-rich ecosystem. *Biological Conservation* 40, 301-314.

Willems, J. H. & Bobbink, R. (1990) Spatial processes in the succession of chalk grassland on old fields in The Netherlands. *Spatial processes in plant communities* (eds F. Krahulec, A. D. Q. Agnew, S. Agnew & J. H. Willems. pp. 237-249. Academia Prague, Prague.

Yoshida, S. (1981) *Fundamentals of rice crop science.* International rice research Institute, Los Banos, Philippines.

Zhang, J., Cheng, G., Yu, F., Kräuchi, N. & Li, M. H. (2008) Intensity and importance of competition for a grass (*Festuca rubra*) and a legume (*Trifolium pratense*) vary with environmental changes. *Journal of Integrative Plant Biology* 50, 1570-579.

In: Grassland Biodiversity and Conservation ...
Editors: Pierre Mariotte and Paul Kardol

ISBN: 978-1-61761-619-8

Chapter 8

GRASSLANDS IN SILVOPASTORAL MOUNTAIN ECOSYSTEMS

Alexandre Buttler*

Ecole Polytechnique Fédérale de Lausanne, Lausanne, Switzerland
Swiss Federal Institute WSL, Site Lausanne, Switzerland
Laboratoire de Chrono-Environnement, UFR ST,
Besançon, France

ABSTRACT

The structure and biodiversity of silvopastoral ecosystems, also called "wooded-pastures", is the result of a long tradition of multipurpose extensive land-use. The patterned vegetation consists of semi-natural grasslands found with isolated trees or intermingled with patches of forest or clumped trees. The main interactions among herbaceous vegetation, trees and shrubs, soil and cattle activities in mountain wooded-pastures of the Jura Mountains are synthesized, and their sensitivity to land-use and climate change is addressed in this chapter.

Cattle grazing plays a major role in vegetation dynamics of wooded pastures through inducing three types of disturbances: herbage removal, dung deposition, and trampling. At large scales, the selection of grazing areas by cattle depends on various factors such as herbage quality and quantity, water availability, relief, slope, natural and artificial barriers, herd social interactions and learned behaviour. At small scales there is a high complexity in species response to disturbances by cattle, which varies according to the vegetation type, induced by the various light conditions in relation to tree density. Sensitivity of species to gaps induced by trampling is critical, and a complex dynamic of recolonisation is key to understanding plant species coexistence in pastures. This involves, for example, the persistence of subordinate plant species, which are recognized as having a functional importance in ecosystems, especially in the context of perturbation and climate change.

Global warming and associated environmental changes are predicted to have a strong impact on mountain ecosystems over the course of this century. Plant communities

* Corresponding author: Alexandre Buttler, E-mail: alexandre.buttler@epfl.ch.

transplanted in mesocosms to warmer and drier climate experienced a marked loss in species richness.

Another major finding from this transplantation experiment, was that across land use types, intensively-used unwooded pastures were consistently more responsive to climate change than their counterparts from extensively-used wooded-pastures of either sparsely or densely wooded type. Simulations show that in extensive grazing systems, it is not only the generally lower stocking density, but also the resulting grassland-forest mosaic that contributes to the robust provision of forage. The apparent advantage in terms of productivity of unwooded pastures diminishes with drought, and wooded pastures can become a forage source of similar importance. Such formerly unproductive patches show a clear potential to compensate for the breakdown of productivity within open grasslands. Thus, across the landscape, spatial variation itself stimulates ecosystem resilience and resistance to drought.

Wooded-pastures are a good example of a landscape where high biodiversity can coexist together with an extensive land-use. So, the presence of many transitional zones (ecotones) in a highly heterogeneous and fluctuating environment, creates optimal conditions for biodiversity.

Conservation of these ecosystems, including adaptive management in the face of land-use and climate change, will require a broad approach and integrated management tools with a transdisciplinary and consensual perspective.

Keywords: Semi-natural grasslands, climate change, land-use, functional diversity, warming, drought, ecosystem functioning, grazing

INTRODUCTION

The structure and biodiversity of silvopastoral ecosystems, also called "wooded-pastures", is the result of a long tradition of multipurpose extensive land-uses (Buttler et al., 2009; Etienne 1996). In the high range of the Jura Mountains, which are situated at the border between Switzerland and France, forests were cleared during the Middle Ages (pollen diagrams - Sjögren 2005, 2006).

Logging for firewood and charcoal burning reduced the cover of old beech-fir forests (potential natural vegetation) and grazing of various livestock suppressed beech (*Fagus sylvatica*) and silver fir (*Abies alba*), leaving mostly regeneration patches of Norway spruce (*Picea abies*). Thus, this patterned vegetation shows semi-natural grasslands, which are often found with isolated trees or intermingled with patches of forest or clumped trees (Figure 1).

The vegetation of wooded-pastures is very diverse, and four main cover types can be recognized in a typical wooded-pasture (Gallandat et al., 1995; Vittoz 1998): unwooded pastures (open pastures with less than 1% tree cover), sparsely wooded pastures (tree cover between 1 and 20%, trees are mainly scattered in a grassland matrix), densely wooded pastures (tree cover between 20 and 70%, trees are aggregated in thickets) and grazed forests (closed forests with more than 70% tree cover).

In the Jura Mountains, wooded-pastures occur mainly at an elevation between 800 m and 1400 m (Gillet and Gallandat 1996a). At 1200 m, mean day temperature is below 0°C for more than 60 days per year, and the ground is generally snow-covered from December untill April. As a consequence, the climate limits cattle management to the summer period, i.e. from end of May to end of September. Cattle herds are mainly composed of heifers, but dairy cows can be seen on many pastures and in some areas horses are the main livestock type.

Figure 1. Wooded-pasture in the Swiss Jura Mountains. Photo: J.-D. Gallandat.

Livestock density ranges from 0.5 to 1.5 adult bovine units per hectare (Gillet and Gallandat 1996a), which is low compared to intensive grazing systems in the lowlands. Two grazing systems are applied in wooded-pastures; (1) free range: the animals spend the whole summer season roaming freely through the pastures; (2) grazing rotation: the pasture is subdivided into paddocks and the animals circulate from one paddock to another according to a variable rotation period (between 2 and 7 rotations per grazing season, corresponding to a stay in each paddock of 10 to 80 days). Wooded-pastures are more and more managed according to the rotation grazing system, with the aim of optimizing the utilization of resources.

The use of these semi-natural ecosystems is critical because their existence and survival depend on a subtle equilibrium between management and site conditions, aiming to provide various ecological goods and services, including grass and wood production, biodiversity, landscape aesthetics and recreational values. From a socio-economic point of view, the main users of wooded-pastures are farmers, even if in some regions revenues generated from forestry activities may be quite significant. Generally, foresters intervene mainly to remove dying trees and to check on natural re-growth. Thus, the landscape structure results from a balance between counteracting ecological processes, namely livestock browsing and tree regeneration. A mosaic of pastures, isolated trees and woodlands is therefore very sensitive to changes in the management of grazing and forest. Both land-use extensification and intensification can lead to a simplification of the mosaic patterns and concomitant loss of biodiversity via the development of either closed forests or treeless pastures. Considering the great decline of this type of heterogeneous landscape in Europe during the last century (Gillet and Gallandat 1996a; Gallandat et al., 1995), the question of adapted and integrated management is important, and becomes even more critical considering the present changes in agricultural practices in mountain regions, as a result of the transformation of this sector's economy (Huber et al., 2012, 2013; Briner et al., 2012), or due to climate change (Buttler et al., 2012a, 2012b; Gavazov et al., 2013). In this chapter the main interactions among herbaceous vegetation, trees and shrubs, soil and cattle activities in mountain wooded-pastures of the Swiss Jura are synthesized, as summarized in Figure 2. Based on recent literature, patterns and processes in these very heterogeneous and spatio-temporal fluctuating environments are highlighted, and their sensitivity to land-use and climate change is addressed.

PLANT DIVERSITY AND ENVIRONMENTAL HETEROGENEITY

A large-scale survey (Gallandat et al., 1995) revealed that one sixth of the Swiss vascular flora (total Swiss Flora has about 3000 species) was observed in the wooded-pastures of the Swiss Jura Mountains. Moreover, Vittoz (1998) listed 554 vascular plant species occurring within a 70 km^2 area. Plant biodiversity is also very high at fine scale, up to 40 species per square meter (Buttler et al., 2009). Several studies have pointed out the influence of trees on the plant diversity of grasslands (e.g. Hubert et al., 1982; Zobel et al., 1994). Trees in grasslands may decrease wind speed and evaporation, protect against erosion, improve the water regime, but may also decrease light availability, which reduces the quality and the quantity of herbage, increases litter accumulation, and ultimately this will affect soil properties. All these influences may affect plant biodiversity (e.g. Heikkinen 1996).

The origin of the multi-scale high plant species richness in the Jura wooded-pastures may thus be multiple and boldly originate from the interplay between biotic and abiotic factors (Figure 2). First, soil variability is important; Jura soils range from shallow calcareous to deep, acidic and silt soils (Havlicek et al., 1998), with sometimes a very fine scale heterogeneity, which induces a fine mosaic of various vegetation types (Gobat et al., 1989).

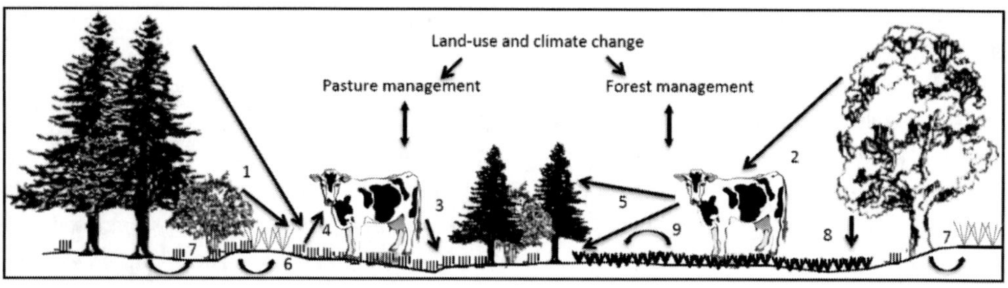

Figure 2. Conceptual diagram synthesizing interactions in mountain wooded-pasture in a highly heterogeneous and spatio-temporal fluctuating environment. 1: Effects of trees and shrubs cover on microclimate (e.g. light, air temperature and humidity, wind, soil temperature and moisture), and herbaceous vegetation (e.g. biodiversity, patchiness, growth). 2: Effect of trees on cattle behaviour (shelter, movements). 3: Effects of cattle activities such as grazing, trampling, dunging and gap creation on vegetation (e.g. species turnover, weeds) and soils (e.g. nutrient transfer, soil biological activity). 4: Effects of herb quality and quantity, water availability, relief, slope, natural and artificial barriers on cattle behaviour (e.g. selectivity, movements). 5: Effect of browsing of shrubs and young trees (e.g. regeneration, growth). 6: Effects of soil and microtopographic heterogeneity on the vegetation (e.g. biodiversity). 7: Tree regeneration. 8: Litter fall, rain interception, through fall and stem flow, and effects on the soil. 9: Plant ecological processes (competition, facilitation, dominant and subordinate species, reproduction, dispersal, resistance, resilience) related to species traits.

Secondly, a complex mosaic of trees, shrubs and open grasslands creates various microclimates favouring different habitats and species assemblages. Thirdly, cattle activities can change plant community composition (Kohler 2004a; Kohler et al., 2004b) and impact soil biological properties (Kohler et al., 2005).

Finally, plant species richness is related to topographic complexity (Dufour et al., 2006a). Wooded-pastures are for above-mentioned reasons a good example of a landscape where high biodiversity can coexist together with an extensive land-use. So, the presence of many transitional zones (ecotones) in a highly heterogeneous and fluctuating environment, creates optimal conditions for biodiversity.

The relationships between several diversity indices and tree cover in subalpine wooded-pastures (1500 to 1720 m a.s.l.) with *Larix decidua* were analysed at two spatial scales (Gillet et al., 1999), (i) phytocoenoses (50 m × 50 m plots) and (ii) herbaceous synusia (1 m^2 quadrats). Diversity indices had a non-linear relationship to *Larix* cover and species richness was more influenced than evenness by *Larix* cover. At the phytocoenosis scale, the optimum for species richness and richness of synusial communities in the herb layer was near 25% tree cover (Figure 3). At the synusial scale, species richness of the herb layer showed an optimum when the mean distance to the four nearest *Larix* trees was ca. 30 m, or if the potential number of sunshine hours between April and September was close to 1250 h, which represents about 7 hours of sunshine per day (Figure 3).

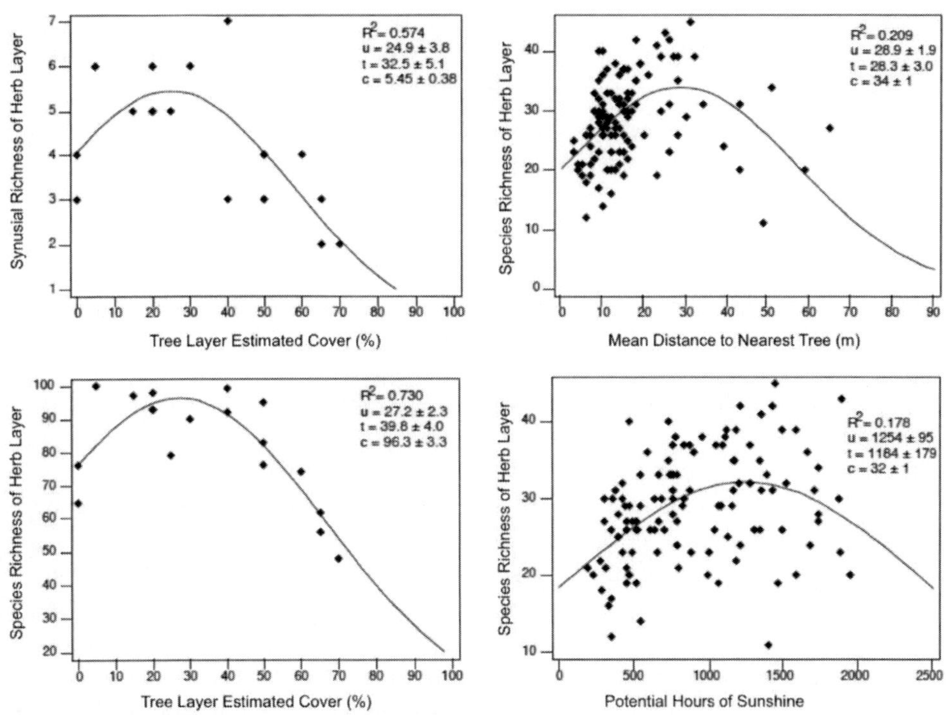

Figure 3. Species and synusial community richness of the herb layer along a gradient of tree cover at phytoceonosis scale (graphs on the left); species richness of the herb layer along a gradient of mean distance to nearest tree and potential hours of sunshine at synusial scale (graphs on the right). Adapted from Gillet et al., 1999.

These results confirm the conclusions of other studies in *Picea*-dominated wooded-pastures of the Jura mountains (Gillet and Gallandat 1996a,b). At the phytocoenosis scale, the presence of isolated or clumped trees in a pasture increases the number of potential habitats, which in turn promotes alpha- and beta-diversity (Whittaker 1975). At the synusial scale, the influence of shade on species richness is critical, and therefore also the position with respect to the tree crown.

Among the external factors, which can influence vegetation diversity, altitude, slope and aspect are 'the usual suspects', together with soil physico-chemical properties and management intensity (e.g. grazing intensity, fertilization, logging, tree regeneration).

Microtopography and soil heterogeneity are other driving forces for biodiversity. Dufour et al. (2006a) found evidence that local species diversity is related to spatial environmental heterogeneity. Biodiversity is therefore expected to be influenced by two components of spatial heterogeneity: 1) the variability of environmental conditions, which will affect the number of habitats, and 2) the spatial configuration of habitats ("roughness"), which will affect the rates of ecological processes, such as dispersal or competition. This was tested using field data collected in a wooded-pasture in the Swiss Jura mountains (1300-1400 m a.s.l.) using a four-level hierarchical sampling design for assessing the scale effects ($cell_{size}$: plots and nested subplots of 100 m × 100 m, 50 m × 50 m, 25 m × 25 m and 12.5 m ×12.5 m) and for linking plant species richness assessed in cells (mean richness of all 1 m^2 quadrats included in the plots at a given $cell_{size}$) and quadrat scales (total richness in the plots at a given $cell_{size}$) to topographic features calculated with a digital elevation model and tree cover. Richness generally increased with increasing environmental variability and decreasing spatial dependence (increasing "roughness"). However, the relationship was not simple: it changed with cell size and the level (quadrat or cell) at which richness was measured; an interaction between variability and roughness added further complexity.

Results were consistent with an interpretation related to both niche-based (i.e. competition) and dispersal-based ecological processes, as has been suggested in other contexts (Tilman 2004; Lortie et al., 2004) and are summarized in Figure 4 (Dufour et al., 2006a). When variability increases, environmental conditions are more diverse, more habitats are present, and more species can coexist (Statzner and Moss 2004; Sabo et al., 2005). In contrast to variability, an increase in roughness does not increase the number of habitat types. Rather, increasing roughness decreases the area of individual habitats, reduces connectivity between areas with similar habitats, so increasing habitat isolation (Figure 4).

The underlying mechanisms may differ for cell and quadrat richness. On the one hand, increasing roughness (i.e. decreasing spatial aggregation), may increase species turnover through ecological equivalency (Shmida and Wilson 1985) where habitat types are well intermixed. Some ecologically equivalent species may then inhabit different areas of the same habitat type because of delayed competitive exclusion (Hurtt and Pacala 1995). As a consequence, two samples in the same habitat type, but different area, will be on average more different in species composition than two samples of the same habitat type in a spatially aggregated landscape. On the other hand, as increasing roughness is likely to increase the mass effect (the continual immigration of poor competitors from dissimilar neighbouring habitats), roughness tends to homogenize the species composition and decrease species turnover. Therefore, the effect of roughness on species turnover should depend on the relative strength of ecological equivalency and mass effect.

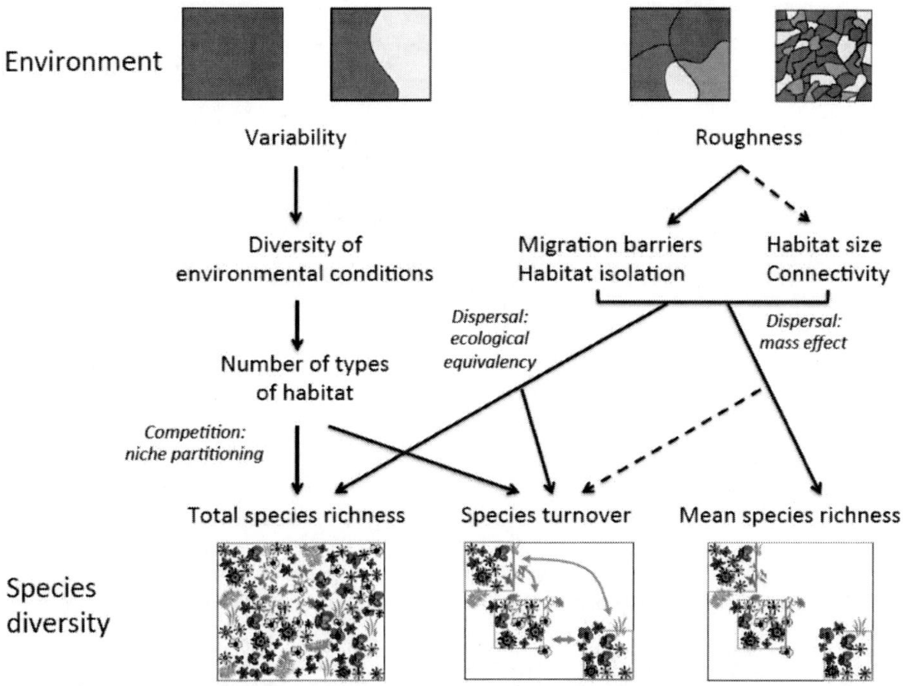

Figure 4. Hypothesized effects of the two components of environmental heterogeneity, variability and spatial configuration (measured as a roughness), on three types of species diversity, total species richness, species turnover and mean species richness. Total species richness represents quadrat richness (equivalent to gamma diversity), mean species richness represents mean cell richness (equivalent to alpha diversity), and species turnover is equivalent to beta diversity. Solid lines represent positive effects, dotted lines represent negative effects. Adapted from Dufour 2006b.

Environmental variables and biological processes are the main factors invoked to explain spatial variation in species composition, but the scale at which they act remains still unclear. In the work by Dufour (2006b), based on the same plots as before (Dufour 2006a) and using vegetation from the $1m^2$ quadrats, variation in species composition at different spatial scales was predicted from a set of environmental variables measured at the same scale. Results are synthetized in Figure 5. Dufour (2006b) show a spatial hierarchy among the environmental variables: topography was the most important predictor of the spatial pattern of species composition at the coarsest scales, whereas insolation appeared to have a large effect at finer scales. The "pure spatial" fraction of species composition was related to species traits in order to define the possible underlying biological processes and their scale of influence. Species traits, such as dispersal characteristics and C-S-R life strategies, had a complex relationship with scale, and this enables expansion of the hypotheses traditionally proposed to explain fine-grained vegetation patterns. Albeit, environmental heterogeneity remains difficult to quantify because it has multiple dimensions. The hypothesis that environmental heterogeneity is a major driver of species richness unifies several theories of biodiversity, and has the potential to be developed into a management tool. The scale-dependence of effects suggests a change in the relative importance of mechanisms, which will need to be taken into account when scaling up to larger landscapes.

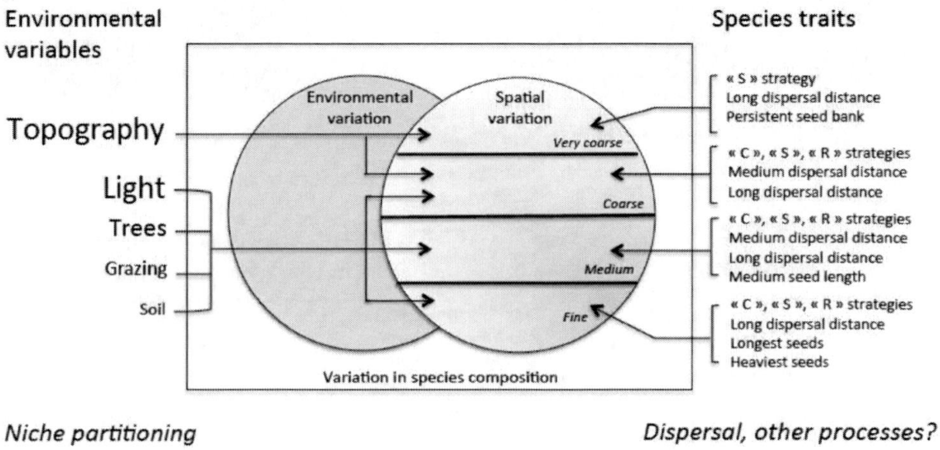

Figure 5. Relationships between the variation in species composition and two sets of explanatory variables, i.e. environmental variables and species traits, at four spatial scales. The rectangle represents the total variation in species composition, the grey circles on the left schematizes the environmental fraction of the variation and the grey circles on the right the spatial fraction of the variation. Environmental variables that are more important at a given scale are written with bigger letters. Redrawn from Dufour 2006b.

Impact of Cattle Activity on Herb Layer

During the summer season, cattle grazing plays a major role in vegetation dynamics of wooded pastures through inducing three types of disturbances: herbage removal, dung deposition and trampling. Herbage removal – or grazing *sensu stricto* – is the main biotic factor affecting herbaceous vegetation structure and dynamics in pastures (Rook et al., 2004).

The effect of herbage removal on plants is principally the loss of aboveground biomass and consequently a change in competition for light between species (Grime 2001). In addition, by selecting for species, grazing also changes the competitive balance between species, and can cause a competitive release for some species. Furthermore, herbage removal induces an exportation of biomass, linked with a local loss of nutrients. In turn, dunging locally increases soil fertility and influences vegetation productivity and composition of herbaceous or dwarf-shrub communities (Bakker and Olff 2003; Gillet et al., 2010).

Trampling affects the vegetation through detaching or destroying plant material by hoof action and soil compaction, leading to hydrological changes (Abdelmagid et al., 1987). Trampling can also create gaps and thus drives community dynamics, facilitating regeneration and invasions (Kohler et al., 2006b). Finally, large herbivores are key drivers of nutrient cycling (Schrama et al., 2012). In extensively grazed summer pastures, these disturbances occur moderately. At large scales, the selection of grazing areas by cattle depends on various factors such as herbage quality and quantity, water availability, relief, slope, natural and artificial barriers, herd social interactions and learned behaviour (Rice et al., 1983).

In order to assess spatial and seasonal variations of cattle activities at a large scale, Kohler et al. (2006a) surveyed the intensity of herbage removal, dung deposition and trampling in an entire paddock in the Swiss Jura mountains. There was a lack of positive relationships between cattle activities (e.g. dunging, trampling and grazing), suggesting different spatial habitat use for each activity. Spatial patterns weakened throughout the season for grazing and trampling, whereas dunging patterns changed less between grazing periods. Multivariate analysis showed that cattle activities depended on different environmental factors within the paddock, and that the importance of management-induced, versus natural structures, was highest for herbage removal. The distance to the wire fence was also important in explaining grazing patterns. Patterns of dung deposition were mainly negatively correlated with slope and rock outcrops, and trampling patterns were positively correlated with tree cover and negatively correlated with vegetation openness and fodder potential. Dunging patterns were dominated by large-scale, herbage removal dominated by medium-scale, and trampling by fine-scale spatial structures (Kohler et al., 2006a). Systematic differences in the spatial and seasonal patterns of cattle activities may result in complex interactions with vegetation, involving feedback effects through nutrient shifts, with strong implications for ecosystem management.

An experimental approach was chosen to explore the effects of cattle activities on vegetation dynamics at a fine scale (Kohler et al., 2004b). The three cattle activities, herbage removal, dung deposition and trampling were simulated in three experiments in pastures, at 1200 m a.s.l., using fenced areas comprising the following treatments: (1) repeated mowing - three levels; (2) intensive trampling - two levels; (3) manuring with a liquid mixture of dung and urine - three levels. Additionally 6 plots outside the fenced areas represented reference plots with normal cattle pasturing. Vegetation response was recorded twice a year; in spring and in autumn.

With respect to short-term responses, differences in species richness between records when comparing treatments were neither significant in autumn, nor in spring of the second year. Nevertheless, fertilized plots, when compared to all other plots, showed a significantly lower species-richness in the manured treatment in autumn, but this difference disappeared in the second spring. In the first year's autumn, there was a clear seasonal community shift and significant differences induced by treatments. Thus, at the considered time scale, treatments did not affect the qualitative species composition, but they affected abundance of established species. Significant vegetation differences between treatments were observed in autumn of the first year and in spring of the second year (Figure 6). In autumn, treatments were clustered into four groups: (1) intensive mowing, (2) treatments combining trampling, manuring and mowing, (3) trampling, (4) abandonment and manuring. In spring of the second year, this trend persisted but only abandonment and intensive mowing remained significantly different from each other, indicating that herbage removal is the most determinant factor for maintaining typical pasture vegetation at a fine scale. It is noteworthy that seasonal treatment effects were only partially carried over to the next spring, showing an unexpected resilience of the plant community, probably due to life-history traits and competitive release following climatic disturbance in winter. In successive years, the cattle activity over the summer period may change spatially, starting from a vegetation mosaic that is not entirely the result of the patchiness created by the former pasturing season. This is in contrast to what has been reported in temperate grasslands without lasting snow cover (Bakker et al., 1984).

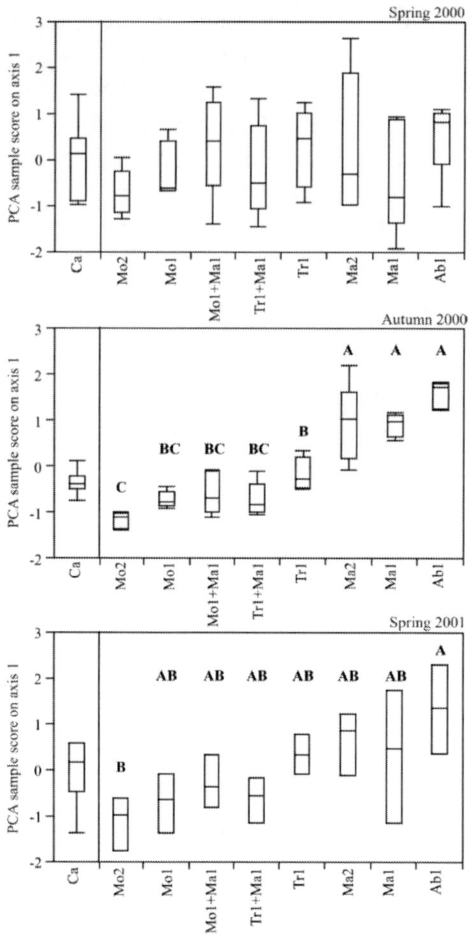

Ca: control with natural grazing by cattle, Mo1: mowing once a month with a lawn mower and a cutting height of 30 mm, Mo2: mowing twice a month, Tr1: trampling once a month, 1000 footsteps per m^2 with ca. 70 kg per footstep of 0.0035 per m2, representing a mean pressure of 20000 kg per m^2, Ma1: manuring 0.5 L per m^2 (equivalent to normal cattle activity), Ma2: manuring 2 L per m^2 (equivalent to intensive cattle activity), Ab1: abandonment.

Figure 6. Coordinates on the first ordination axis of principal component analysis (PCA) of vegetation records in a grazed meadow in spring (before treatment) and autumn of the first year and in spring of the following year, grouped by treatments. For autumn 2000 and spring 2001, significant differences ($P < 0.05$) with the Tukey-Kramer HSD are indicated with different characters (for spring 2000 there were no significant differences). Plots with cattle activity were excluded from this analysis because they were not included in the experimental design. Kohler et al., 2004b.

Furthermore, before the herds settle on the pasture, the vegetation is growing quickly, without cattle disturbance, which triggers temporary competition processes that are different from those under heavy cattle pressure. Processes acting during winter and early spring are therefore key factors controlling community dynamics in mountain pastures. In the context of global change, such factors may change in the future, for example as a consequence of shorter duration of snow cover and milder temperatures.

The long-term response is represented in Figure 7, which includes the grazed meadow used in Figure 6, and two other meadows so as to represent a gradient of exposure to light.

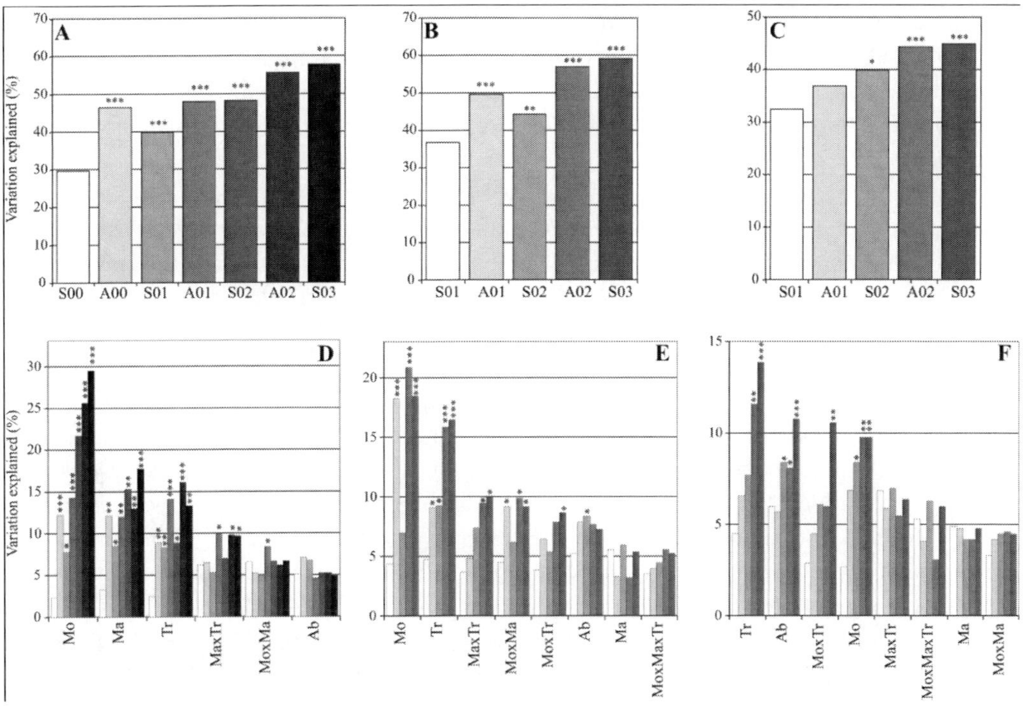

Asterisks show significant result of the Monte Carlo permutation test (999 permutations): ***: P = 0.001; **: P < 0.01; *: P < 0.05.

Figure 7. Evolution of variation explained by mowing (Mo), trampling (Tr), manuring (Ma), abandonment (Ab) and their interactions during the experiments in separated Redundancy analyses (RDAs) using species records at each sampling date in a grazed meadow in full light (A and D), a grazed meadow slightly shaded by trees (B and E) and a grazed meadow more shaded at the forest edge (C and F). Above: variation explained by the full model (all factors and their interactions; S: spring; A: autumn; 00-01-02-03: years). Below: variation explained by each treatment or interaction separately. Different grey scales represent the different recording sessions as in (A), (B) and (C). Kohler 2004a.

In grazed meadows in full light or slightly shaded (Figure 7a,b), treatment effects were already significant in the first autumn, but the effect decreased by the following spring, in absence of treatments in early spring, as described before. In both meadows, when considering all factors and interactions together, the resilience did not appear the following years, and the explained variation increased throughout the experiment. Succession is generally described as directional changes in species composition (Platt and Connell 2003). Indeed, this was the case in meadows at the forest edge (Figure 7c, f), where the resilience appeared neither in the global model, nor in each treatment separately. Vegetation response to the treatments was weaker than in grazed meadows in the full light, showing higher resistance to changes in this community with low light availability. The importance of each treatment for explaining the species variation varied also depending on the community. In both meadows under full light or slightly shaded (d, e), repeated mowing had the strongest effect, followed by manuring and trampling when the meadow was in the full light (d), and by trampling and the interaction with manuring in the slightly shaded meadow (e).

This ranking was different in the meadow at forest edge (f) with trampling as the most important factor, followed by abandonment, and the interaction between mowing and trampling. For the three community types together, light conditions explained about 33%, and treatment about 14% of the variation in the third year of the experiments. These findings highlight the importance of scattered or clumped trees in wooded-pastures, since light condition, in interaction with cattle activities, engenders various spatial responses of the vegetation.

In the grazed meadow in full light, intensive manuring and abandonment showed the largest positive deviations from the reference plots, which had natural cattle activity (Kohler et al., 2004a). Manuring, trampling and the combined treatment showed a similar picture, but with a less pronounced trend, and were opposed to mowing treatments. Species with positive weight (e.g. *Agrostis capillaris* and *Dactylis glomerata)* are expected to increase with intensive manuring and abandonment. At the opposite end, repeated mowing promotes species like *Hieracium pilosella* or *Luzula campestris*. The slightly shaded grazed meadow showed a similar pattern whereas the more shaded grazed meadow at the forest edge was different. There, the contrast was highest between abandoned plots characterized by species such as *Luzula sylvatica* and *Geranium sylvaticum*, opposed to treatments with trampling where *Agrostis* increased, while this species was linked to intensive manuring and abandonment in the two other meadows. Concerning species richness, all species appearing in the plots belonged to the original communities, and therefore were adapted to regular cattle activity. In grazed meadows, mowing was the main factor acting positively on species richness, and intensive manuring and abandonment had a strong negative effect. In the more shaded meadow, trampling acted negatively on species richness but manuring and abandonment positively. This fact can be explained by a different starting position of the community along the "humped-backed" curve, describing the relation between biomass and species richness (Grime 1973a, 1973b; Rapson et al., 1997). Indeed, the lower biomass in this community and therefore the lower competition between species, allowed for additional niches to be occupied. This high complexity in species response to disturbances and their interactions becomes even higher if one considers combined effects of gap creation by heavy trampling. In an exclosed pasture, four controlled treatments simulating cattle activity (repeated mowing twice a month, trampling once a month, manuring once a month and untreated control – see caption to Figure 6) were applied on 2 x 2 m plots where one artificial gap of 60 x 60 cm was created in the centre, and the vegetation surveyed on a grid of cells (Kohler et al., 2006b). At the level of functional traits, differences between the gap areas and references over time accounted for 68% of the variance and were highly significant, but treatments showed only a very weak effect. Over time, the centre and the edge of the gap became more and more similar to the undisturbed area around the gap (Figure 8). Colonizers generally had very light seeds (< 0.2 mg), unspecialized seed dispersal, persistent seed bank and high vegetative spread. Among the species present, *Agrostis capillaris, Veronica serpyllifolia* and *Campanula rotundifolia* combined these four traits. Zoochorous species were more common in intact vegetation. Our results showed that a persistent seed bank was clearly associated with colonizers but, as reported from perennial grasslands elsewhere (Rapp and Rabinowitz 1985; Milberg 1993; Kotanen 1997; Mariott et al., 1997), rapid vegetative spread was also a characteristic trait of species present in gaps. The importance of this trait was not surprising considering the high proportion of species using this strategy in perennial grasslands (Klimes et al., 1997; Macek and Leps 2003).

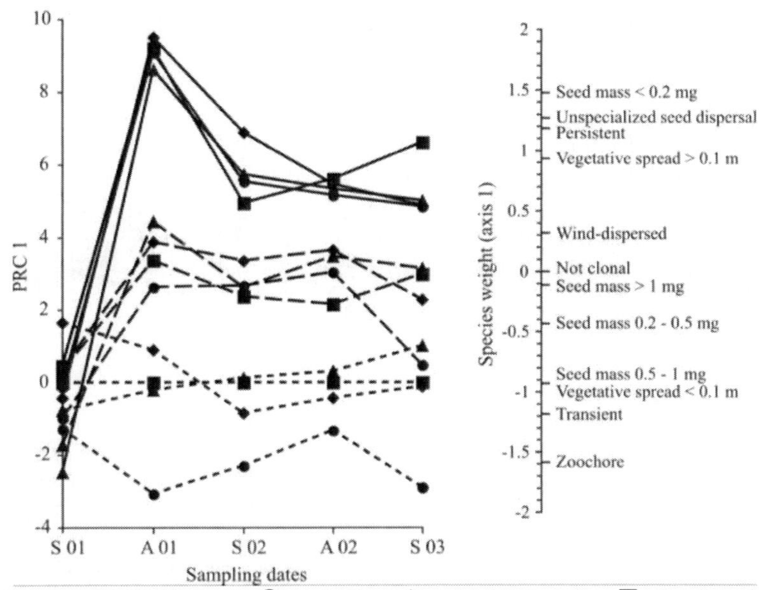

S: spring; A: autumn; 01-02-03: years; ●: trampling; ▲: repeated mowing; ■: abandoned; ♦: manuring. Axis on the left corresponds to the canonical coefficient, axis on the right corresponds to the traits weight.

Figure 8. Principal Response Curve (PRC) diagram and species weights on axis 1 for the functional traits matrix. Vegetation around the gap in abandoned plot was used as reference. Continuous line: centre of the gap; large dashed lines: edge of the gap; fine dashed lines: around the gap. Kohler et al., 2006b.

Species are first filtered by gap environment according to their regenerative attributes, and the filtering role of dung deposition, herbage removal or trampling by cattle is not of primary importance, but may moderate vegetation response. These complex dynamics are probably key to understanding plant species coexistence in pastures.

A cell-grid method was used to survey seasonal changes in four herbaceous communities on a mountain wooded-pasture that was pastured by heifers in a rotational grazing system (Kohler et al., 2004c). In each cell, the dominance of the plants, the grazing intensity of each species and the occurrence of disturbance by trampling were recorded. Seasonal changes in species composition were very strong, and scale-dependant (Figure 9). The mean number of species per cell did not show any obvious trend, whereas the cumulative richness increased. Maximum values for the cumulative number of species per cell increased up to about half of the total cumulative number of species per plot. The lawn pattern appeared as the most constant with a small difference between the number and cumulative number of species. At this fine scale, the four vegetation types showed a similar behaviour. By contrast, at plot scale, the richness varied with a maximum in June for the temporary refused meadow patches, the underwood and the oligotrophic lawn, and in September for the eutrophic grazed meadow. In the temporary refused meadow, the underwood and the lawn, the cumulative richness increased dramatically in June and more gently after.

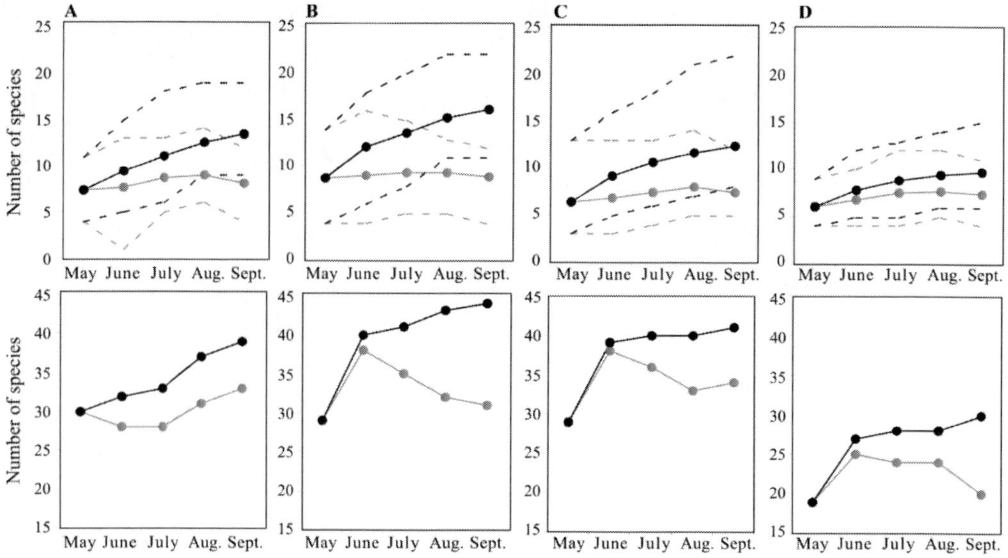

Figure 9. Species richness (grey lines) and cumulative species richness (black lines) in four vegetation types (A: eutrophic grazed meadow, B: temporary refused meadow patches, C: underwood herb community, D: oligotrophic lawn) from May to September. At 1 dm^2 scale, the cells (above), mean (solid lines), and maximum and minimum (dashed lines) are represented. At 1 m^2 scale, the plot (below), absolute values are given. Kohler et al., 2004c.

Classification of the species according to their occurrence in the plots in casual (species appearing only once), pulsating (species disappearing and appearing newly during the season) and persistent (species remaining at least through two successive months, and which are not pulsating), indicated that most of the species were persistent at the plot scale. Nevertheless, at the cell scale, almost all species could behave as persistent, pulsating or casual. In particular, less-frequent species showed pulsating or casual behaviours (Kohler et al., 2004c). From a community point of view, there was no correlation between the mean species turnover and the grazing intensity or the trampling occurrence, but at the population level, there were some positive correlations between persistent and pulsating species and the mean grazed species cover, and negative correlations with casual species. Changes at the plot scale were mainly driven by a seasonal shift, whereas changes at the cell scale suggested strong small-scale dynamics of species, as observed by Herben et al. (1993), and Pärtel and Zobel (1995). Despite changes at the cell scale, the spatial structure of the community at plot scale (autocorrelation pattern) did not change (Kohler et al., 2004c). The high proportion of species with clonal reproduction in grasslands (Tamm et al., 2002; Klimes et al., 1997) points to a high possibility for adult plants to colonise. Overall, the fact that many species showed a constant mean frequency during the season, while dynamically active suggests a carousel at this time scale (see Palmer and Rusch 2001). Another result, which confirms this mobility, is the mean number of species per cell (1 dm^2) which did not show any trend over time despite high fluctuation at the plot scale (1 m^2). This is in agreement with the Niche-limited Carousel Model of Wilson et al. (1995), which proposes that species are mobile in the community but can emigrate to a microsite only if a species disappears.

In conclusion, dynamics and internal species turnover of the community at fine scales and short time periods seem more driven by internal characteristics of the community rather than by disturbances induced by cattle. The latter plays a role at larger scales (more than 1 m^2) by maintaining the composition and the structure of the different communities (see Kohler et al., 2004b).

Dung deposition by cattle, together with grazing and trampling, can be considered as one of the important factors driving vegetation dynamics in pastures. Gillet et al. (2010) investigated at 10-cm and 1-month resolution the plant community dynamics induced by dung deposition in two plant communities (a mesotrophic and an oligotrophic grassland) in a wooded-pasture of the Swiss Jura Mountains. Vegetation around dung pats was subjected to two opposing fertilizer and grazing gradients, which induced changes in vegetation at fine scales and over the short term. The canopy was higher near the dung pat only one week after dung deposition. At final sampling, the tallest vegetation was found near the edge of the dung pat and vegetation size decreased linearly up to 40-50 cm distance (Figure 10).

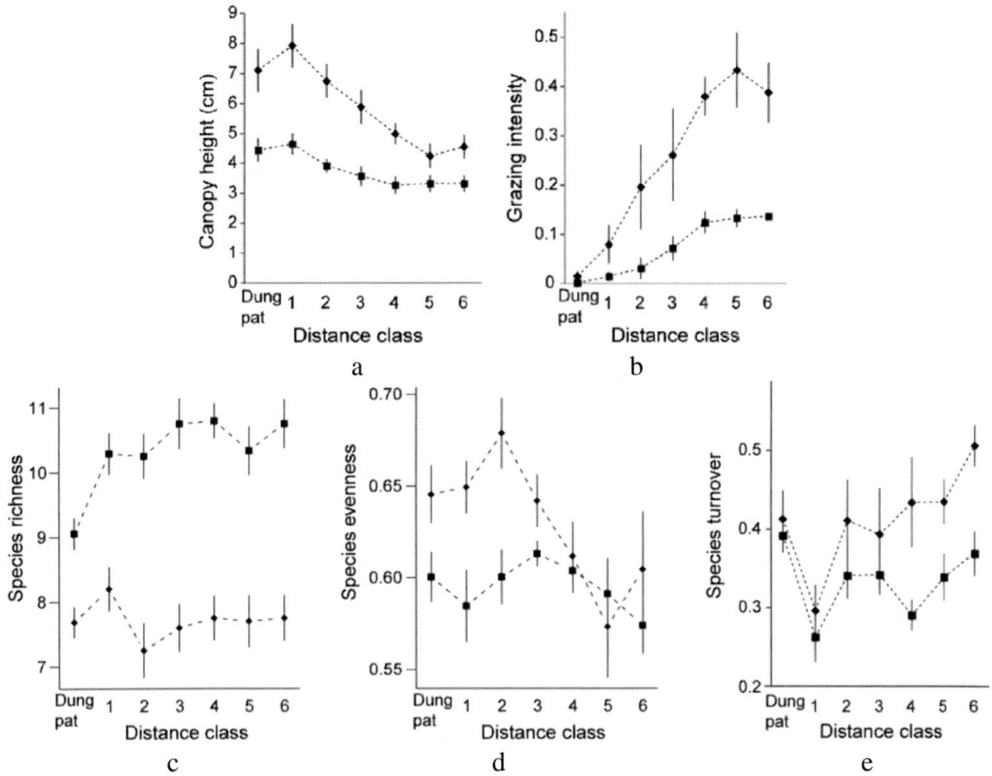

-◆- MG after 14 weeks.

-●-OG after 12 weeks.

Diamonds: grazed mesotrophic grassland (MG); squares: grazed oligotrophic limestone grassland (OG).

Figure 10. (a) Mean canopy height, (b) grazing intensity, (c) species richness, (d) species evenness and (e) temporal species turnover (in reference to the first observation), in relation to distance to the dung pat for the last observation: 14 weeks after dung deposition for MG and 12 weeks for OG. Segments represent standard error of the mean. Gillet et al., 2010.

There was a significant positive correlation between grazing intensity and distance for all sampling dates in both plant communities, with a lowest grazing intensity near the dung pat. The distance to the dung pat also significantly explained the variation in the species assemblage data after 5 to 7 weeks. An elevated number of species near the dung pat, as compared to further away, was observed at the first observation in both vegetation types, but this effect disappeared during the season indicating a transient increase of species number. The turnover rates were always significantly lower near the dung pat in both vegetation types, suggesting a seasonal stabilizing effect of dung on the plant composition of their direct surroundings (0–10 cm) likely due to cattle avoidance. Since dung pats are dropped every year in different locations, they create a shifting mosaic of nutrient availability and grazing intensity, inducing micro-successions in plant communities. Only very few new species appeared and the few new species recorded were common species of the studied pastures. This is in line with findings of a long-term experiment (Dai 2000), where changes in relative abundance of species were observed, but not an establishment of typical vegetation linked to dung pats.

At the landscape scale, cattle have favoured dunging areas, associated with ruminating activity (Kohler et al., 2006a). In areas with high dung-pat density, plant species, which react positively to high nutrient soil content, will be more abundant and this pattern may be relatively stable over time. In contrast, at very fine scales, dung pats are dropped every year in different locations. These spatial changes create a shifting mosaic of nutrient availability and grazing intensity. Plant communities react rapidly to these changes and micro-successions may be induced at seasonal scale. These micro-successions are likely to be temporary and cyclic phenomena affecting vegetation in pasture ecosystems at fine scales, but could become directional and propagate to larger scales in certain circumstances, such as changes in local grazing pressure. Therefore, dung deposition, together with other cattle activities such as selective grazing, could play an important role in grassland ecosystems by promoting species turnover and habitat diversity.

Role of Subordinate Plant Species

Many studies tried to classify species in plant communities and to demonstrate the importance of functional groups (Wardle et al., 2003; MacLaren and Turkington 2010), functional traits (Lavorel et al., 2011) and keystone species (Lyons and Schwartz 2001; Boeken and Shachak 2006) for ecosystem functioning. In his attempt to define the functional role of species, Whittaker (1965) recognized that the simplest way to classify the components of biodiversity is to order species according to their relative abundance or productivity. Plant classifications from dominance-diversity profiles distinguish three groups of species, which may vary in functional traits: dominant, subordinate and transient species (Grime 1979, 1987, 1998; Olff and Bakker 1998; Gibson et al., 1999; Mariotte et al., 2013a, 2013b,c,d). Dominant species are generally few in number, taller and account for a high proportion of the total community biomass, while subordinate species are smaller, grow under the canopy of dominants and account for a low proportion of the total community biomass. While dominant and subordinate species are frequent, transient species generally do not persist over years and appear locally through soil seed banks or seed rain.

According to the "mass ratio hypothesis", the functioning of ecosystems is suggested to be mainly driven by the most dominant species, or species group, and to be proportionate to the primary production of these species (Grime 1998). However, an increasing number of studies attempted to highlight the role of subordinate plant species and recognized the functional importance of this species-group, especially in the context of perturbation and climate change (e.g. insurance hypothesis, Yachi and Loreau 1999). Indeed, it is expected that more diverse communities would have a higher chance of containing species that are well-adapted to perturbations (e.g. warming, drought, flooding), and thus could compensate for the biomass loss of less adapted species. Mariotte et al. (2013a) investigated how subordinate and dominant species in semi-natural mountain grasslands relate to productivity. They showed that species diversity was clearly linked to productivity (Figure 11) and best explained by a polynomial function. These results confirm the diversity–productivity dependency, observed for semi-natural grasslands (Grime 1973b; Al-Mufti et al., 1977; Grace et al., 2007; Adler et al., 2011).

The maximum diversity occurred at intermediate productivity and this supports the humpback model of Grime (1973b, 1987), where species diversity is predicted to be higher at intermediate productivity and fertility. Interestingly, the contribution of species-groups differed along the diversity–productivity relationship. The contribution of dominant species to the total diversity remained similar along the productivity gradient while subordinate species followed a polynomial distribution along the productivity gradient and highly contributed to the total species diversity. Grime (1987) proposed that low productivity sites favour transient species, intermediate productivity favours subordinate species and high productive sites favour dominant species.

It is suggested that in semi-natural mountain grasslands, large herbivores may stabilize the proportion of dominant species along the productivity gradient, whereas subordinate species are favoured at intermediate level of grazing. Because subordinate species contribute greatly to the diversity of semi-natural grasslands, this has direct implications for the way grasslands should be managed, and highlights the importance of maintaining intermediate levels of pasturing to preserve biodiversity. An integrated management scheme, which considers local vegetation properties and allows for various grazing intensities and possibly different livestock breeds could help to improve the biodiversity of grasslands.

Mariotte et al. (2013d) undertook an aboveground plant removal experiment in a mountain grassland of the Swiss Jura Mountains to test the impact of subordinate plant species on soil microbial communities and ecosystem functioning. Plant community composition did not show significant shifts following removal treatments after 2 years. Nevertheless, the removal treatments significantly influenced bacterial communities, assessed by means of T-RFLP profiles of soil samples, with three distinct communities being discriminated according to the treatments, no biomass removal control, removal of all subordinate species, and partial biomass removal of dominant species to amounts equal to those removed in the subordinate removal plots. Removal of subordinates had a greater impact on bacterial communities than partial removal of dominants. Arbuscular mycorrhizal fungi (AMF) communities differed only marginally between the treatments. AMF are obligate symbionts because they completely depend on host plants for carbon (Johnson et al., 2002) and plant removal is expected to reduce AMF biomass in soil and to shift AM fungal community composition.

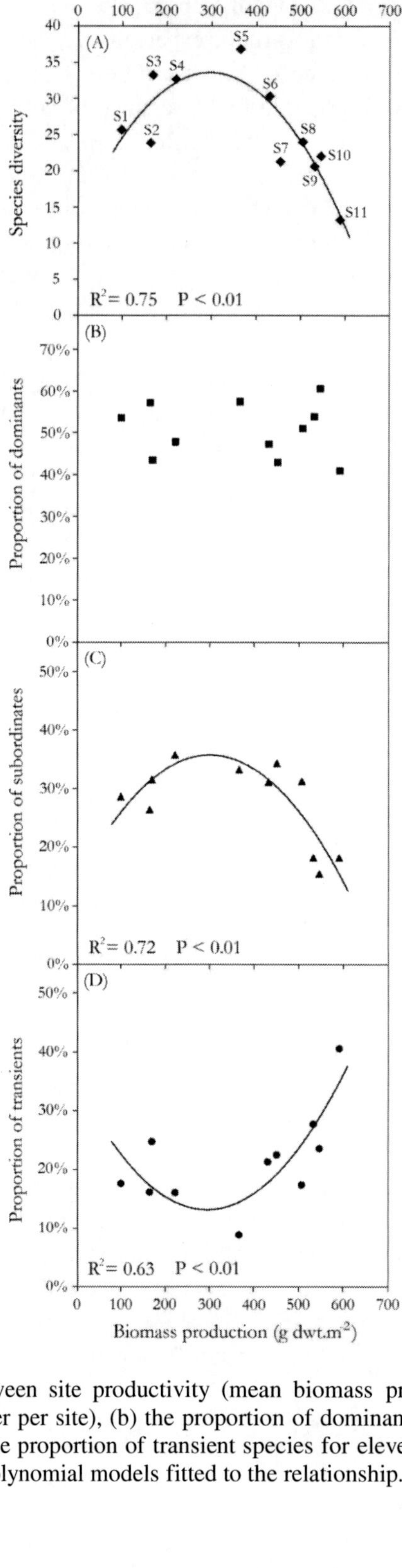

Figure 11. Relationships between site productivity (mean biomass production) and (a) the species diversity (mean species number per site), (b) the proportion of dominant species, (c) the proportion of subordinate species, and (d) the proportion of transient species for eleven semi-natural grassland sites. Curves represent significant polynomial models fitted to the relationship. Mariotte et al., 2013a.

Considering that AMF are generalists and may colonize different plant species, mycorrhizal communities are very resilient to changes in vegetation composition (Wardle et al., 1999; Johnson et al., 2004; Urcelay et al., 2009) and, in the short term, the effect of subordinate species removal might have a relatively small influence. It was suggested that the direct (i.e. biomass) and indirect (i.e. disturbance) effects of biomass removal treatments rapidly affect AMF communities, whereas the species effect of removal of subordinates might become visible only in the longer term through changes in above and belowground plant competition and/or facilitation (Mariotte et al., 2012).

With respect to ecosystem processes, during the first year of the experiment, no changes in aboveground biomass between the treatments were observed, but significant effects occurred in the second year and increased in the third year, with lower aboveground biomass production where subordinates were removed (Mariotte et al., 2013d).

Also, litter decomposition rate, soil respiration rate and soil dissolved inorganic nitrogen decreased in the subordinate removal plots. Thus, the removal of subordinate species in this experiment, and consequent modification of the quality of litter inputs, might explain the observed changes in the bacterial community composition. This seems to have had cascading effects on litter decomposition, soil respiration and fertility. Collectively, these findings indicate that not only the loss of species diversity (random loss), but also the loss of plant functional groups (i.e. subordinate), which represent non-random plant species, can impact ecosystem functioning in wooded pastures.

IMPACT OF CLIMATE CHANGE

Silvopastoral systems of the Swiss Jura Mountains serve as a traditional source of forage and timber in the subalpine vegetation belt, but their sensitivity to historic land-use and climate change puts their future sustainability at stake.

Global warming and associated environmental changes are predicted to have a strong impact on mountain ecosystems this century (IPCC 2007). The observed trend in Swiss mountains during the 20th century for a 1.5°C rise in mean annual air temperature (Beniston et al., 1997) is paralleled by increased instances of extreme summer temperature maxima and prolonged droughts (Schär et al., 2004; Beniston 2009). Throughout the current century, these tendencies are predicted to become more pronounced, and Northern hemisphere temperate mountains will experience the most intensive temperature rise, with a rate of warming typically two to three times higher (range +2.8°C to +5.3°C) than that recorded over the 20th century (Nogues-Bravo et al., 2007). Additionally, predicted hot spells and a 30% diminishing precipitation during the growing season will intensify the dry periods in Central Europe (Beniston 2007, CH2011). These will bring about changes in summer soil moisture availability, plant phenology, and growing season length, which would ultimately have repercussions on ecosystem distribution and function (Parry 2000, FOEN/FSO 2011).

In their work, Gavazov et al. (2014) used an altitudinal gradient to evaluate the effect of year-round warmer and drier climate on semi-natural subalpine grasslands of the Swiss Jura Mountains. Turf mesocosms from wooded-pastures were transplanted downslope, from 1350 m a.s.l. (control site), down to 1010 m a.s.l and 570 m a.s.l. respectively, testing for their resistance to three intensities of climate change.

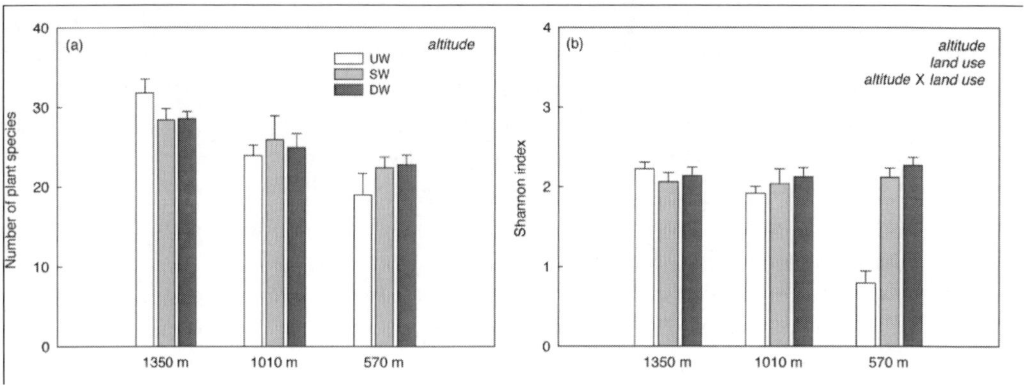

Figure 12. Plant species (a) richness and (b) Shannon diversity of the vegetation communities in the three land use types: unwooded pasture (UW), sparsely wooded pasture (SW), and densely wooded pasture (DW). Significant ($P < 0.05$) factors (in italics) and interactions are shown for each response variable. Gavazov et al., 2014.

Each mesocosm contained one of the following three land use types: unwooded pasture, sparsely-wooded pasture and densely-wooded pasture, spanning a gradient from high to low management intensity. Plant communities transplanted in mesocosms to warmer and drier climate experienced a marked loss in species richness (Figure 12a). The number of species decreased by up to 30 %, with a significant effect at both transplantation altitudes: 1010 m a.s.l. and 570 m a.s.l., compared to the control site at 1350 m a.s.l. Species richness among the land use types was equal, and did not show a significant interaction with the transplantation treatment. Most prominent losses were observed for graminoids and leguminous species. In contrast to species richness, Shannon diversity index (Figure 12b) varied significantly both across altitude and land use types, with a significant interaction between the two factors. This pattern was mainly driven by the significant decrease in the Shannon index of unwooded pasture mesocosms transplanted to 570 m a.s.l.

Total herbaceous above-ground biomass (AGB) across all three land use types decreased down to 45 % of that of the control site following the transplantation to warmer and drier climate, with a significant effect at the lowest site at 570 m a.s.l., and this was associated with the strongest deficiency in soil moisture content during the plant growing season. Land use exerted a strong effect too, and interacted significantly with the transplantation treatment. This was manifest by a significant effect of climate change on unwooded pasture and on sparsely wooded pasture vegetation at the lowest altitude, but not on the densely wooded pasture type (Figure 13a). Different plant growth forms (PGFs) reacted with different intensities and direction in response to the experimental climate change (Figure 13b,c,d). Graminoids experienced a consistent decrease in biomass in both unwooded and sparsely wooded pastures, and contributed substantially to the overall decrease in the total aboveground biomass in these land use types. Inherent differences in the morphology, growth and ecophysiology between graminoids and forbs could have driven the dichotomous response of these growth forms, for example with respect to belowground resource storage and leaf bud preformation (Wookey et al., 2009), leaf turnover, and water use efficiency (Signarbieux and Feller 2012).

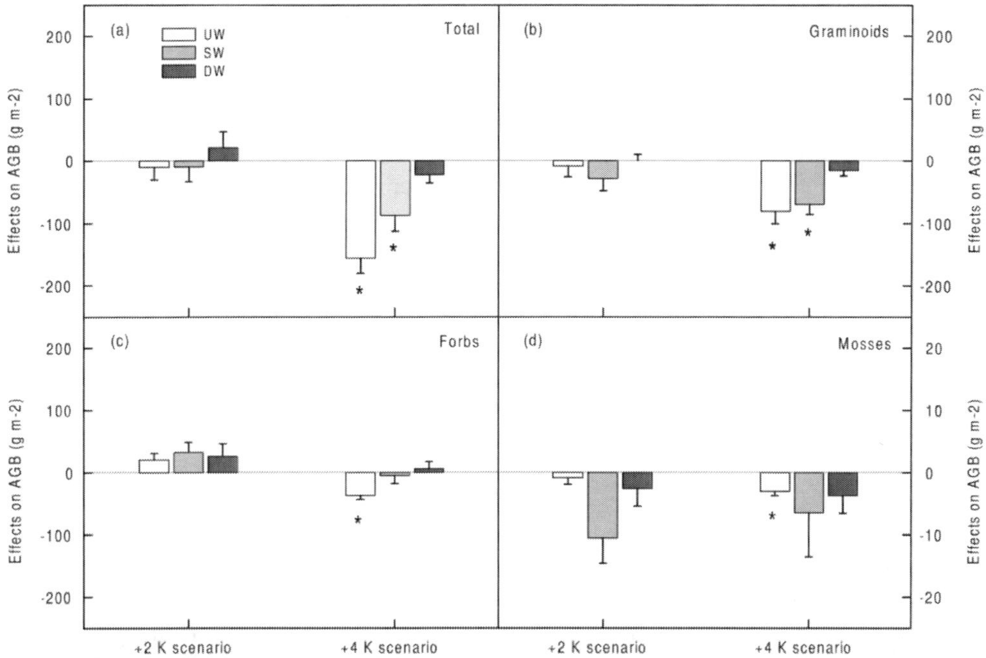

The three land use types are: unwooded (UW), sparsely wooded (SW), and densely wooded (DW) pastures. Asterisks (*) indicate significant differences (from controls at 1350 m a.s.l.) at the $P < 0.05$ threshold.

Figure 13. Effect of transplantation to warmer and drier climate at 1010 m a.s.l. (+2 K scenario) and 570 m a.s.l. (+4 K scenario) on aboveground biomass (AGB) production. Presented data are for (a) total biomass, and biomass of (b) graminoids, (c) forbs, and (d) mosses (y-axis differently scaled). Gavazov et al., 2014.

An advanced phenological development of 30 days was observed, due to shorter snow cover duration and warmer soils in spring, as demonstrated in Inouye (2008) and Wipf and Rixen (2010). Furthermore, a mid-season senescence due to drought stress (NDVI measures), was observed in unwooded pastures, while no such effects were recorded for the other land use types.

Overall, significant deviations from control conditions in plant aboveground biomass, richness, diversity and phenology were observed primarily for the communities exposed to the intensive climate change treatment (+4°C scenario), while there was little to no effect for those exposed to moderate climate change treatment (+2°C scenario). Another major finding from this experimental manipulation was that across land use types, intensively-used unwooded pastures were consistently more responsive to climate change than their counterparts from extensively-used wooded pastures of either sparsely or densely wooded type. These experimental findings corroborate modelling evidence (see Engler et al., 2011) for a strong dependence of mountain plant performance and floristic composition on the severity of drought episodes during projected climate warming this century. At the landscape scale, forests and wooded pastures, in contrast to grasslands, could exhibit a more conservative water use through a limited evaporation and hence cope better with long-lasting heat and drought stress.

According to the insurance hypothesis (Yachi and Loreau, 1999), more diverse plant communities are more likely to be resistant to drought. In their work, Mariotte et al. (2013c) conducted a rainout-shelter summer drought experiment in a mountain Swiss Jura grassland, in combination with the removal of subordinate species, in order to test the plant community resilience. Plant community composition was determined after the drought and based on biomass measurements made each year before and after the drought event (July-September), resistance, recovery and resilience of the plant community for each combination of treatments was determined.

Plant community resistance was significantly affected by the drought and removal treatments, and importantly, there was a significant drought × removal interaction effect. In watered plots, plant communities showed about 60% of regrowth, whilst there were no significant effects of the removal treatments. In drought plots, plant communities showed about 20% of regrowth in no removal and dominant removal plots, whereas community resistance was significantly lower in plots where subordinates were removed, with only 2.5% of the biomass re-growing after the drought treatment. Resistance of dominant and transient species against drought was similar to those of the whole plant community, but lower for dominant than for transient species. In contrast to dominant and transient species, subordinate species showed significantly stronger resistance in drought plots than in control plots. The proportional biomass of species groups shifted in response to the drought treatment: dominant species decreased from 60% to 25%, whilst subordinate species increased from 2% to 12%. Plant community recovery was not affected by the drought and removal treatments, but plant community resilience significantly decreased by the removal of subordinate species. Plant community composition (subordinate species excluded) at the end of the experiment was significantly affected by the drought and removal treatments.

As suggested by Wang et al., 2007, the resistance of plant community is strongly related to the pre-drought biomass in the community. Mariotte et al. (2013c) demonstrated that post-drought biomass decreased with increasing pre-drought biomass (Figure 14a) and a positive correlation between the pre-drought biomass of subordinate species and post-drought community biomass appeared (Figure 14b). Together, these results demonstrate that the plant community was more resistant and produced more biomass after drought when containing high biomass of subordinate species. Their work demonstrates that, independent of plant diversity, the presence of drought-resistant subordinate species increases plant community insurance against drought and, hence, is important for the functioning of grassland ecosystems. Overall, the proportion of dominant species decreased in response to drought, whereas the proportion of subordinate and transient species increased. These findings may be explained by reduced competitiveness of dominant species, which were strongly negatively affected by the drought, and/or by physiological adaptations of subordinates to drought such as better water use efficiency (WUE) or a deeper rooting system.

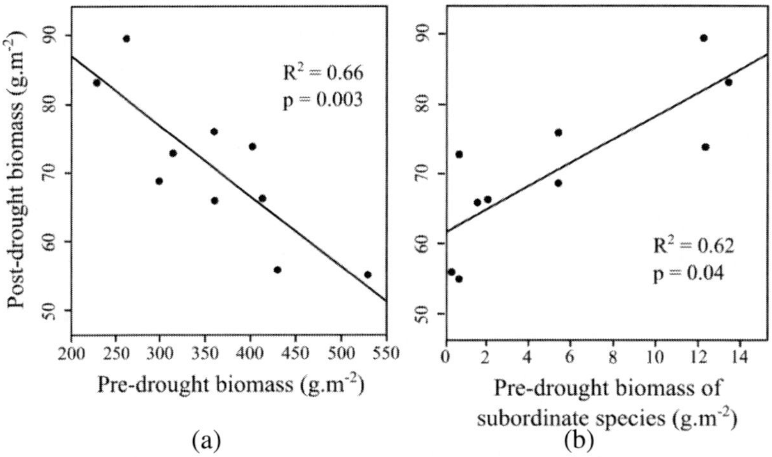

Figure 14. Linear relationships between post-drought community biomass (September 2010) and (a) pre-drought total biomass and (b) biomass of subordinate plant species (July 2010). Regressions include drought plots without plant removal and with partial removal of dominant biomass. Mariotte et al., 2013c.

Ecosystem Modelling

The turf transplantation experimental results of Gavazov et al. (2013) were implemented into a spatially explicit model of wooded pastures dynamics and by this means extrapolated to the landscape scale. The ecosystem dynamic simulation model of wooded-pastures, WoodPaM, was developed by Gillet et al. (2002) and Gillet (2008) to investigate the successional dynamics of wooded pastures in the Swiss Jura Mountains.

During simulations, selective foraging by cattle causes local impacts on vegetation (i.e., grazing, browsing, dunging, and trampling). Those, in turn, and together with geomorphologic and climatic conditions, promote or hinder forest development, and by this means drive the development of the landscape structure.

Forage production of a wooded-pasture in the Swiss Jura Mountain was simulated along regionalized climate change scenarios (observed for 1950–2000 AD and simulated for 2000–2050 AD annual temperatures and vegetation growing season temperatures) for two scenarios from the Intergovernmental Panel on Climate Change *Special Report Emissions Scenarios* (IPCC-SRES 2000): B2 and A1FI (Gavazov et al., 2013). Simulated time series of Global Utilisation Rates (GUR) were calculated. For the modelling, the spin-up period ended around year 1980, and realistic utilization rates were simulated for an extensive (EWP: ~70%), moderate (MWP: ~80%), and intense pastures (IWP: ~100%) (see Figure 15).

Overall, for current stocking densities, simulations of extensive pastures (sparsely and densely wooded) showed a continuous provision of sufficient forage, while in intense pastures (unwooded), scarcity of forage was indicated. The contribution of landscape heterogeneity to ecosystem process resilience is exemplified through the comparison of a pair of simulated years with hot (2021) and cool (2022) temperatures from the simulation continuum of IPCC-SRES climate change scenario B2 (Figure 15). Within the treeless intensive pasture in the hot year 2021, forage production decreased with drought causing an increase of 12.5% in Global Utilization Rate. Similar drought impacts on forage production were recognizable within the grasslands with more trees (moderate and extensive wooded pasture); however, the utilization rate in the extensive pastures, which is a true forest-grassland mosaic, increased by only 6.7% in GUR. Overall, this indicates a buffering potential of wooded pastures.

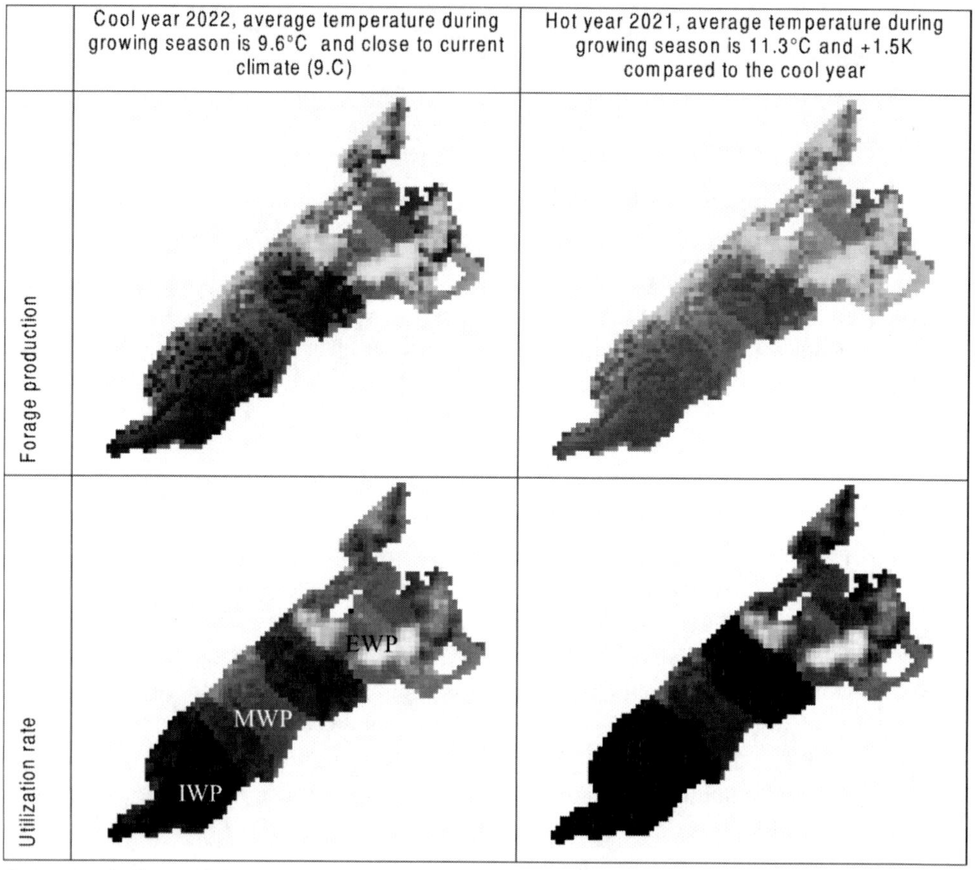

Darker tones indicate higher values of the respective parameters (Gavazov et al., 2013). IWP: treeless intensive wooded pasture; MWP: moderate wooded pasture; EWP: extensive wooded pasture.

Figure 15. Output maps of forage production and local utilization rate (percentage consumed forage) in two arbitrarily chosen years – a hot one (2021) and a cool one (2022), based on the moderate climate change scenario B2 with contrasting average temperatures in the vegetation period (Δ 1.5 °C). Gavazov et al., 2013.

This approach of incorporating experimental results into a process-based simulation model of wooded pastures merits more than a simple extrapolation across spatial scales (from plot to landscape levels). Since biomass production of grasslands forms represents the base of WoodPaM simulations, climate change impacts on primary production mechanistically determine forage availability for cattle, which in turn shapes landscape structure via selective grazing behavior (Kohler et al., 2006a). This simulation period (2000-2050) is probably short enough to detect climatic stress while assuming no plant species adaptation (development of drought-adapted ecotypes), or significant community change through immigration, or to produce any substantial shift in the tree layer and thus the landscape structure.

Nevertheless, one should bear in mind that in the long run, impacts of climate change on grassland productivity could produce feedback on landscape structure due to adaptive grazing behaviour of cattle (Smit et al., 2007; Vandenberghe et al., 2007, 2009). In dry years, depleted forage resources in open pastures would provoke cattle grazing in the forest understory, where drought impacts were found to be smaller and forage would still be available.

Forest regeneration would therefore be put at stake, and landscape structural dynamics could be expected to shift towards a more open landscape. These simulations showed that in extensive grazing systems, it was not only the generally lower stocking density but also the resulting grassland-forest mosaic that contributed to the robust provision of forage.

Consequently, the apparent advantage in terms of productivity of unwooded pastures diminished with drought, and wooded pastures became a forage source of similar importance. Such formerly unproductive patches show a clear potential to compensate for the breakdown of productivity within open grasslands. Across the landscape, spatial variation itself stimulates ecosystem resilience and resistance to drought.

Godfree et al. (2011) showed that naturally occurring extreme climatic events such as drought can be mitigated through the protective role of heterogeneous environments. This concept is shared by Foley et al. (2005) who advocate the maintenance of a diverse portfolio of ecosystem services by a single ecosystem, such that sustainable land use strategies could be implemented for both short- and long-term needs. In the case of subalpine wooded pastures, this can be accomplished through the preservation of the mosaic landscape.

It is, nevertheless, acknowledged that decisions about the future of such cultural landscapes come from stakeholders rather than scientists, and that such a stability requires both maintenance of historically established agricultural practices (Küster 2004) and adaptive management, including change in policies (Huber et al., 2013).

CONCLUSION

The multiple factors acting on grasslands in mountain ecosystems, and in particular the interplay between biotic and abiotic factors are key to understanding the high plant species richness and coexistence of habitats. Pastures constitute a patchwork of micro-successions at various successional stages, depending on the major and changing constraints in relation to cattle activity, and thus of management scheme. Not only species diversity, but also plant functional groups are important for ecosystem functioning in wooded pastures, particularly for facing climate change. Both functional diversity of species and diversity of habitats will grant wooded-pasture higher resilience in a changing climate.

Despite the fact that environmental heterogeneity remains difficult to quantify because it has multiple dimensions, the hypothesis that environmental heterogeneity is a major driver of species richness unifies several theories of biodiversity and has the potential to be developed into a management tool.

The scale-dependence of effects suggests a change in the relative importance of mechanisms, which will need to be taken into account when scaling up to larger landscapes.

For sustaining the typical heterogeneous mountain wooded pasture landscape, it is important to integrate ecological patterns and processes, e.g. cattle behaviour as the main driving force affecting vegetation dynamics, as well as socio-economic processes, e.g. farming, logging, expectations and wishes of users and stakeholders or alternative land-use for wood energy. Considering trade-offs is essential if one aims at meeting the requirements of sustainable management schemes of wooded-pasture landscape, including their various ecosystem goods and services.

Thus, integrated management planning of wooded-pastures requires an intensive collaboration between agronomists, foresters, ecologists and economists (e.g. Barbezat et Boquet 2008). New challenges with land-use and climate change call for adaptive management and this can only be achieved by a transdisciplinary and consensual approach.

ACKNOWLEDGMENTS

We are grateful to former students, scientific collaborators and colleagues who contributed greatly with their work and publications to this synthesis, in particular: F. Gillet, B. Murisier, J.-D. Gallandat, J.-M. Gobat, A. Dufour, F. Kohler, M.-A. Progin, C. Vandenberghe, P. Mariotte, F. Gilgen, T. Spiegelberger, K. Gavazov, F. Hagedorn, P. Kardol. A. Peringer and S. Siehoff. Many thanks also to L. Bragazza, B. Robroek, Rob Mills, P. Mariotte and P. Kardol for their careful revision and useful comments.

REFERENCES

Abdelmagid, A. H., Trlica, M. J. and Hart, R. H. 1987. Soil and vegetation responses to simulated trampling, *J. Range Manage.* 40: 303-306.

Adler, P. B., Seebloom, E. W., Borer, E. T., Hillebrand, H., Hautier, Y., Hector, A., et al. 2011. Productivity is a poor predictor of plant species richness. *Science* 333: 1750–1752.

Al-Mufti, M. M., Sydes, C. L., Furness, S. B., Grime, J. P., and Band, S. R. 1977. A quantitative analysis of shoot phenology and dominance in herbaceous vegetation. *Journal of Ecology* 65: 759-791.

Bakker, J. P., de Leeuw, J. and van Wieren, S. E. 1984. Micro- patterns in grassland vegetation created and sustained by sheep-grazing. *Vegetatio* 55: 153-161.

Bakker, E. S. and Olff, H. 2003. Impact of different-sized herbivores on recruitment opportunities for subordinate herbs in grassland. *J. Veg. Sci.* 14: 465-474.

Barbezat, V. and Boquet, J.-F. (Eds) 2008. *Gestion intégrée des paysages sylvo-pastoraux de l'Arc jurassien – Manuel' (Hankbook).* Conférence TransJurassienne. La Chaux-de-Fonds, Besançon. p. 160 et 1 CD-ROM.

Beniston, M., H. F. Diaz and R. S. Bradley. 1997. Climatic change at high elevation sites: an overview. *Climatic Change* 36: 233–251.

Beniston, M., D. B. Stephenson, O. B., Christensen, C. A. T., Ferro, C., Frei, S., Goyette, K., et al. 2007. Future extreme events in European climate: an exploration of regional climate model projections. *Climatic Change* 81: 71–95.

Beniston, M. 2009. Decadal-scale changes in the tails of probability distribution functions of climate variables in Switzerland. *International Journal of Climatology* 29: 1362–1368.

Boeken, B. and Shachak, M., 2006. Linking community and ecosystem processes: the role of minor species. *Ecosystems* 9: 119–127.

Briner, S., Elkin, C., Huber, R., and Grêt-Regamey, A. 2012. Assessing the impacts of economic and climate changes on land-use in mountain regions: A spatial dynamic modeling approach, *Agriculture, Ecosystem and Environment* 149: 50–63.

Buttler, A., Gavazov, K., Peringer, A., Siehoff, S., Mariotte, P., Wettstein, J.-B., et al. 2012a. Erhaltung der Wytweiden im Jura: Klimatische und agrarpolitische Herausforderungen. *Agrarforschung* 3, no. 7-8: 346-353.

Buttler, A., Spiegelberger, T., Chételat, J., Kalbermatten, M., Lannas, K., Peringer, A., et al. 2012b. Evolution récente et future des paysages sylvo-pastoraux du Parc Jurassien Vaudois. *Schweiz. Z. Forstwes*, 163: 469-480.

Buttler, A., F. Kohler and F. Gillet 2009. The Swiss mountain wooded pastures: patterns and processes, In: A. Rigueiro-Rodriguez, J. McAdam and M. R. Mosquera-Losada (eds). *Agroforestry in Europe: Current Status and Future Prospects*, Springer, New York, pp. 377-396.

CH2011. 2011. *Swiss climate change scenarios CH2011.* C2SM, MeteoSwiss, ETH, NCCR Climate, and OcCC, Zurich, Switzerland.

Dai, X. 2000. Impact of cattle dung deposition on the distribution pattern of plant species in an alvar limestone grassland. *Journal of Vegetation Science* 11: 715-724.

Dufour, A., Gadallah, Z., Wagner, H., Guisan, A., and Buttler, A. 2006a. Plant species richness and environmental heterogeneity in a mountain landscape: effects of variability and spatial configuration. *Ecography* 29: 573-584.

Dufour, A. 2006b. *Multiscale assessment of plant diversity in wooded-pastures of the Jura mountains.* Thèse UFR Sciences et Techniques, Université Franche Comté, Besançon, 111p.

Engler, R., et al. 2011. 21st century climate change threatens mountain flora unequally across Europe. *Global Change Biology* 17: 2330-2341.

Etienne, M. 1996. *Western European silvopastoral systems.* INRA, Paris.

FOEN/FSO. 2011. Federal Office for the Environment and Federal Statistical Office. *Environment Switzerland 2011.* Bern and Neuchâtel, Switzerland.

Foley, J. A., R., DeFries, G. P., Asner, C., Barford, G., Bonan, S. R., Carpenter, F. S., et al. 2005. Global consequences of land use. *Science* 309: 570–574.

Gallandat, J. D., Gillet, F., Havlicek, E., and Perrenoud, A. 1995. *Pâtubois: typologie et systémique phytoécologiques des pâturages boisés du Jura suisse.* Laboratoire d'écologie végétale, Université de Neuchâtel, Neuchâtel, Switzerland.

Gavazov, K., Peringer, A., Gillet, F., Siehoff, S., Buttler, A., and Spiegelberger, T. 2013. Dynamics of forage production in pasture-woodlands of the Swiss Jura Mountains under projected climate change scenarios. *Ecology and Society* 18, no. 1, p. 38.

Gavazov, K., T. Spiegelberger and A. Buttler 2014. Transplantation of subalpine wood-pasture turfs along a natural climatic gradient reveals lower resistance of unwooded pastures to climate change compared to wooded ones. *Oecologia*, DOI:10.1007/s00442-013-2839-9.

Gobat, J. M., Duckert, O. and Gallandat, J. D. 1989. Quelques relations microtopographie-sols-végétation dans les pelouses pseudo alpine du Jura suisse: exemples d'un système naturel et d'un système anthropisé. *Bulletin de la société neuchâteloise des sciences naturelles* 112: 5-17.

Gibson, D. J., Ely, J. S. and Collins, S. L. 1999. The core-satellite species hypothesis provides a theoretical basis for Grime's classification of dominant, subordinate, and transient species. *Journal of Ecology* 87: 1064-1067.

Gillet, F. and Gallandat, J. D. 1996a. Wooded pastures of the Jura mountains, In: Etienne, M. (ed.), *Western european silvopastoral systems*. INRA, Paris.

Gillet, F. and Gallandat, J. D. 1996b. Integrated synusial phytosociology: some notes on a new, multiscalar approach to vegetation analysis. *J. Veg. Sci.* 7: 13-18.

Gillet, F., O. Besson and J.-M. Gobat. 2002. PATUMOD: a compartment model of vegetation dynamics in wooded pastures. *Ecological Modelling* 147: 267–290.

Gillet, F., Kohler, F., Vandenberghe, C., and Buttler, A. 2010. Effect of dung deposition on small-scale patch structure and seasonal vegetation dynamics in mountain pastures. *Agric. Ecosyst. Environ.* 135: 34-41.

Gillet, F., Murisier, B., Buttler, A., Gallandat, J.-D., and Gobat, J. M. 1999. Influence of tree cover on diversity of herbaceous communities in subalpine wooded pastures. *Applied Vegetation Science* 2: 47-54.

Gillet, F. 2008. Modelling vegetation dynamics in heterogeneous pasture-woodland landscapes. *Ecological Modelling* 217: 1-18.

Godfree, R., B., Lepschi, A., Reside, T., Bolger, B., Robertson, D., Marshall and M. Carnegie 2011. Multiscale topoedaphic heterogeneity increases resilience and resistance of a dominant grassland species to extreme drought and climate change. *Global Change Biology* 17: 943–958.

Grace, J. B., Anderson, T. M., Smith, M. D., Seabloom, E., Andelman, S. J., Meche, G., et al. 2007. Does species diversity limit productivity in natural grassland communities? *Ecology Letters* 10: 680–689.

Grime, J. P. 1973a. Competitive exclusion in herbaceous vegetation. *Nature*, 242: 344-347.

Grime, J. P. 1973b. Control of species density in herbaceous vegetation. *Journal of Environmental Management* 1: 151-167.

Grime, J. P. 1979. *Plant strategies and vegetation processes*. John Wiley and Sons, Chichester, UK.

Grime, J. P. 1987. Dominant and subordinate components of plant communities – implications for succession, stability and diversity. *Colonisation, Succession and Stability* (eds A. Gray, P. Edwards and M. Crawley), pp. 413-428. Blackwell, Oxford, UK.

Grime, J. P. 1998. Benefits of plant diversity to ecosystems: immediate, filter and founder effects. *Journal of Ecology* 86: 902-910.

Grime, J. P. 2001. *Plant strategies, vegetation processes, and ecosystem properties.* Wiley, Chichester, UK.

Havlicek, E., Gobat, J. M. and Gillet, F. 1998. Reflections on relationships between vegetation and soil: three examples on allochtonous material in the Jura Mountains. *Ecologie* 4: 535-546.

Heikkinen, R. K. 1996. Predicting patterns of vascular plants species richness with composite variables: a meso-scale study in Finnish Lapland. *Vegetatio* 126: 151-165.

Herben, T., Krahulec, F., Hadincova, V., and Kovaoova, M. 1993. Small-scale spatial dynamics of plant-species in a grassland community over six years. *J. Veg. Sci.* 4: 171–178.

Huber, R., Bebi, P., Briner, S., Bugmann, H., Buttler, A., Grêt-Regamey, A., et al. 2012. MOUNTLAND: Klimawandel und nachhaltige Landnutzung im Berggebiet. *Agrarforschung* 3, no. 7-8: 340-345.

Huber, R., S. Briner, A. Peringer, S. Lauber, R. Seidl, A. Widmer, et al. 2013. Modeling social-ecological feedback effects in the implementation of payments for environmental services in pasture-woodlands. *Ecology and Society* 18, 2: 41.

Hubert, D., Ovalle, C. and Daget, P. 1982. Influence du couvert des arbres sur la végétation d'une pelouse des Causses. *Coll. Phytosoc.* 11: 569-586.

Hurtt, G. C. and Pacala, S. W. 1995. The consequences of recruitment limitation: reconciling chance, history and competitive differences between plants. *J. Theor. Biol.* 176: 1-12.

Inouye, D. W. 2008. Effects of climate change on phenology, frost damage, and floral abundance of montane wildflowers. *Ecology* 89: 353-362.

Intergovernmental Panel on Climate Change (IPCC) 2000. *Special report emission scenarios.* Cambridge University Press, Geneva, Switzerland.

Intergovernmental Panel on Climate Change (IPCC) 2007. *Climate change 2007: synthesis report.* Contribution of Working Groups I, II and III to the Fourth Assessment Report of the Intergovernmental Panel on Climate Change. Geneva, Switzerland.

Johnson, D., Leake, J. R., Ostle, N., Ineson, P., and Read, D. J. 2002. In situ $^{13}CO_2$ pulse-labelling of upland grassland demonstrates a rapid pathway of carbon flux from arbuscular mycorrhizal mycelia to the soil. *New Phytol.* 153: 327–334.

Johnson, D., Vandenkoornhuyse, P. J., Leake, J. R., Gilbert, L., Booth, R. E., Grime, J. P., Young, P. W., and Read, D. J. 2004. Plant communities affect arbuscular mycorrhizal fungal diversity and community compo- sition in grassland microcosms. *New Phytol.* 161: 503–515.

Klimes, L., Klimesova, J., Hendriks, R., and van Groenendael, J. 1997. Clonal plant architecture: a comparative analysis form and function. In: DE Kroon, H. and van Groenendael, J. (eds.), *The ecology and evolution of clonal plants,* Backhuys, Leiden, pp. 1–29.

Kohler, F. 2004a. Influence of grazing, dunging and trampling on short-term dynamics of grasslands in mountain wooded pastures. *PhD thesis*, University of Neuchâtel, Switzerland.

Kohler, F., Gillet, F., Gobat, J.-M., and Buttler, A. 2004b. Seasonal vegetation changes in mountain pastures due to simulated effects of cattle grazing. *Journal of Vegetation Science* 15: 143-150.

Kohler, F., Gillet, F., Progin, M.-A., Gobat, J.-M., and Buttler, A. 2004c. Seasonal dynamics of plant species at fine scale in wooded pastures. *Community Ecology* 5: 7-17.

Kohler, F., Hamelin, J., Gillet, F., Gobat, J. M., and Buttler, A. 2005. Soil microbial community changes in mountain wooded pastures due to simulated effects of cattle grazing. *Plant and soil* 278: 327-340.

Kohler, F., Gillet, F., Reust, S., Wagner, H. H., Gadallah, F., Gobat, J.-M., and Buttler, A. 2006a. Spatial and seasonal patterns of cattle habitat use in a mountain wooded pasture. *Landscape Ecol.* 21: 281-295.

Kohler, F., Gillet, F., Gobat, J.-M., and Buttler, A. 2006b. Effect of cattle activities on gap colonisation in mountain pastures. *Folia Geobot.* 41: 289–304.

Kotanen, P. M. 1997. Effects of gap area and shape on recolonization by grassland plants with differing reproductive strategies. *Canad. J. Bot.* 75: 352–361.

Küster, H. 2004. Cultural landscapes. In: M. Dieterich and J. Straaten, editors. *Cultural landscapes and land use*, pages 1–11, Springer, The Netherlands.

Lavorel, S., Grigulis, K., Lamarque, P., Colace, M. P., Garden, D., Girel, J., et al. 2011. Using plant functional traits to understand the landscape distribution of multiple ecosystem services. *J. Ecol.* 99: 135–147.

Lortie, C. J., Brooker, R. W., Choler, P., Kikvidze, Z., Michalet, R., Pugnaire, F. I., and Callaway, R. M. 2004. Rethinking plant community theory. *Oikos* 107 (2): 433-438.

Lyons, K. G. and Schwartz, M. W. 2001. Rare species loss alters ecosystem function – invasion resistance. *Ecol. Lett.* 4: 358–365.

Macek, P. and Leps, J. 2003. The effect of environmental heterogeneity on clonal behaviour of *Prunella vulgaris* L. *Pl. Ecol.* 168: 31–43.

MacLaren, J. R. and Turkington, R., 2010. Ecosystem properties determined by plant functional group identity. *J. Ecol.* 98: 459–469.

Mariott, C. A., Fisher, J. M., Hood, K. J., and Smith, M. A. 1997. Persistence and colonization of gaps in sown swards of grass and clover under different sward management. *Grass Forage Sci.* 52: 156–166.

Mariotte, P., Buttler, A., Jonhson, D., Thébault, A., and Vandenberghe, C. 2012. Exclusion of root competition increases competitive abilities of subordinate plant species through root-shoot interactions. *Journal of Vegetation Science* 23: 1148-1158.

Mariotte, P., Buttler, A., Kohler, F., Gilgen, A., and Spiegelberger, T. 2013a. How do subordinate and dominant species in semi-natural mountain grasslands relate to productivity and land-use change. *Basic and Applied Ecology* 14: 217-224.

Mariotte, P., Meugnier, C., Johnson, D., Thébault, A., Spiegelberger, T., and Buttler, A. 2013b. Arbuscular mycorrhizal fungi reduce the differences in competitiveness between dominant and subordinate plant species. *Mycorrhiza* 23: 267-277.

Mariotte, P., Vandenberghe, C., Hagedorn, F., Kardol, P., and Buttler, A. 2013c. Subordinate species enhance community insurance to drought in semi-natural grassland ecosystems. *Journal of Ecology* 101: 763-773.

Mariotte, P., Vandenberghe, C., Meugnier, C., Rossi, P., Bardgett, R. D., and Buttler, A. 2013d. Subordinate plant species impact on soil microbial communities and ecosystem functioning in grassland: findings from a removal experiment. *Perspectives in Plant Ecology, Evolution and Systematics* 15: 77-85.

Milberg, P. 1993. Seed bank and seedling emerging after soil disturbance in a wet semi-natural grassland in Sweden. *Ann. Bot. Fenn.* 30: 9–13.

Nogues-Bravo, D., Araujo, M. B., Errea, M. P., and Martinez-Rica, J. P. 2007. Exposure of global mountain systems to climate warming during the 21st Century. *Global Environmental Change-Human and Policy Dimensions* 17: 420-428.

Olff, H. and Bakker, J. P. 1998. Do intrinsically dominant and subordinate species exist? A test statistic for field data. *Applied Vegetation Science* 1: 15-20.

Palmer, M. W. and Rusch, G. M. 2001. How fast is the carousel? Direct indices of species mobility with examples from an Oklahoma grassland. *Journal of Vegetation Science* 12: 305-318.

Parry, M. L. 2000. *Assessment of potential effects and adaptations for climate change in Europe: the Europe ACACIA project*. Jackson Environmental Institute, University of East Anglia, Norwich.

Partel, M. and Zobel, M. 1995. Small-scale dynamics and species richness in successional alvar plant-communities. *Ecography* 18: 83-90.

Platt, W. J. and Connell, J. H. 2003. Natural disturbances and directional replacement of species. *Ecological Monographs* 73: 507-522.

Rapp, J. K. and Rabinowitz, D. 1985. Colonization and establishment of Missouri prairie plants on artificial soil disturbances. I. Dynamics of forb and graminoid seedlings and shoots. *Amer. J. Bot.* 72: 1618–1628.

Rapson, G. L., Thompson, K. and Hodgson, J. G. 1997. The humped relationship between species richness and biomass - testing its sensitivity to sample quadrat size. *Journal of Ecology* 84: 293-295.

Rice, R. W., MacNeil, M. D. and Koong, L. J. 1983. Simulation of herbage/animal interface of grazioso land', In: Lauenroth, W. K., Skogerboe, G. V., Flug, M. (eds) *Analysis of ecological systems: state-of-the-art in ecological modelage*, Developments in Ecological Modelling 5. Elsevier, Amsterdam.

Rook, A. J., Dumont, B., Isselstein, J., Ososro, K., WallisDeVries, M. F., Parente, G., and Mills, J. 2004. Matching type of livestock to desired biodiversity outcomes in pastures - a review. *Biological Conservation* 119: 137-150.

Sabo, J. L., Sponseller, R., Dixon, M., Gade, K., Harms, T., Heffernan, J., Jani, A., Katz, G., Soykan, C., Watts, J., and Welter, J. 2005. Riparian zones increase regional species richness by harboring different, not more species. *Ecology* 86: 56-62.

Schär, C., Vidale, P. L., Luthi, D., Frei, C., Haberli, C., Liniger, M. A., and Appenzeller, C. 2004. The role of increasing temperature variability in European summer heatwaves. *Nature* 427: 332-336.

Schrama, M., Veen, G. F., Bakker, E. S., Ruifrok, J. L., Bakker, J. P., and Olff, H. 2012. An integrated perspective to explain nitrogen mineralization in grazed ecosystems. *Perspectives in Plant Ecology Evolution and Systematics* 15: 32-44.

Shmida, A. and Wilson, M. V. 1985. Biological determinants of species diversity. *J. Biogeogr.* 12: 1-20.

Signarbieux, C. and Feller, U. 2012. Effects of an extended drought period on physiological properties of grassland species in the field. *Journal of Plant Research* 125: 251-261.

Sjögren, P. 2005. Palaeoecological investigations of pasture woodland in Combe des Amburnex, Swiss Jura Mountains. *PhD Thesis*, Universität Bern, Bern.

Sjögren, P. 2006. The development of pasture woodland in the southwest Swiss Jura Mountains over 2000 years, based on three adjacent peat profiles. *Holocene* 16: 210-223.

Smit, C., C. Vandenberghe, J. den Ouden, and H. Mueller- Schäerer. 2007. Nurse plants, tree saplings and grazing pressure: changes in facilitation along a biotic environmental gradient. *Oecologia* 152: 265–273.

Statzner, B. and Moss, B. 2004. Linking ecological function, biodiversity and habitat: a mini-review focusing on older ecological literature. *Basic Appl. Ecol.* 5: 97-106.

Urcelay, C., Diaz, S., Gurvich, D. E., Chapin III, F. S., Cuevas, E., and Dominguez, L. S., 2009. Mycorrhizal community resilience in response to experimental plant functional type removals in a woody ecosystem. *J. Ecol.* 97, 1291–1301.

Tamm, A., Kull, K. and Sammul, M. 2002. Classifying clonal growth forms based on vegetative mobility and ramet longevity: whole community analysis. *Evol. Ecol.* 15: 383–401.

Tilman, D. 2004. Niche tradeoffs, neutrality, and community structure: a stochastic theory of resource competition, invasion, and community assembly. *Proc. Nat. Acad. Sci. US* 101: 10854-10861.

Vandenberghe, C., Freléchoux, F., Moravie, M. A., Gadallah, F., and Buttler, A. 2007. Short-term effects of cattle browsing on tree sapling growth in mountain wooded pastures', *Plant Ecol.* 188: 253-264.

Vandenberghe, C., Smit, C., Pohl, M., Buttler, A., and Freléchoux, F. 2009. Does the strength of facilitation by nurse shrubs depend on grazing resistance of tree saplings. *Basic and Applied Ecology* 10: 427-436.

Vittoz, P. 1998. Flore et végétation du Parc jurassien vaudois: typologie, écologie et dynamique des milieu. *PhD Thesis*, University of Lausanne.

Wang, Y., Yu, S. and Wang, J. 2007. Biomass-dependent susceptibility to drought in experimental grassland communities. *Ecology Letters* 10: 401–410.

Wardle, D. A., Bonner, K., Barker, G. M., Yeates, G. W., Nicholson, K. S., Bardgett, R. D., Watson, R. N., and Ghani, A. 1999. Plant removals in perennial grassland: vegetation dynamics, decomposers, soil biodiversity, and ecosystem properties. *Ecol. Monogr.* 69: 535–568.

Wardle, D. A., Yeates, G. W., Williamson, W., and Bonner, K. I. 2003. The response of a three trophic level soil food web to the identity and diversity of plant species and functional groups. *Oikos* 102: 45–56.

Wilson, J. B., Sykes, M. T. and Peet, R. K. 1995. Time and space in the community structure of a species-rich limestone grassland. *Journal of Vegetation Science* 6: 729-740.

Whittaker, R. H. 1965. Dominance and diversity in land plant communities. *Science* 147: 250-260.

Whittaker, R. H. 1975. *Communities and ecosystems.* Collier MacMillan, London, 2nd ed.

Wipf, S. and Rixen, C. 2010. A review of snow manipulation experiments in Arctic and alpine tundra ecosystems. *Polar Research* 29: 95-109.

Wookey, P. A., Aertz, R., Bardgett, R. D., Baptist, F., Brathen, K., Cornelissen, J. H. C., et al. 2009. Ecosystem feedbacks and cascade processes: understanding their role in the responses of Arctic and alpine ecosystems to environmental change. *Global Change Biology* 15: 1153-1172.

Yachi, S. and Loreau, M. 1999. Biodiversity and ecosystem productivity in a fluctuating environment: the insurance hypothesis. *Proceeding of the National Academy of Science of the US* 96: 1463-1468.

Zobel, K., Zobel, M. and Rosén, E. 1994. An experimental test of diversity maintenance mechanisms, by a special removal experiment in a species-rich wooded meadow. *Folia Geobotanica* 29: 449-457.

In: Grassland Biodiversity and Conservation …
Editors: Pierre Mariotte and Paul Kardol
ISBN: 978-1-61761-619-8

Chapter 9

ALPINE GRASSLANDS: CHANGES OF PLANT SPECIES COMPOSITION AND FUNCTIONAL TYPES AFTER RESTORATION AND SIMULATION OF CLIMATE CHANGES

Francis Isselin-Nondedeu[1,*] and Thomas Spiegelberger[2]
[1]UMR CNRS 7324 CITERES Département d'Aménagement,
Ecole Polytechnique de l'Université François Rabelais (Department of Landscape, Environmental and Urban Planning, Polytechnic School of University), Tours, France
[2]Irstea, UR EMGR Mountain Ecosystems,
St-Martin-d'Hères, France

ABSTRACT

Alpine grasslands are fragile ecosystems formed by plant species that have developed different adaptations to survive and reproduce in harsh environments. The chapter summarizes some of these adaptations and stresses out peculiarities of plant community functioning at high altitude. Vegetation composition and structure at meso- and micro-scales respond to many ecological factors that are strongly linked with slope and topography, such as snow cover, thickness and duration, nutrients and water availability, and wind exposure. However, alpine grasslands are subjected to major changes of land-uses and climate conditions. Along this chapter we discuss the responses of vegetation to these global changes, focusing on how it can affect species and functional composition. Experimentally simulated conditions of increased temperature and reduced precipitation (study case 1) significantly reduced species richness and affected functional groups in different ways. The development of ski areas, mostly with the construction of ski runs, greatly disturbs soil and vegetation. For many alpine grasslands, active restoration is thus required to stabilize soil and help revegetation. As it is described in a second study case, restoration is a long process taking at least a decade. Judging by the composition of the vegetation, many seeded species persisted in the community whereas many local species failed to establish. The cultivation of local

* Corresponding author: Francis Isselin-Nondedeu, email: francis.isselin@univ-tours.fr.

species, in order to use their seeds in the seed mixtures for restoration, is developing in many regions and the first results are promising.

Keywords: Restoration, alpine grasslands, climate change, land-use, skiing industry, functional traits, plant community, treeline

INTRODUCTION

Alpine ecosystems, which encompass the bio-geographical zone above the treeline, are among the species richest ecosystems of the planet (Körner 2002). Despite the fact that the zone above the treeline represents only 3% of the land surface and that 12% of all the people in the world live in mountain areas, numerous hot-spots of biodiversity are located in this zone, containing at least 4% of the world's vascular plant species (Körner & Spehn 2002). Indeed, the alpine environment offers a great diversity of habitats with a combination of various geological substrates, different soil types and topographic situations, past and present changes of climate and microclimate, and sometimes a long history of human land-uses. In Europe, two modifications of land-use occurred simultaneously during the last 60 years: the intensity of mountain agriculture, together with the number of farmers, decreased in some regions, and continue to decrease (Tasser & Tappeiner 2002; Gehrig-Fasel et al. 2007), while in other regions, alpine ecosystems have been altered through agricultural intensification, development of tourism, urbanization, and expansion of ski areas (Wipf et al. 2005).

In this chapter we focus on alpine grasslands, i.e. lands above the treeline that are dominated by grass or grasslike vegetation. We give an overview of the threats on these grasslands by focusing on plant species and functional composition in the context of alpine environments, disturbances, climate change and ecological restoration. In the first part, we present the main characteristics of the alpine grasslands. We then illustrate the effects of land-use change and the impacts of climate change with a first case study. The next part describes restoration goals and techniques for alpine grasslands, with a second case study showing variations of plant composition after the ecological restoration of ski runs.

CHARACTERISTICS OF ALPINE ENVIRONMENTS

Alpine environments are found near and above the treeline, at altitudes ranging from 1100 to 2500 meters in the Rocky Mountains and Scandinavian Alps, at 1500 to 3200 m in the European Alps and up to 4200 m in the Californian mountains (Körner 1999b). This chapter focuses on a belt between 1800 m to 3200 m because (i) in some regions, the natural versus anthropogenic origins of the treeline is still debated (Harsch et al. 2009; Korner & Paulsen 2004), (ii) most of the grasslike vegetation are well developed in this altitudinal range, (iii) this belt is the most exploited by the skiing industry and for grazing activities. Three main ecological factors conditioning plant life in alpine environment have been identified (Troll 1973): i) altitude, which involves variations of factors, such as the air pressure and moisture, temperature, and insolation; ii) continental characteristics governing the precipitation regime and the seasonal temperature difference; and iii) climatic constraints related to latitude, which may superimposed on the altitude. Altitude is the most obvious environmental variable constraining plant life. With elevation, air temperature decreases on average by -0.6°C per 100 m. Diurnal amplitudes of temperature can be high with frost events that may occur during summer, but amplitude narrows with elevation due to high-altitude winds (Linacre 1982). Air moisture content decreases with elevation while precipitation usually increases. From sea level to higher summits, atmospheric pressure decreases with elevation but not linearly. However, between 1000 m and 3000 m, atmospheric pressure drops by 100 hPa per 1000 m elevation. Similarly, partial pressure of O_2 and CO_2 at 1000 m is reduced by 20% compared to sea level, and by 30% at 3000 m (Körner 1999a). Total solar radiation increases by 8-10% per 1000 m of elevation, and UV-radiation by around 10% per 1000 m (Blumthaler et al. 1997).

FUNCTIONAL TRAITS OF ALPINE PLANT SPECIES

Plants adapted to high mountain conditions show a multitude of strategies, reflecting the ecological conditions of habitats and micro-habitats. Here, we focus on the main adaptive features related to the aforementioned environmental variables in order to highlight peculiarities of the plants, which constitute alpine grasslands. The books of Larcher (2003) and Körner (1999b) provide a detailed review of alpine plants ecophysiology. Actually, Körner (1999b) considers that cold temperatures represent the first environmental filter that plants have to pass to become "alpine". Depending on their locations, plants living in the same range of high altitude do not experience cold to the same extent. Patterns of snow distribution and duration create micro-environments, where temperature regimes aboveground and belowground are different (Sohlberg & Bliss 1984; Bowman 1992; Fisk et al. 1998; Freppaz et al. 2012). The effects of snow on plant community composition and functioning will be discussed in the next section. During winter, plants living under snow are protected from lower temperatures, strong winds and desiccation because of the dampening effect of snow. This effect is relatively high for low density and important thickness of snow cover. Even though plants living under long lasting snow cover do not necessarily require adaptations to extreme cold, they developed many other adaptations, especially because long-lasting snow cover involves a short growing season. While some species keep their photosynthetic tissues functional even under snow (e.g. *Soldanella sp.*), the photosynthetic organs of most other plants die. However, just after snow melting, plants rapidly reallocate nutrients from roots to create new photosynthetic active tissues (Chapin et al., 1990; Onipchenko et al., 1998; Choler 2005b; Monson et al., 2006). In this habitat, many species complete their life cycle by combining rapid resource acquisition, fast growth and early flowering (Kudo et al., 1999; Lluent et al., 2013).

The protective effect of snowpack takes place where the snow can accumulate, such as in depressions, areas protected from winds and shadowed places, but alpine plant species have also developed strategies to cope with cold environments. By their growth form, plants can modulate the microclimate, for example, by keeping their stems or branches close to the ground. Prostrate and dwarfs plants limit exposure to cold winds and benefit from soil heat accumulation (Bliss 1971). The cushion form is another efficient way to protect the plant from desiccation, to heat the whole plant and to retain the heat (Molenda et al., 2012).

Compact rosettes are also very common in alpine grasslands, the basal leaves lying on the ground are protected while the other leaves receive solar radiation and heat the center of the rosette; temperature differences between the core of the rosette and the air can reach 25°C (Körner 1999b). Many sedges and grasses form large tussocks where the dry leaves provide protection from cold and desiccation by remaining attached at the periphery and inside the tussock (Monteiro et al., 2011).

FUNCTIONING AND DYNAMICS OF ALPINE GRASSLANDS

Spatial Heterogeneity and Topography

Contrary to most low elevation grasslands, alpine grasslands exhibit high spatial heterogeneity encompassing diverse types of plant communities. Indeed, variation in meso- and micro-topography, slope angle, slope exposures and shapes, soil types, nutrient availability, snow cover and duration, and disturbance regimes, generate high diversity of habitats and micro-habitats, and therefore of plant community types. Environmental heterogeneity, at different scales, is known to be related to species diversity and functional diversity (Snyder & Chesson 2004; de Bello et al., 2013), but see also Rose & Malanson (2012) who did not find any positive relationship between species diversity and topographic heterogeneity at fine scale (1 m²). Abiotic and biotic disturbances are among the main factors that generate environmental heterogeneity and may even be more important for structuring plant communities than resource competition (Pierce et al., 2007; le Roux et al., 2013). Related to alpine climate conditions, spatial variability of soil freeze-thaw cycles creates patches with different nutrient levels related to contrasted productivity (Freppaz et al., 2007; Kreyling et al., 2010) and gaps for plant colonization (Urbanska 1997b; Cichini et al., 2011).

Figure 1. Left to right – A- alpine grassland dominated by grass species (2100 m); B- an alpine grassland dominated by sedges, developped on a steep slope, serving as "reference ecosystem" for the ski runs – C- showing large areas of bare soils 15 years after restoration (2300m, parent material made of gypsum – la Plagne, French-Alps) – photo F. Isselin-Nondedeu.

In alpine environments, because of the prevalence of slopes (Figure 1), movements of materials play an important role in structuring plant communities. Therefore, in addition to biotic disturbances, such as grazing and trampling (by cows, gophers, marmots), abiotic disturbances, such as landslides and soil slides, soil pedoturbation, and avalanches, play a major role in structuring plant communities. Shapes of the ground, like depressions and mounds, from bottom to upslope, determine a meso-topographic gradient, which underpins a complex ecological gradient, sometimes at scales below 10 meters. For instance changes in nutrient availability (Bowman 1992; Fisk et al., 1998; Boyce et al., 2005), changes of snow cover duration and soil temperature (Billings & Bliss 1959; Körner 1999b) were all linked to small-scale changes of species composition along meso-topographic gradients (Sohlberg & Bliss 1984; Onipchenko et al., 1998: Choler 2005b). Within a single grassland, there may be many vegetation patches that function differently. For example, when comparing species that dominate the vegetation along a topographic gradient, from bottom to the top, Oksanen et al. (1992) highlighted a shifts from competitive (C, sensu Grime (1977a)) to stress-tolerant (S) species. Other studies completed this functional approach by studying leaf-traits. Since leaves are crucial for plant functioning, variations in leaf traits reflect species adaptations and their tolerance to the spatial variability of ecological factors. It has been shown that deciduous species with short leaf life-span are frequent in protected habitats, with long periods of snow cover and with generally high soil nutrient content (Kudo et al., 1999), whereas communities exposed to wind, often located on poorer soils, are dominated by species with long life-span and tough leaves (Kikuzawa & Kudo 1995; Kudo et al., 2001). Leaf traits related to photosynthetic activity, resource acquisition and conservation, such as leaf mass per area, leaf nitrogen concentration and specific leaf area (SLA) show more variation along meso-topographic gradients than along elevation gradients (Kudo et al., 1999; Choler 2005a). Mean values of specific leaf area of the dominant species can be multiplied by five when comparing vegetation growing up-slope and bottom-slope, only distant from 8-10 meters, with extreme values of 40 cm^2g^{-1} for *Arctostaphyllos uva-ursi* and 400 $cm^2.g^{-1}$ for *Viola biflora* (Isselin-Nondedeu & Bédécarrats 2004).

Through plant-soil feedback effects, traits of the standing plants can affect ecosystem processes (Lavorel & Garnier 2002; Mariotte et al., 2013). Leaf-traits and composition affect the soil microbial community and therefore litter decomposition, which in turn may affect plant community composition (Cornwell et al., 2008). Because most of the alpine species are slow-growing stress tolerant species, they have high leaf dry matter content (LDMC), high leaf C/N ratio and high tannin concentration, which slowdown nitrogen mineralization and decrease nutrient availability (Meier & Bowman 2008). Moreover, by analyzing the links between functional composition of microbial and plant communities in subalpine grasslands, a recent study showed that a shift from soil communities dominated by fungi to communities dominated by bacteria occurred with increasing dominance of plant species with high LDMC and C/N ratio (Grigulis et al., 2013). Besides competition for resource, root traits and morphology also play an important role in ecosystem processes. For example, establishment of species with high root tensile resistance increase slope stability, limit runoff and may increase colonization by other species (Pohl et al., 2011). The main links between habitat factors and plant community composition are synthesized in a conceptual model presented in Figure 2.

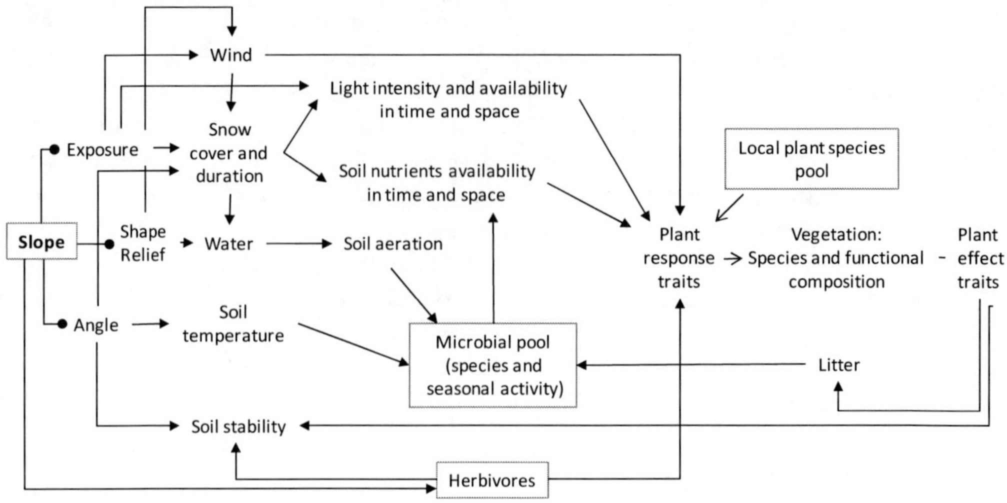

Figure 2. Conceptual model of vegetation control by principal habitat factors linked to slope and topography. The general exposure of slope, facing North, South, East or West, but also the angle influence exposition to solar radiation, which in turn influence soil temperature regime and snow duration along the year. Slope relief (convex-shape or concave-shape) mediates exposure to wind – strong winds blow out snow in the convex part, creating accumulation along the slope and in the concave part. Snow duration therefore differs depending on the relief, involving a variability of photosynthetic active radiations (PAR) in space and time. Plants growing where snow accumulates have less day of PAR along the year than other plants. Water runs downslope so as with soluble nutrients. Water and soil temperature select bacterial and fungal communities that decompose plant organic matter. Unsaturated soils provide aerobic conditions favoring soil N-mineralization. Species assemblage is realized locally with respect to environmental conditions and biotic interactions. Within the standing vegetation, some plant traits may in turn affect ecosystem processes (litter decomposition, soil stabilization by root systems). See section 1.3 for details.

Facilitation and Nurse Effects

By contrast to low elevation grasslands, the dynamics and organization of plant communities in alpine grasslands take place under stressful conditions. Community composition at a given time and at a given place is the result of a filtering process: the species of a regional and local pool being filtered in or out depending on their functional traits and tolerance with respect to abiotic and biotic filters (Keddy 1992; Diaz et al., 1998; de Bello et al., 2013). The results of such community assembly processes are visible at first sight when analyzing distribution patterns of species and plant traits along meso-topographic gradients (see the previous paragraph). However, the contribution of abiotic factors or biotic interactions in driving community assembly and dynamics is still debating. Nonetheless, for alpine ecosystems, numerous studies have found support for the stress-gradient hypothesis. According to this hypothesis, the frequency and importance of positive relationships increase with increasing levels of stress (Callaway et al., 2002).

The general mechanism involved in facilitative interactions is related to modification of abiotic factors: the facilitator improves the micro-environment around it, in which other plants, the beneficiaries, can grow. By improving temperature regime, soil moisture, nutrient content and availability, cushion plants are very effective facilitators creating microhabitats for others plants, soil fauna and pollinators (Arroyo et al., 2003; Molenda et al., 2012, Reid & Lortie 2012; Dvorsky et al., 2013). Facilitative effects are linked to species growth form, but since growth form changes with environmental conditions, the intensity of facilitation for the same species may vary in space (Schob et al., 2013). Most of the facilitative interactions seem also not stable in time, because when the facilitator facilitates "too much", the beneficiaries may become competitors (Olofsson 2004). Nonetheless, even though facilitators may be outcompeted, they play a crucial role in the successional dynamics of alpine grasslands. At high altitude, the regeneration phase is risky: sexual reproduction can be rare, in contrast with clonal propagation, and therefore the production of seeds (Bliss 1971; de Witte et al., 2012) and the seed dispersal distances may be limited (Urbanska et al., 1998; de Witte et al., 2012). In addition, seeds may not stay in bare areas or steep slopes because of the recurrent strong winds and rainfalls (Chambers & McMahon 1994; Chambers 1995). Moreover, seeds may not germinate because of unsuitable conditions, and when emerged, 60-95% of the seedlings generally die from summer frosts or desiccation (Stöcklin & Bäumler 1996; Erschbamer et al., 2001). Success of plant establishment for initiating alpine grasslands succession is strongly dependent on the availability of safe-sites, i.e. microsites that provide good conditions for germination and seedlings establishment (Urbanska 1997b; Elmarsdottir et al., 2003, Austrheim & Eriksson 2003; Frei et al., 2012). In these conditions, even micro-topographic soil features play an important role in the beginning of succession by entrapping seeds (Chambers et al., 1991; Chambers & McMahon 1994; Isselin-Nondedeu & Bédécarrats 2007) – see for example Figure 3. Similarly, standing plants can entrap propagules within their canopy (Bullock & Moy 2004; Dona & Galen 2007; Haussmann et al., 2010) or inside small mounds of soil deposited upslope of the plants (Isselin-Nondedeu & Bédécarrats 2007) as a prerequisite for further facilitative effects.

Figure 3. A hoof imprint in the soil entrapping several seeds with pappus. Such microtopography can act as safe-site and play an important role in regeneration of species at high altitude – photo F. Isselin-Nondedeu.

Alpine Grasslands under Changes

Land-uses and Disturbances

Mountains have always attracted human societies, and humans constantly modified these ecosystems through its land-uses. In a socio-ecological system perspective, the nature and intensity of the relationship have changed many times during the recent history. In Europe, the extent of alpine grasslands has been favored since the Holocene by burning trees, pushing down the treeline, and by hay cutting (Novak et al., 2010; Hejcman et al., 2013). Keeping high-altitude open spaces started with the seasonal transhumance (i.e. when shepherds lead their cattle at high altitude in the mountain during the whole summer), dating from 5500-4000 years BP (Oeggl et al., 2009). The progressive development of sedentary agriculture, until the 19th century, involved intensification in the valleys and the maintenance of pastoral activities in the upper zones, mainly for dairy products. Major changes of plant community structure and composition started to affect alpine grasslands during the industrial revolution. Then, several phenomena occurred. First, the mechanization of agriculture related to changes of the economic system that shifted from local to global scale with an increase of productivity (Stoate et al., 2009), concomitant with agriculture and farming intensification, led to the abandonment of unproductive areas. Moreover, migrations of population from farmlands to urban areas involved a decreasing number of farmers and consequently the abandonment of large areas of land (Tasser & Tappeiner 2002; Simmons 2005). Since 1960, the use of alpine grasslands by agriculture globally declined, but at the same time, the tourism industry considerably increased, especially with the development of the skiing industry. Continuous changes of land-use through time affected the structure and composition of the grasslands leading to shifts of ecosystem states and transformations of alpine ecosystems (Figure 4)

Ski areas can be seen as hot-spots of environmental deterioration because many types of disturbances and degradations are concentrated in these areas. Since ski resorts are entirely new cities or sprawlings of ancient hamlets, environmental issues due to urbanization are now seen at high altitude (e.g., pollution, management of water, waste treatment and management, roads and water runoff). Then, disturbances associated with skiing include construction and maintenance of ski runs, ski lifts and access roads (Bayfield 1996; Urbanska 1997a; Roux-Fouillet et al., 2011), which all damage vegetation and soil biota, enhance soil erosion and landscape fragmentation (Geneletti 2008). The development of ski areas involves the use of heavy machinery for the construction of ski runs, topographic adjustment and the compaction of the snow to facilitate recreational skiing. Such earth works alter nutrient turnover rates and their pathways, within and between the various biotic or abiotic ecosystem compartments. As a consequence of the altered nutrient cycle, vegetation composition may also change.

These changes in nutrient cycles do not only affect today's ecosystem functioning, but may also result in long-term legacy affecting ecosystem processes. Several experiments (Spiegelberger et al., 2006, 2010; Hejcman et al., 2007a,b; Klaudisova et al., 2009) have shown that small changes in key elements of the ecosystems, such as pH, may have a long-lasting effect, which is generally visible in the vegetation (Spiegelberger et al., 2006, 2010), but appears even more prominently in soil. Thus, from a general perspective, it can be concluded that small-scale human and natural perturbations cause long-lasting patterns in nutrient cycling and hence influence belowground and aboveground community composition in mountain grasslands (Spiegelberger et al., 2006). Long-lasting legacies of human activities might be particularly common in mountain ecosystems, which are characterized by low rates of nutrient cycling (DeAngelis et al., 1989). From the more specific view of sustainable ecosystem and soil management, the widely used practice of soil liming, related the development of ski runs, might have long-lasting effects on plant and soil microbial biodiversity as well as on the chemical composition of soils (soil pH, Ca and Al content), at least for several decades or even centuries.

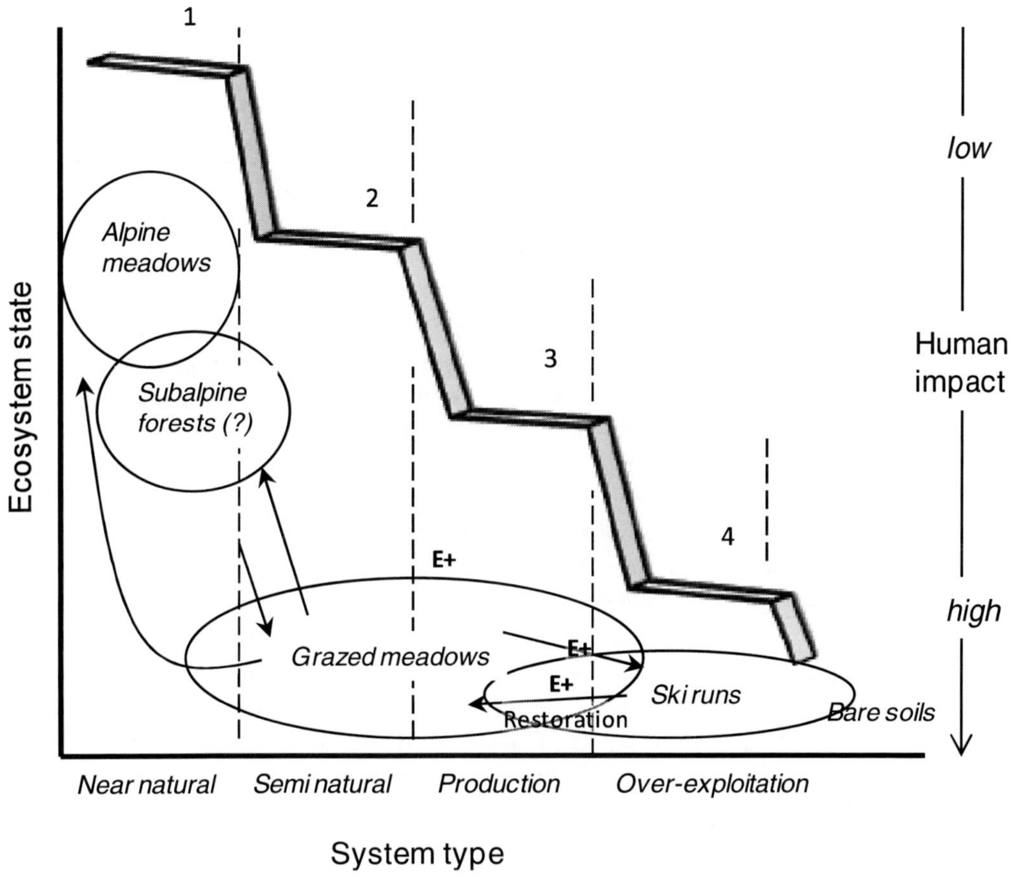

Figure 4. Shifts of ecosystem states and transformations of alpine ecosystems (adapted and modified from van Andel & Aronson (2005). E+ indicates where energy is required to maintain or transform the system. For instance, mostly metabolic energy (cattle) is required to maintain alpine meadow, otherwise the system changes of state (to an ungrazed meadows or maybe to a subalpine forest). Also, energy is needed to transform grazed meadows into ski runs, mostly by burning fossil fuels for heavy machinery. Attempts to restore the system, from ski runs to a reference ecological state, also require energy. In this case, the use of fossil energy should be reduced as far as possible. The arrow on the right indicated. A release of grazing pressure is observable from the center "grazed meadows" to the top left, whereas to the right the semi-natural system of grazed meadows is transformed to over-exploited systems by increasing human impact.

Study Case 1: Modifications of Community Composition under Climate Changes

Mountain grasslands are known to be highly sensitive to land-use changes (see above), but this vulnerability may even increase when climate changes occur at the same time (Huber et al., 2013). However, this is always difficult to estimate the relative impacts of climate change, which can subsequently be separated into increased CO_2 and temperature and altered precipitation (quantity and timing), from the perturbations associated to land-use change. In various experiments presented below, we focus either on the combined effects of precipitation and temperature changes or only in the shift in precipitation regime.

Methods

An extended description of the study site and methods used in these experiments can be found in Gavazov et al. (2013) and in Chapter 8 of this book for the combined precipitation-temperature experiment, and in Jung et al. (2014) for the effects of altered precipitation on community composition. Briefly, we transplanted mesocoms from the Swiss Jura Mountain crest (1350 m a.s.l.) at lower altitudes to assess the vulnerability of plant communities (i.e. diversity and composition) to the combined effects of temperature increase and reduction in precipitation. Plots were transplanted at three experimental sites: Combe des Amburnex (1350 m a.s.l.), Saint-George (1010 m a.s.l.), and Arboretum d'Aubonne (570 m a.s.l.). The site at the Jura Mountain crest (1350 m a.s.l.) served as a control site with a mean annual rainfall of ca.1750 mm, and a mean annual temperature of 4.5°C. The remaining two sites were chosen to represent a combination of annual temperature increase of +2 K and a precipitation decrease of -20% at 1010 m, and of +4 K and -40% at 570 m. At each site, fifteen mesocosms were transplanted to a common garden following a completely randomized design. In the second experiment, rainout shelters were set up at a similar altitude as at the Swiss Jura Mountain crest but in the Northern French Alps to intercept a precipitation quantity of 115 mm corresponding to 62% of the June-July average rainfall (1952-2009 period). The intercepted rain quantity corresponds to a 30-year return period of droughts at the site based on precipitation records from 1952-2009. Consequently, the average volumetric soil water content during the experiment went down to 6% ± 3% in the drought plots, against 17% ± 4% in the control plots.

Results and Discussion

After two years, the combined effects of temperature increase and reduction in precipitation (Gavazov et al., 2013) resulted in a significant reduction of plant diversity in the mesocoms ($F_{2,36} = 15.2$, $P < 0.001$), with species loss reaching 30% in some cases. The species groups mostly affected were graminoïds and legumes, but some forbs such as *Caltha palustris* or *Polygonum bistorta* were also no longer present in the lowest mesocosms, which underwent an increase in temperature by about 5°C and a reduction in precipitation by about 40%. The reduction in precipitation alone, simulated in the Northern French Alps (Jung et al., 2014), increased the differences in species relative abundance between drought and control plots (Bray-Curtis dissimilarity higher than the 95% percentile of a null distribution), but remained similar for species presence-absence (Bray-Curtis dissimilarity <95% percentile of a null distribution), with the exception of *Ranunculus montanus*, which totally disappear in drought plots.

Conclusion

The comparison of the two experiments showed that in both cases and rapidly after the perturbation, plant species composition of mountains ecosystems shifted following climate change related to increased temperature and reduced precipitation. However, changes in plant composition were less pronounced than expected as plant communities counter-balanced the perturbation induced by these extreme effects through intraspecific variability. Significant intraspecific trait responses to drought were observed for three out of four functional traits measured: leaf dry matter content, leaf nitrogen concentration and leaf carbon concentration while specific leaf area did remain similar. These findings indicate a key role of intraspecific variability to a short-term drought event, thus providing evidence for the importance of intraspecific variability in the functional response of plant communities to temporal environmental events such as droughts. Therefore intraspecific variability can be a more important driver of the short-term functional response than changes in the vegetation composition (Jung et al., 2014). These experiments give important findings about the resistance of plant community to extreme climatic event. The next step would be to extend these short-term experimental approaches by simulating, in long-term field experiments, the climatic conditions forecasted for the future in order to estimate changes in vegetation composition for the coming decades.

ECOLOGICAL RESTORATION ABOVE THE TIMBERLINE

Restoration Objectives

Setting restoration goals is of prior importance in order to assist the recovery of a degraded ecosystem. Soil stabilization is the first and critical step that will place the ecosystem into the good dynamic trajectory. Construction of ski runs and associated earth works remove or destabilize soils. These areas, used for skiing activities, are situated in steep slopes and soils are particularly prone to erosion. Under a set of environmental conditions, dynamic interactions between plants, soils, and the bedrocks maintain the soil at equilibrium on the slopes. The first target is then to stabilize the soil, and sometimes to recreate soil when it has been removed or destroyed. This first step is already a methodological and technical challenge (see next part) because the restored ecosystem should be similar to an "undisturbed" reference ecosystem in terms of structure, species composition and diversity, biological types and functional groups. Evidences are accumulating that increases of species diversity and functional diversity improve alpine grasslands functioning (Chapin & Körner 1996, Klanderud & Totland 2008; Grigulis et al., 2013), nutrient cycling (Rixen et al., 2008; Ashton et al., 2010; Onipchenko et al., 2012; Song et al., 2012,), soil processes and stability (Pohl et al., 2009; Pohl et al., 2012), ecosystem resilience (Spiegelberger et al., 2010; de Witte et al., 2012; Blanco-Pastor et al., 2013) and resistance to invasion (Quiroz et al., 2011;

Milbau et al., 2013). Variation of functional traits and genetic variability within populations is of great importance (Bilton et al., 2010; Manel et al., 2012) and should be taken into account. Including all these parameters in restoration projects was not of prime concern 40 years ago (Granjou et al., 2010). Revegetation due to earth movements was first launched in the 1970's in order to fight ground erosion; then was associated the idea of "turning the mountain green again" to heal the landscape and improve the scenic view. This was encouraged by a societal demand due to the development of summer tourism but also by dairy farmers for whom large areas of bare soils could represent a loss of earnings (Granjou et al., 2010). Therefore, in the early years, the expression "revegetation" was used instead of "restoration". The idea was basically to create a vegetation cover, as fast as possible, ignoring plant species diversity and composition and the importance of functional diversity and traits. In later years, projects evolved by using new approaches, diversifying seed mixtures and more recently by using local species to gain in biodiversity and sustainability.

Dealing with Characteristics of Alpine Environments and Alpine Species: Constraints for Restoration

The harshness of the alpine environment makes difficult the reconstitution of the ecosystem after some disturbances related to ski runs construction or resurfacing. In some situations, disturbances put back the ecosystem at the beginning of a primary succession (parent materials outcrop). All ecological factors linked to altitude act as strong filters, which constrain plant life. Many species of conventional seed mixtures do not tolerate such conditions and either not germinate, suffer during the seedling stage or survive only during one growing season. Combination of steep slopes with frequent intense rainfall events can increase runoff, soil loss and also the loss of seeds that disperse away from the bare site (Isselin-Nondedeu et al., 2006).

Adaptative features of plants to alpine conditions may limit the speed and success of the restoration, and limit the recolonization process. First, many alpine species show a limited dispersal in space, and only colonize sites that are very close to the parent plants and to the parent standing vegetation (Chambers & McMahon 1994; Urbanska & Fattorini 2000; Dirnbock et al., 2004). Germination is also a complicated phase since many alpine species exhibit different types of dormancy (Urbanska & Schütz 1986). As indicated previously, human intervention is thus required to assist the natural recovery, which then increase the resilience and vegetation cover of degraded sites. A number of techniques are available and can be divided in three main groups according to the objectives and the ecological state of the degraded sites: (1) rehabilitation of soil and soil protection, (2) seeding or planting, and (3) addition of nutrients.

Restoration Techniques: From Large Scale to Site-specific Restoration

The natural topography of the slopes is sometimes completely reorganized in degraded sites and the soil can be altered to such an extent that the re-establishment of vegetation, even under conventional restoration techniques, is very difficult. When it is possible, topsoil should be saved in order to be returned to the degraded zone. If not, trials of topsoil reconstitution have been made mainly by adding organic matter, compost mixtures on the remaining substrate, or by transferring soils from other sites (Barni et al., 2007). For small areas and steep slopes, soil stabilization can be done with several bio-engineering techniques using fiber-mats, biodegradable mats or geotextiles.

The most conventional revegetation methods are hand-sowing for small areas and hydro-seeding for larger areas. On steeper slopes and at high altitudes, seeds and topsoil must be protected with mulch (straw, geotextiles, bio-degradable nets). In the hydro-seeding method, a mixture containing seeds, fertilizer, and different adjuvant substances (fixating substances, fungicides) are sprayed over the soil surface. Straw-mulching is a simple and not expensive technique, which is very efficient in cold environment. The straw layer creates a micro-climate favorable to seed germination and seedlings by enhancing moisture at the air-soil interface and by protecting from frost heaving (Dinger 1983; Price et al., 1998).

While all these methods currently use commercial seed mixtures, made of species and cultivars, which are not part of the local species pool, alternative methods were tested to restore local species. The issues behind the development of these methods are the following: some restoration failures are due to the irrelevant composition of the seed mixtures, that could be corrected by seeding plant species functionally adapted to the local environment or by planting local species (Chambers 1997); and cross-breeding between cultivars and native species may lead to genetic pollution of local genetic pools (Malaval et al., 2010; Schroder & Prasse 2013). Turf transplant of local species from a donor site was successfully applied in small areas, with each transplanted turf functioning as an island which propagate seeds and asexual fragments (Fattorini 2001a; Bay & Ebersole 2006) – see Aradottir (2012) for further details on turf transplant methods. Hay transfer and hay-mulch seeding are other methods used to introduce local species in the degraded sites. The donor grassland is mowed when species are mature and plant materials (stems, straw, fruits, seeds…) are dried and spread over the site (Krautzer & Graiss 2009). The donor grassland should therefore correspond as close as possible to the degraded site.

Figure 5. Seeding of local species on experimental plots at the La Plagne ski resort – photo S. Labonne.

Another option consist in harvesting seeds of local species, which are then seeded in the restoration sites before snow fall in autumn or after snow melt in the spring (Figure 5). Such approach is currently under investigation (2013) in a subalpine pasture (2100m a.s.l.) of the La Plagne ski resort (French Alps), where seeds have been harvested with 4 different machines: a hand-hold brush strippers, a vehicle-tracked brush stripper, a hand-held vacuum collector and lawn-mower like vacuum collector. Seeds of the collected species were then seeded in October 2013 at three study sites situated between 2100 and 2200 m according to five treatments: locally collected seed with a density of 15 grams/m², locally collected seed with a density of 20 g/m², seeds collected in a local hay barn, seeds from commercial seeds suppliers and natural recolonisation.

Study Case 2: Alpine Grasslands Composition After Restoration

Considering that the aim of restoration is to assist ecosystem to recover from a perturbation in order to approach the state of a semi-natural or natural reference ecosystem, it generally takes at least 40 years to restore alpine grasslands (Chambers 1997; Rydgren et al., 2011). Combinations of harsh climatic conditions, deeply altered soils, seed limitation, functional characteristics of alpine plants, and sometimes the use of inappropriate seed mixtures or restoration techniques explain the slow speed of restoration. Theoretically, construction of ski runs and slope resurfacing initiate a secondary succession, but without intervention, soil erosion processes often occur and impede plant colonization. Therefore, seed mixtures are sowed to rapidly stabilize the soil and to obtain a plant cover allowing recolonization and establishment of local species. It is then expected that the abundance of seeded species declines after restoration while the community composition becomes similar to those observed in reference communities in the same habitat conditions (slope, mesotopography, exposure, altitude…). However, the number of successful restorations declines with increasing altitude (Chambers 1997). Soil nutrient content, as well as texture, structure and chemical composition play an important role in the success of restoration especially where soils have been deeply disturbed and the parent materials outcrop (Ozenda 1985; Michalet et al., 2002). Even though ecological dynamics are very slow above the treeline, long-term monitoring studies are very rare and most of all are based on diachronic monitoring (Bayfield 1996; Fattorini 2001b). To our knowledge, there is no study comparing vegetation composition on ski runs, from one to several decades after the restoration with different parent materials but in the same bio-climatic zone. The objective of this study was to analyze the species composition of restored ski runs by comparing a set of sites, restored from one to forty-five years ago, on three different parent materials.

Methods

The study was conducted in the ski resort of la Plagne in the northern French Alps (45°33' N, 06°40' E). The ski area covers about 3500 ha, ranging from 1250 m to 3250 m. Most of the grasslands used for ski activities are located in the subalpine zone. Within the ski area, three main bedrocks are found and constitute three main zones with different soil properties and vegetation: a zone dominated by gypsum and associated evaporitic stones (carbonated breccias), a zone with quartzite, and a third one with green shales (shales, ophiolite, serpentinite). Soils formed on gypsum range from lithosols on crests to rendzinas.

Well-developed cambisols on concave and flat areas, have high pH (6.5-8) and high concentrations of Ca^{2+} and K^{+}. Soils derived from green shales, ophiolite, serpentinite complex are mostly rankers on crests, podsols and cambisols in other reliefs, with a more acid pH (5-6.5). Soils on quartzite evolved towards podsols or acid and shallow soils (pH 4.5-5.5) with high organic matter content (Gensac 1990; Isselin-Nondedeu & Bedecarrats 2013). Mean annual precipitation and temperature at the ski resort are 1165 mm and 2.7°C, respectively (MeteoFrance). The first important construction of ski runs started in 1961, followed by the first works for soil stabilization. Sixty-one plots (200 m²) on north-facing ski runs were surveyed: 24 plots in the gypsum zone, 17 in the quartzite zone and 20 in the green-shales zone. We selected ski runs for which we had enough information about management and restoration. Indeed, information is scarce, especially for the oldest ski runs, and most often, one ski run can be divided in several parts that differ in terms of disturbances, restoration techniques, and time since the restoration. Information were collected from the ski resort databases (reports, GIS of the ski area), from technical reports (Dinger 1983; Perrussel 1986; Bédécarrats 1998) and from previous survey (Isselin-Nondedeu 2004). Restoration was done mainly by hand sowing and hydro-seeding with the following standard seed mixture: Ray-grass (*Lolium perene*) 10-20%, Red fescues (*Festuca rubra*) 20-50%, Timothy grass (*Phleum pratense*) 10-20%, Fabaceae (mainly *Trifolium repens*, *T. pratense*, *Lotus corniculatus*, *Anthyllis vulneraria*) 10-30%, *Achillea millefolium* and/or *Dactylis glomerata* and/or *Festuca ovina* 0-10%. In order to perform diachronic analysis, ski runs were grouped into 6 classes of age: (1) 1 to 4 years after the restoration (n = 15), (2) 5 to 8 years (n = 16), (3) 9 to 15 years (n = 12), (4) 20 to 25 years (n = 3), (5) 28 to 30 years (n = 6), and (6) 35 to 45 years (n = 9). We also selected natural and undisturbed sites as "references" of alpine grasslands (see an example in Figure 1), which were comparable in altitude, slope angle, and parent materials: 12 plots on gypsum, 6 on quartzite, and 9 on green-shales. All reference sites are extensively grazed during summer by dairy cows. In each ski run, vegetation composition was sampled on homogenous plots, with flat regular slopes of 15-20°, located in the center of the ski slope. The line-point intercept method was used by setting three lines of 10 meters perpendicular to the slope, separated from each other by 10 meters. This method is well adapted for sites with low vegetation cover and with patches of bare soil, and allows to rapidly capture the variability of the site. A metal rod (1 mm diameter) was dropped every 10 cm along in each line and all plant species in contact with the rod were identified and recorded. The relative frequency of the species *i* was calculated by dividing the number of hit of species *i* by the numbers of hits of all plant species. Percent of vegetation cover was visually estimated at the plot scale (200 m²). We applied the same method for the reference sites. Most of the data were collected during summer 2005, except for 4 ski runs of the first age class, 1 of the fifth age class, and 5 of the sixth age class where the data were collected in august 2013. Canonical correspondence analysis (CCA) was performed with species frequencies of all plots, and with altitude, parent material, and age class as plot variables. To compare plant functional types between the sites, species were grouped according the Grime's CSR strategies (Grime 1977b; Pierce et al., 2007).

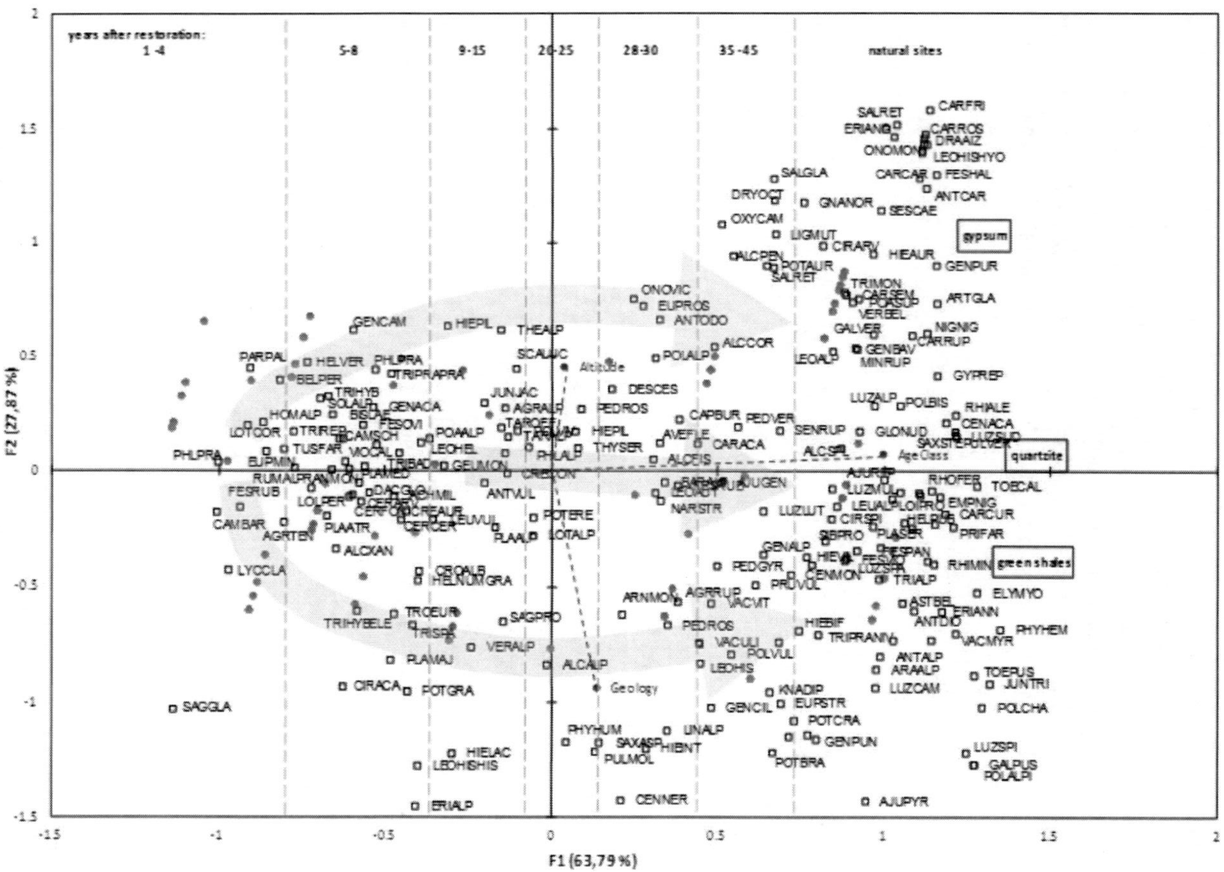

Figure 6. The CCA ordination bi-plot of the first two axes showing the distribution of the different species (some species abbreviations have been removed for more visibility see table in appendix 1 for species names) and plot variables (altitude, geology represented by the main parent materials, age classes) of the ski runs. Age classes are detailed on the graph, showing the range of years after the restoration of the ski runs. The parent materials are listed on the right where the natural reference sites appear.

Results and Discussion

Results of the CCA analysis (Figure 6) showed that most of the inertia in the species data was explained by the age class variable (axis 1; more than 63%) and the geology (axis 2; more than 27%). The results identified distinct groups: the first group is composed mainly by seeded and pioneer species, which can be found on the ski runs from 1 to 8 years after they have been restored and in every parent material. The seeded species *Phleum pratense (cultivars), Festuca rubra, Lotus corniculatus, Lolium perenne, Trifolium repens* were among the most frequent species, with *Cerastium sp., Campanula scheuchzeri, Ranunculus montanum, Tussilago farfara, Geum montanum* as local species (species names are given in the Appendix 1).

For every restored plots one to 5 years after the restoration, the vegetation cover was between 65-80% (Figure 7), which was satisfactory to limit soil surface runoff (Tasser et al., 2003). Thereafter, the vegetation cover is generally lower than the plots of the first age class, traducing the loss of the majority of the seeded species. These young restored grasslands are dominated by plants of the CSR – CR – R groups (sensu Grime 1977a; Figure 8). Others studies found that high soil nutrient content and soil disturbances associated with all activities related to skiing selected the most competing, fast-growing, even ruderals species (Guleryuz et al., 2008; Barrantes et al., 2013). If the seeded species decreased over time, some persisted after the restoration or are the dominant species (in term of biomass or cover) of some of the ski runs. This was the case for *Phleum pratense* (cultivars climax), *Festuca rubra*, and *Deschampsia caespitosa* at lower altitude. Even though the seeded species failed to reproduce by seeds in some years, they may be very good competitors, resprout over time, spread laterally, and thus present many traits allowing for their persistence in sites managed by human (Prach & Pysek 1999, Schroder & Prasse 2013). This issue has already been pointed out by others studies in mountain environments (Bayfield 1996; Isselin-Nondedeu & Bedecarrats 2009; Hagen & Evju 2013).

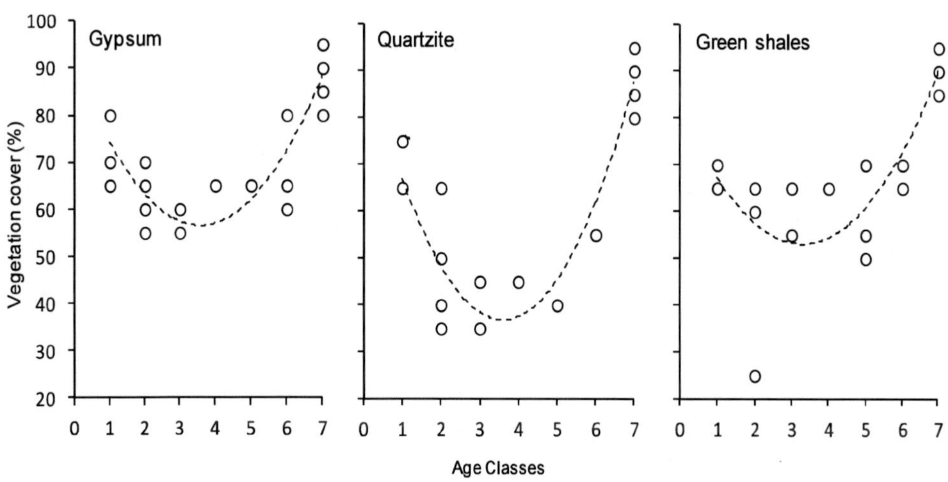

Figure 7. Vegetation covers at the plot scale on the different ski runs and in the three main parent materials of the ski area. Numbers on the X-axis are the age class of each ski run: 1- 1 to 4 years after the restoration, 2- 5 to 8 years, 3- 9 to 15 years, 4- 20 to 25 years, 5- 28 to 30 years, 6- 35 to 45 years, 7 being the natural sites.

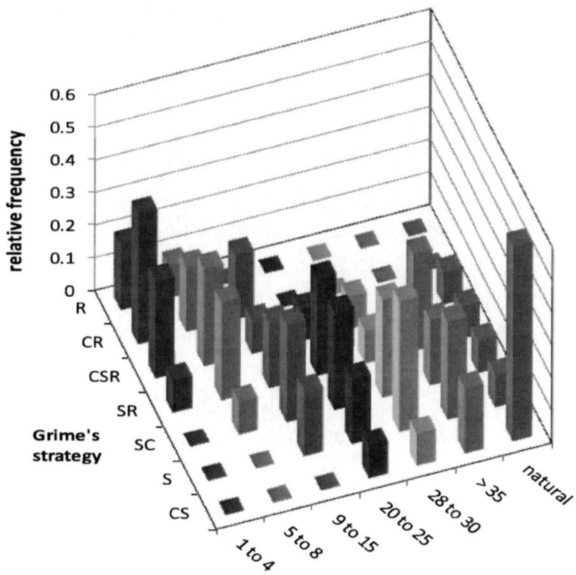

Figure 8. Distribution of CSR strategy types of the plant species found in the ski runs according to their age class after restoration.

A shift toward more stress tolerant species (CS, S) is visible starting from ski runs of 5 to 8 years after the restoration, which is consistent to the functional composition of the alpine grasslands (Oksanen & Ranta 1992). According to their species composition 10 years after the restoration, grasslands tend to be separated in three groups reflecting progressive influences of the parent materials. Chemical composition, pH and porosity of parent materials and soils are environmental filters that shape community composition by acting on seedlings (Wenk & Dawson 2007; Isselin-Nondedeu & Bedecarrats 2013), and on adult plants (Michalet et al., 2002; Arnesen et al., 2007).

The reference alpine grasslands were located on the right part of the ordination graphic, with those associated with gypsum clearly separate from the two others (Figure 1). Most of the species on the top-right corner of the graphic were either not found on the restored ski runs, or only at very low frequency (*Nigritella nigra, Antennaria carpatica, the different species of Carex, Sesleria caerulea, Salix retusa*). Similarly, many species of the reference communities on green shales, and particularly on quartzite, were not found in the restored ski runs. This finding is consistent with the experiment of Grabherr (2003), who found that communities on siliceous substrates took more than 25 years to recover partially. The oldest ski runs on quartzite (more than 40 years after restoration) have not the physiognomy of the vegetation of natural sites. They should have evolved to heaths, or at least to mixed vegetation with patches of sedges and grasses and patches dominated by subshrubs and prostrate shrubs (*Rhododendron ferrugineum, Loiseuleria procumbens, Empetrum nigrum*). Vegetation cover (Figure 7) is lowest in each restored ski run on quartzite comparing to the other types of soil. This might be explained by the low reconstitution of soil on acid parent material, such as quartzite, which impedes plant development. Moreover, subshrubs species, which develop on quartzite, are slow-growing plants and the disturbances involved by ski activities, like snow compaction, snow-grooming machines crossing, and strokes by skis during the end of the winter season, impede their growth. Thermal conductivity increases with an increase of snow compaction, leading to higher temperature fluctuation and therefore higher risk for plants to suffer from cold. Snow compaction occurs naturally when snow accumulates during winter but can be artificially emphasized by the repeated actions of snow-grooming machines, especially on ski runs. As we observed in our sites, this tends to select non-ligneous species and fast-growing species like snow-bed specialists that can tolerate long period of snow cover. The high frequency of pioneer species, C and CR species as for example *Poa alpina, Silene rupestris, Rumex acetosella, Polygonum viviparum, Cardamine resedifolia, Achillea millefolium, Taraxacum officinale, Polygonum bistorta, and Dactylis glomerata*, 10 to 12 years after the restoration was attributed to the continuous soil alterations and vegetation disturbances related to skiing activities (Barni et al., 2007) - see also similar results for other ski areas in the Alps (Rixen et al., 2003; Roux-Fouillet et al., 2011).

Conclusion

The results of this study show that even 40 years after the restoration, grasslands on the ski runs are not similar to the natural grasslands in term of plant composition. The vegetation composition and structure after the restoration works could have been closer to the natural sites in absence of all the recurrent disturbances due to the skiing activities. The results also pointed out that the soil, linked to the parent material strongly influences the dynamic trajectory of the restored grasslands, and that quartzite may hinder the restoration process.

PERSPECTIVES

Past human activities have already reduced the capacity of ecosystems to cope with environmental changes (Scheffer et al., 2001). The resilience of ecosystems may be an essential factor underlying the sustainable production of natural resources and ecosystem services faced with uncertainty (Gunderson 2000). Ecological resilience seems to depend on the taxonomical and/or functional diversity of the ecosystems, its material or nutrient fluxes and, on the connectivity within the system.

Biological diversity appears to play a substantial role in ecosystem resilience and in sustaining desirable ecosystem states in face of changes (Peterson et al., 1998). Ecological redundancy, meaning that multiple species or functional types (Leps et al., 1982; MacGillivray et al., 1995) contribute to provide similar functions to the ecological system and therefore guarantee the provision of ecosystems services even in the face of changes (Naeem 1998), seems therefore important to be maintained. A high diversity of functional groups in a dynamic ecosystem undergoing change, together with a high species diversity within these groups, seems important in helping to maintain ecosystem services (Luck et al., 2003) in general and more particularly in grasslands (Tilman & Downing 1994). Such a variability in the responses of individuals has been demonstrated in a drought experiment on the High Plateaux of the Vercors, where intra-specific variability over-proportionally contributes to the high resistance to environmental change and the resilience of those subalpine pastures to extreme drought event (Jung et al., 2014).

Besides ecological diversity, theoretical and experimental work suggest that resilience is further affected by fluxes of material, such as nutrient cycling (DeAngelis et al., 1989; Moore et al., 1993). The 'resilience – productivity hypothesis' posits that the time required by an ecosystem to regain equilibrium after a disturbance is similar to the turnover time of nutrients in the system (DeAngelis et al., 1989; Moore et al., 1993). This suggests that resilience may be particularly low in ecosystems with harsh environmental conditions, like alpine and arctic ecosystems. Consequently, resilience to perturbations and in particular to changes in key ecosystem properties (pH or phosphorus) of grassland ecosystems under harsh environmental conditions may be particularly low (Spiegelberger et al., 2006) and may even be irreversible (Semelová et al., 2008). While intensive forms of land-use are essential for humans, they have contributed to the transformation of ecosystem patterns and processes (Foley et al., 2005). In particular, human activities have altered the rate, pathways and efficiency of the movement of nutrients within and between the various biotic or abiotic ecosystem compartments. These changes in nutrient cycles do not only affect today's ecosystem functioning but may also result in long-term legacy effects on ecosystem processes, thereby changing the resilience of ecosystems and their adaptive capacity to sustain ecosystem services in the face of uncertainty and global change (Folke et al., 2002; Elmqvist et al., 2003).

Moreover, in this context, we can wonder whether global change can bring an opportunity to improve restoration success. Studies in Europe (Mosimann 1983 in (Peratoner 2006)) as well as our observations showed that alpine grassland restoration was satisfactory up to 1600-1800 m. Between 1600 and 2200 m the number of successful revegetated ski runs decreased rapidly, and above 2200 m many restoration attempts failed, letting large areas of bare soil. The climatic constraints become progressively unfavourable for germination, growth and establishment of the seeded plants. However, with an increase of temperatures in the mountain landscape due to global warming (Parolo & Rossi 2008) (see Study case 1), it could be possible that the growth and survival of species used in the seed mixtures increases. Further researches should be conducted (1) to improve restoration of alpine grasslands in the context of climate change, for example, by studying interactions between local and seeded species and whether they are or not altered by the modifications of climate (2) to develop the seed production of local species. This is also important not to be focused only on the sites to be restored, but to have a broad vision of what happens in the landscape. Decline of plant populations or degradation of other communities in the landscape, reduces seed availability and diversity and landscape fragmentation decrease connectivity and therefore the dispersal of seeds between non-degraded and degraded grasslands. Sustainable management of alpine grasslands should therefore take into account for all perturbations associated with on-going land-use and climate change and remains a hard challenge for the future.

ACKNOWLEDGMENT

A part of the case studies was carried out on the long term research site Zone Atelier Alpes (LTSER platform Central French Alps), a member LTER France. TS acknowledges support through the Interreg ALCOTRA project n° 219 «Alp'Grain». FIN & TS thank the managers of La Plagne ski resort.

APPENDIX 1

Table of abbreviations and species names found in the vegetation surveys

ACHMIL	*Achillea millefolium (seeded)*	GALPUS	*Galium pusillum*	PHYORB	*Phyteuma orbiculare*
AGRALP	*Agrostis alpina*	GALVER	*Galium verum*	PLAALP	*Plantago alpina*
AGRRUP	*Agrostis rupestris*	GENACA	*Gentiana acaule*	PLAATR	*Plantago atrata*
AGRTEN	*Agrostis tenuis*	GENALP	*Gentiana alpina*	PLAMAJ	*Plantago major*
AJUGEN	*Ajuga geneviensis*	GENBAV	*Gentiana bavarica*	PLAMED	*Plantago media*
AJUPYR	*Ajuga pyramidalis*	GENCAM	*Gentiana campestris*	PLASER	*Plantago serpentina*
AJUREP	*Ajuga reptans*	GENCIL	*Gentiana ciliata*	POAALP	*Poa alpina*
ALCALP	*Alchemilla alpina*	GENPUN	*Gentiana punctata*	POASUP	*Poa supina*
ALCCOR	*Alchemilla coriacea*	GENPUR	*Gentiana purpurea*	POLALP	*Polygala alpestris*
ALCFIS	*Alchemilla fissa*	GENVER	*Gentiana verna*	POLALP	*Polygala alpina*
ALCPEN	*Alchemilla pentaphyllea*	GEUMON	*Geum montanum*	POLCHA	*Polygala chamaebuxus*
ALCSPL	*Alchemilla splendens*	GLOCOR	*Globularia cordifolia*	POLVUL	*Polygala vulgaris*
ALCXAN	*Alchemilla xanthochlora*	GLONUD	*Globularia nudicaulis*	POLBIS	*Polygonum bistorta*
ALNVIR	*Alnus viridis*	GNANOR	*Gnaphalium norvegicum*	POLVIV	*Polygonum viviparum*
ANTCAR	*Antennaria carpatica*	GYPREP	*Gypsophila repens*	POTAUR	*Potentilla aurea*
ANTDIO	*Antennaria dioica*	HELNUMGRA	*Helianthemum nummularium ssp grandiflorum*	POTBRA	*Potentilla brauneana*
ANTALP	*Anthoxanthum alpinum*	HELPUB	*Helicotrichon pubescens*	POTCRA	*Potentilla crantzi*
ANTODO	*Anthoxanthum odoratum*	HELVER	*Helicotrichon versicolor*	POTERE	*Potentilla erecta*
ANTVUL	*Anthyllis vulneraria*	HIEAUR	*Hieracium aurantiacum*	POTGRA	*Potentilla grandiflora*
ARAALP	*Arabis alpina*	HIEBIF	*Hieracium bifidum*	POTNOR	*Potentilla norvegica*
ARNMON	*Arnica montana*	HIEINT	*Hieraciumintybaceum*	PRIFAR	*Primula farinosa*
ARTGLA	*Artemisia glacialis*	HIELAC	*Hieracium lactucella*	PRIGLU	*Primula glutinosa*
ASTBEL	*Aster bellidiastrum*	HIEPIL	*Hieracium pilosella*	PRUVUL	*Prunella vulgaris*
AVEFLE	*Avenella flexuosa*	HIEPIL	*Hieracium pilosum*	PULMOL	*Pulmonaria mollis*
BARALP	*Bartsia alpina*	HIEVIL	*Hieracium villosum*	PULALP	*Pulsatilla alpina*
BELPER	*Bellis perenis*	HIPCOM	*Hippocrepis comosa*	PULVER	*Pulsatilla vernalis*
BISLAE	*Biscutella laevigata*	HOMALP	*Homogyne alpina*	RANMON	*Ranunculus montanus*
BRIMED	*Briza media*	HUPSEL	*Huperzia selago*	RHIALE	*Rhinanthus alectorolophus*
BUPRAN	*Bupleurum ranunculoides*	JUNJAC	*Juncus jacquinii*	RHIMIN	*Rhinanthus minor*
CAMBAR	*Campanula barbata*	JUNTRI	*Juncus trifidus*	RHOFER	*Rhododendron ferrugineum*
CAMROT	*Campanula rotundifolia*	KNADIP	*Knautia dipsacifolia*	RUMALP	*Rumex alpina*
CAMSCH	*Campanula scheuchzeri*	LEOAUT	*Leontodon autumnalis*	SAGGLA	*Sagina glabra*
CAPBUR	*Capsella bursa-pastoris*	LEOHEL	*Leontodon helveticus*	SAGPRO	*Sagina procumbens*
CARCUR	*Carex curvula*	LEOHIS	*Leontodon hispidus*	SAGSAG	*Sagina saginoides*
CARFRI	*Carex frigida*	LEOHISHIS	*Leontodon hispidus ssp hispidus*	SALGLA	*Salix glabra*
CARROS	*Carex rosae*	LEOHISHYO	*Leontodon hispidus ssp hyoseroides*	SALRET	*Salix reticulata*
CARRUP	*Carex rupestris*	LEOALP	*Leontopodium alpinum*	SALRET	*Salix retusa*
CARSEM	*Carex sempervirens*	LEUALP	*Leucanthemopsis alpina*	SAPPUM	*Saponaria pumila*
CARACA	*Carlina acaulis*	LEUVUL	*Leucanthemum vulgare*	SAXASP	*Saxifraga aspera*
CARCAR	*Carum carvi*	LIGMUT	*Ligusticum mutellina*	SAXSTE	*Saxifraga stellaris*
CENACA	*Centaurea acaule*	LIGMUT	*Ligusticum mutellinoides*	SCALUC	*Scabiosa lucida*
CENMON	*Centaurea montana*	LINALP	*Linaria alpina*	SEMMON	*Sempervirum montanum*
CENNER	*Centaurea nervosa*	LOIPRO	*Loiseleuria procumbens*	SENRUP	*Senecio rupestris*
CERARV	*Cerastium arvense*	LOLPER	*Lolium perene (seeded)*	SESCAE	*Sesleria caerulea*
CERCER	*Cerastium cerastiodes*	LOTALP	*Lotus alpinus*	SIBPRO	*Sibbaldia procumbens*
CERFON	*Cerastium fontanum*	LOTCOR	*Lotus corniculatus*	SOLALP	*Soldanella alpina*
CIRACA	*Cirsium acaule*	LUZALP	*Luzula alpinopilosa*	SOLPUS	*Soldanella pusilla*
CIRARV	*Cirsium arvense*	LUZCAM	*Luzula campestris*	TARALP	*Taraxacum alpinum*
CIRSPI	*Cirsium spinossissinum*	LUZLUT	*Luzula lutea*	TAROFF	*Taraxacum officinale*
CREAUR	*Crepis aurea*	LUZMUL	*Luzula multiflora*	THEALP	*Thesium alpinum*
CRECON	*Crepis conyzifolia*	LUZSPA	*Luzula spadicea*	THYSER	*Thymus serpyllum*
CREVESTAR	*Crepis vesicaria ssp taraxacifolia*	LUZSPI	*Luzula spicata*	TOECAL	*Toefeldia calyculata*
CROALB	*Crocus albiflorus*	LUZSUD	*Luzula sudetica*	TOEPUS	*Toefieldia pusilla*
DACGLO	*Dactylis glomerata (seeded)*	LYCCLA	*Lycopodium clavatum*	TRIALP	*Trifolium alpinum*
DESCES	*Deschampsia cespitosa*	MEUATH	*Meum athamanticum*	TRIBAD	*Trifolium badium*
DRAAIZ	*Draba aizoides*	MINRUP	*Minuartia rupestris*	TRIHYB	*Trifolium hybridum*
DRYOCT	*Dryas octopetala*	MYOALP	*Myosotis alpestris*	TRIHYBELE	*Trifolium hybridum ssp elegans*
ELYMYO	*Elyna myosuroides*	NARSTR	*Nardus stricta*	TRIMON	*Trifolium montanum*
EMPNIG	*Empetrum nigrum*	NIGNIG	*Nigritella nigra*	TRIPRANIV	*Trifolium pratense ssp nivale*
ERIALP	*Erigeron alpinus*	ONOMON	*Onobrychis montana*	TRIPRAPRA	*Trifolium pratense ssp pratense*
ERIANN	*Erigeron annuus*	ONOVIC	*Onobrychis viciifolia*	TRIREP	*Trifolium repens (seeded)*
ERIANG	*Eriophorum angustifolium*	OXYCAM	*Oxytropiscampestris*	TRISPA	*Trifolium spadiceum*
EUPCYP	*Euphorbia cyparissias*	PARPAL	*Parnassia palustris*	TRIFLA	*Trisetum flavescens*

Appendix 1. (Continued)

EUPMIN	*Euphrasia minima*	PEDGYR	*Pedicularis gyroflexa*	TROEUR	*Trollius europaeus*
EUPROS	*Euphrasia rostkoviana*	PEDROS	*Pedicularis rostrato-capitata*	TUSFAR	*Tussilago farfara*
EUPSTR	*Euphrasia stricta*	PEDROS	*Pedicularis rostrato-spicata*	VACMYR	*Vaccinium myrtillus*
FESHAL	*Festuca agg. halleri*	PEDVER	*Pedicularis verticillata*	VACULI	*Vaccinium uliginosum*
FESOVI	*Festuca ovina*	PHLALP	*Phleum alpinum*	VACVIT	*Vaccinium vitis-idae*
FESPAN	*Festuca paniculata*	PHLPRA	*Phleum pratense*	VERALP	*Veronica alpina*
FESRUB	*Festuca rubra*	PHLPRA	*Phleum pratense (seeded)*	VERBEL	*Veronica bellidioides*
FESRUB	*Festuca rubra (seeded)*	PHYHEM	*Phyteuma hemisphaericum*	VERSER	*Veronica serpyllifolia*
FESVIO	*Festuca violacea*	PHYHUM	*Phyteuma humile*	VIOCAL	*Viola calcarata*

REFERENCES

Aradottir, A. L. (2012) Turf transplants for restoration of alpine vegetation: does size matter? *Journal of Applied Ecology,* 49, 439-446.

Arnesen, G., Beck, P. S. A. & Engelskjon, T. (2007) Soil acidity, content of carbonates, and available phosphorus are the soil factors best correlated with alpine vegetation: Evidence from troms, North Norway. *Arctic Antarctic and Alpine Research,* 39, 189-199.

Arroyo, M. T. K., Cavieres, L. A., Penaloza, A. & Arroyo-Kalin, M. A. (2003) Positive associations between the cushion plant Azorella monantha (Apiaceae) and alpine plant species in the Chilean Patagonian Andes. *Plant Ecology,* 169, 121-129.

Ashton, I. W., Miller, A. E., Bowman, W. D. & Suding, K. N. (2010) Niche complementarity due to plasticity in resource use: plant partitioning of chemical N forms. *Ecology,* 91, 3252-3260.

Austrheim, G. & Eriksson, O. (2003) Recruitment and life-history traits of sparse plant species in subalpine grasslands. *Canadian Journal of Botany,* 81, 171-182.

Barni, E., Freppaz, M. & Siniscalco, C. (2007) Interactions between vegetation, roots, and soil stability in restored high-altitude ski runs in the Alps. *Arctic Antarctic and Alpine Research,* 39, 25-33.

Barrantes, O., Reiné, R. & Ferrer, C. (2013) Changes in Land Use of Pyrenean Mountain Pastures — Ski Runs and Livestock Management — Between 1972 and 2005 and the Effects on Subalpine Grasslands. *Arctic, Antarctic, and Alpine Research,* 45, 318-329.

Bay, R. F. & Ebersole, J. J. (2006) Success of turf transplants in restoring alpine trails, Colorado, USA. *Arctic Antarctic and Alpine Research,* 38, 173-178.

Bayfield, N. G. (1996) Long-term changes in colonization of bulldozed ski pistes at Cairn Gorm, Scotland. *Journal of Applied Ecology,* 33, 1359-1365.

Bédécarrats, A. (1998) Etude de la dynamique de reconquête par les espèces natives des groupements supra-forestiers des espaces dégradés en altitude et végétalisés artificiellement, le cas de la Plagne (Savoie) / Study of recolonization by native species of degraded high-altitude zones, la Plagne (Savoie, France). *Recréer la Nature*, pp. 41. CNRS, CEMAGREF.

Billings, W. D. & Bliss, L. C. (1959) An alpine snowbank environment and its effect on vegetation plant development and productivity. *Ecology,* 40, 388-397.

Bilton, M. C., Whitlock, R., Grime, J. P., Marion, G. & Pakeman, R. J. (2010) Intraspecific trait variation in grassland plant species reveals fine-scale strategy trade-offs and size differentiation that underpins performance in ecological communities. *Botany-Botanique,* 88, 939-952.

Blanco-Pastor, J. L., Fernandez-Mazuecos, M. & Vargas, P. (2013) Past and future demographic dynamics of alpine species: limited genetic consequences despite dramatic range contraction in a plant from the Spanish Sierra Nevada. *Molecular Ecology,* 22, 4177-4195.

Bliss, L. C. (1971) Arctic and alpine plant life cycles. *Annual Review of Ecology and Systematics.,* 2, 405-438.

Blumthaler, M., Ambach, W. & Ellinger, R. (1997) Increase in solar UV radiation with altitude. *Journal of Photochemistry and Photobiology B-Biology,* 39, 130-134.

Bowman, W. D. (1992) Inputs and storage of nitrogen in winter snowpack in an alpine ecosystem. *Arctic and Alpine Research,* 24, 211-215.

Boyce, R. L., Clark, R. & Dawson, C. (2005) Factors determining alpine species distribution on Goliath Peak, Front Range, Colorado, U.S.A. *Arctic, Antarctic & Alpine Research,* 37, 88-96.

Bullock, J. M. & Moy, I. L. (2004) Plants as seed traps: inter-specific interference with dispersal. *Acta Oecologia,* 25, 35-41.

Callaway, R. M., Brooker, R. W., Choler, P., Kikvidze, Z., Lortie, C. J., Michalet, R., et al., (2002) Positive interactions among alpine plants increase with stress. *Nature,* 417, 844-848.

Chambers, J. C. (1995) Relationships between seed fates and seedling establishment in an Alpine ecosystem. *Ecology,* 76, 2124-2133.

Chambers, J. C. (1997) Restoring alpine ecosystems in western United States: environmental constraints, disturbance characteristics and restoration success. *Restoration ecology and sustainable development* (eds K. M. Urbanska, N. R. Webb & P. J. Edwards), pp. 161-187. Cambridge university press, Cambridge.

Chambers, J. C. & McMahon, J. A. (1994) A day in the life of a seed: movements and fates of seeds and their implications for natural and managed systems. *Annual Review of Ecology and Systematics.,* 25, 263-292.

Chambers, J. C., McMahon, J. A. & Brown, R. W. (1991) Seed entrapment in alpine ecosystems: effects of soil particle size and diaspore morphology. *Ecology,* 72, 1677-1688.

Chapin, F. S. & Körner, C. (1996) Arctic and alpine biodiversity : patterns, causes and ecosystem consequences. *Functional roles of biodiversity: a global perspective* (ed J. W. S. Ltd), pp. 8-32.

Chapin, F. S. I., Schulze, E.-D. & Mooney, H. (1990) The ecology and economics of storage in plants. *Annual review of ecology and systematics*, 423-447.

Choler, P. (2005a) Consistent shifts in Alpine plant traits along a mesotopographical gradient. *Arctic Antarctic and Alpine Research,* 37, 444-453.

Choler, P. (2005b) Shifts in species' trait-combinations along snowmelt and disturbance gradients in the alpine tundra of the south-western Alps, France. *Arctic, Antarctic & Alpine Research,* 37, 444-453.

Cichini, K., Schwienbacher, E., Marcante, S., Seeber, G. U. H. & Erschbamer, B. (2011) Colonization of experimentally created gaps along an alpine successional gradient. *Plant Ecology,* 212, 1613-1627.

Cornwell, W. K., Cornelissen, J. H. C., Amatangelo, K., Dorrepaal, E., Eviner, V. T., Godoy, O., et al., (2008) Plant species traits are the predominant control on litter decomposition rates within biomes worldwide. *Ecology Letters,* 11, 1065-1071.

de Bello, F., Lavorel, S., Lavergne, S., Albert, C. H., Boulangeat, I., Mazel, F., et al., (2013) Hierarchical effects of environmental filters on the functional structure of plant communities: a case study in the French Alps. *Ecography,* 36, 393-402.

de Witte, L. C., Armbruster, G. F. J., Gielly, L., Taberlet, P. & Stocklin, J. (2012) AFLP markers reveal high clonal diversity and extreme longevity in four key arctic-alpine species. *Molecular Ecology,* 21, 1081-1097.

DeAngelis, D. L., Bartell, S. M. & Brenkert, A. L. (1989) Effects of nutrient recycling and food-chain length on resilience. *American Naturalist*, 134, 778–805.

Diaz, S., Cabido, M. & Casanoves, F. (1998) Plant functional traits and environmental filters at a regional scale. *Journal of Vegetation Science,* 9, 113-122.

Dinger, F. (1983) Revégétalisation des pistes de ski: le point sur l'expérimentation actuelle. pp. 83. Cemagref, Grenoble.

Dirnbock, T., Dullinger, S. & Dirnbock, T. (2004) Habitat distribution models, spatial autocorrelation, functional traits and dispersal capacity of alpine plant species. *Journal of Vegetation Science*, 77-84.

Dona, A. J. & Galen, C. (2007) Nurse effects of alpine willows (Salix) enhance over-winter survival at the upper range limit of fireweed, Chamerion angustifolium. *Arctic Antarctic and Alpine Research,* 39, 57-64.

Dvorsky, M., Dolezal, J., Kopecky, M., Chlumska, Z., Janatkova, K., Altman, J., et al., (2013) Testing the Stress-Gradient Hypothesis at the Roof of the World: Effects of the Cushion Plant Thylacospermum caespitosum on Species Assemblages. *Plos One,* 8.

Elmarsdottir, A., Trlica, M. J. & Aradottir, A. L. (2003) Microsite availability and establishment of native species on degraded and reclaimed sites. *Journal of Applied Ecology,* 40, 815-823.

Elmqvist, T., Folke, C., Nyström, M., Peterson, G., Bengtsson, J., Walker, B. & Norberg, J. (2003) Response diversity, ecosystem change, and resilience. *Frontiers in Ecology and the Environment*, 1, 488–494.

Erschbamer, B., Kneringer, E. & Niederfriniger Schlag, R. (2001) Seed rain, soil seed bank, seedling recruitment, and survival of seedlings on a glacier foreland in the Central Alps. *Flora,* 197, 304-312.

Fattorini, M. (2001a) Establishment of transplants on machine-graded ski runs above timberline in the Swiss Alps. *Restoration Ecology,* 9, 119-126.

Fattorini, M. (2001b) Establishment of Transplants on Machine-Graded Ski Runs Above Timberline in the Swiss Alps. *Restor. Ecology,* 9, 119-126.

Fisk, M. C., Schmidt, S. K. & Seastedt, T. R. (1998) Topographic patterns of above- and belowground production and nitrogen cycling in Alpine tundra. *Ecology,* 79, 2253-2266.

Foley, J. A., DeFries, R., Asner, G. P., Barford, C., Bonan, G., Carpenter, S. R., et al., (2005) Global Consequences of Land Use. *Science*, 309, 570–574.

Folke, C., Carpenter, S., Elmqvist, T., Gunderson, L., Holling, C. S. & Walker, B. (2002) Resilience and Sustainable Development: Building Adaptive Capacity in a World of Transformations. *AMBIO: A Journal of the Human Environment*, 31, 437–440.

Frei, E. S., Scheepens, J. F. & Stoecklin, J. (2012) Dispersal and microsite limitation of a rare alpine plant. *Plant Ecology,* 213, 395-406.

Freppaz, M., Williams, B. L., Edwards, A. C., Scalenghe, R. & Zanini, E. (2007) Simulating soil freeze/thaw cycles typical of winter alpine conditions: Implications for N and P availability. *Applied Soil Ecology,* 35, 247-255.

Freppaz, M., Williams, M. W., Seastedt, T. & Filippa, G. (2012) Response of soil organic and inorganic nutrients in alpine soils to a 16-year factorial snow and N-fertilization experiment, Colorado Front Range, USA. *Applied Soil Ecology,* 62, 131-141.

Gehrig-Fasel, J., Guisan, A. & Zimmermann, N. E. (2007) Tree line shifts in the Swiss Alps: climate change or land abandonment? *Journal of vegetation science*, 571-582.

Geneletti, D. (2008) Impact assessment of proposed ski areas: A GIS approach integrating biological, physical and landscape indicators. *Environmental Impact Assessment Review,* 28, 116-130.

Gensac, P. (1990) Plant and soil groups in the Alpine grasslands of the Vanoise Massif, French Alps. *Arctic & Alpine Research,* 22, 195-201.

German federation for soil-bioengineering, HBLFA Raumberg-Gumpenstein.

Grabherr, G. (2003) Alpine vegetation dynamics and climate change - a synthesis of long-term studies and observations. *Alpine biodiversity in Europe* (eds L. Nagy, G. Grabherr, C. Körner & D. B. A. Thompson). Springer, Berlin, Heidelberg, New-York.

Granjou, C., Gaucherand, S. & Chanteloup, E. (2010) De la réparation à la restauration. *Journal of Alpine Research | Revue de géographie alpine.*

Grigulis, K., Lavorel, S., Krainer, U., Legay, N., Baxendale, C., Dumont, M., et al., (2013) Relative contributions of plant traits and soil microbial properties to mountain grassland ecosystem services. *Journal of Ecology,* 101, 47-57.

Grime, J. P. (1977a) Evidence for the existence of three primary strategies in plants and its relevance to ecological and evolutionary theory. *American Naturalist,* 111, 1169-1194.

Grime, J. P. (1977b) *Plant strategies and vegetation processes.* John Wiley & Sons, Chichester, UK.

Guleryuz, G., Titrek, E. & Arslan, H. (2008) Nitrogen mineralization in the ruderal sub-alpine communities in Mount Uludag, Turkey. *European Journal of Soil Biology,* 44, 408-418.

Gunderson, L. H. (2000) Ecological resilience - in theory and application. *Annual Review of Ecology and Systematics*, 31, 425–439.

Hagen, D. & Evju, M. (2013) Using Short-Term Monitoring Data to Achieve Goals in a Large-Scale Restoration. *Ecology and Society,* 18.

Harsch, M. A., Hulme, P. E., McGlone, M. S. & Duncan, R. P. (2009) Are treelines advancing? A global meta-analysis of treeline response to climate warming. *Ecology Letters,* 12, 1040-1049.

Haussmann, N. S., McGeoch, M. A. & Boelhouwers, J. C. (2010) Contrasting nurse plants and nurse rocks: The spatial distribution of seedlings of two sub-Antarctic species. *Acta Oecologica-International Journal of Ecology,* 36, 299-305.

Hejcman, M., Hejcmanova, P., Pavlu, V. & Benes, J. (2013) Origin and history of grasslands in Central Europe - a review. *Grass and Forage Science,* 68, 345-363.

Hejcman, M., Klaudisova, M., Scheliberg, R. & Honsova, D. (2007a) The Rengen Grassland Experiment: Plant species composition after 64 years of fertilizer application. *Agriculture, Ecosystems and Environment*, 122, 259–266.

Hejcman, M., Klaudisova, M., Stursa, J., Pavlu, V., Schellberg, J., Hejcmanova, P., et al., (2007b) Revisiting a 37 years abandoned fertilizer experiment on Nardus grassland in the Czech Republic. *Agriculture, Ecosystems and Environment*, 118, 231–236.

Huber, R., Rigling, A., Bebi, P., Brand, F. S., Briner, S., Buttler, A., et al., (2013) Sustainable Land Use in Mountain Regions Under Global Change: Synthesis Across Scales and Disciplines. *Ecology and Society*, 18.

Isselin-Nondedeu, F. & Bédécarrats, A. (2004) Variations des traits fonctionnels des plantes le long d'un gradient mésotopographique (La Plagne, Alpes du Nord)/ Plant functional traits variation along a mesotopographical gradient in the subalpine meadow in the Northern French Alps. *Journées d'Ecologie Fonctionnelle / Meeting of Functional Ecology*, pp. 60. CNRS, Rennes, France.

Isselin-Nondedeu, F. & Bédécarrats, A. (2007) Soil microtopographies shaped by plants and cattle facilitate seed bank formation on alpine ski trails. *Ecological Engineering,* in Press.

Isselin-Nondedeu, F. & Bedecarrats, A. (2009) Assessing the dominance of Phleum pratense cv. climax, a species commonly used for ski trail restoration. *Applied Vegetation Science,* 12, 155-165.

Isselin-Nondedeu, F. & Bedecarrats, A. (2013) Germination and Seedling Responses of Subalpine Plants to Different Soil Substrates. *Folia Geobotanica,* 48, 39-53.

Isselin-Nondedeu, F., Rey, F. & Bédécarrats, A. (2006) Contributions of vegetation cover and cattle hoof prints towards seed runoff control on ski pistes. *Ecological engineering,* 27, 193-201.

Jung, V., Albert, C. H., Violle, C., Kunstler, G., Loucougaray, G. & Spiegelberger, T. (2014) Intraspecific trait variability mediates the response of subalpine grassland communities to extreme drought events. *Journal of Ecology*, 102, 45–53.

Keddy, P. A. (1992) Assembly and response rules. *Journal of Vegetation Science,* 3, 157-164.

Kikuzawa, K. & Kudo, G. (1995) Effects of the length of the snow-free period on leaf-longevity in alpine shrubs: a cost-benefit model. *Oikos,* 73, 214-220.

Klanderud, K. & Totland, O. (2008) Diversity-Stability Relationships of an Alpine Plant Community under Simulated Environmental Change. *Arctic Antarctic and Alpine Research,* 40, 679-684.

Klaudisova, M., Hejcman, M. & Pavlu, V. (2009) Long-term residual effect of short-term fertilizer application on Ca, N and P concentrations in grasses Nardus stricta L. and Avenella flexuosa L. *Nutrient Cycling in Agroecosystems*, 85, 187–193.

Körner, C. (1999a) Alpine climate. *Alpine plant life: functional plant ecology of high mountain ecosystems* (ed Springer-Verlag), pp. 21-30. Springer, Berlin.

Körner, C. (1999b) *Alpine plant life: functional plant ecology of high mountain ecosystems.* Springer, Berlin.

Körner, C. (2002) Mountain biodiversity, its causes and function: an overview. *Mountain biodiversity, a global assesment* (eds C. Körner & E. M. Spehn), pp. 336. The Parthenon Publishing Group, London.

Korner, C. & Paulsen, J. (2004) A world-wide study of high altitude treeline temperatures. *Journal of Biogeography,* 31, 713-732.

Körner, C. & Spehn, E. (2002) *Mountain biodiversity – A Global Assessment.* Parthenon Publishers, New York, London.

Krautzer, B. & Graiss, W. (2009) Establishment of semi-natural litter meadows at a Natura 2000 site in the Enns valley. *International Workshop of the SALVERE-Project 2009* (ed A. R. a. E. C. Raumberg-Gumpenstein), pp. 25-26. Agricultural Research and Education Centre Raumberg-Gumpenstein, Raumberg-Gumpenstein.

Kreyling, J., Beierkuhnlein, C. & Jentsch, A. (2010) Effects of soil freeze-thaw cycles differ between experimental plant communities. *Basic and Applied Ecology,* 11, 65-75.

Kudo, G., Molau, U. & Wada, N. (2001) Leaf-trait variation of tundra plants along a climatic gradient: an integration of responses in evergreen and deciduous species. *Arctic, Antarctic & Alpine Research,* 33.

Kudo, G., Nordenhall, U. & Molau, U. (1999) Effects of snowmelt timing on leaf traits, leaf production, and shoot growth of alpine plants: comparisons along a snowmelt gradient in northern Sweden. *Ecoscience,* 6, 439-450.

Larcher, W. (2003) *Physiological plant ecology.* Springer, Berlin, Heidelberg, New-York.

Lavorel, S. & Garnier, E. (2002) Predicting changes in community composition and ecosystem functioning from plant traits: revisiting the Holy Grail. *Functional Ecology,* 16, 545-556.

le Roux, P. C., Virtanen, R. & Luoto, M. (2013) Geomorphological disturbance is necessary for predicting fine-scale species distributions. *Ecography,* 36, 800-808.

Leps, J., Osbornova-Kosinova, J. & Rejmanek, M. (1982) Community stability, complexity and species life-history strategies. *Vegetatio*, 50, 53–63.

Linacre, E. (1982) The effect of altitude on the daily range of temperature. *Journal of Climatology,* 2, 375-382.

Lluent, A., Anadon-Rosell, A., Ninot, J. M., Grau, O. & Carrillo, E. (2013) Phenology and seed setting success of snowbed plant species in contrasting snowmelt regimes in the Central Pyrenees. *Flora - Morphology, Distribution, Functional Ecology of Plants,* 208, 220-231.

Luck, G. W., Daily, G. C. & Ehrlich, P. R. (2003) Population diversity and ecosystem services. *Trends in Ecology and Evolution*, 18, 331–336.

MacGillivray, C. W., Grime, J. P. & The Integrated Screening Programme (ISP) Team. (1995) Testing predictions of the resistance and resilience of vegetation subjected to extreme events. *Functional Ecology*, 9, 640–649.

Malaval, S., Lauga, B., Regnault-Roger, C. & Largier, G. (2010) Combined definition of seed transfer guidelines for ecological restoration in the French Pyrenees. *Applied Vegetation Science,* 13, 113-124.

Manel, S., Gugerli, F., Thuiller, W., Alvarez, N., Legendre, P., Holderegger, R., et al., (2012) Broad-scale adaptive genetic variation in alpine plants is driven by temperature and precipitation. *Molecular Ecology,* 21, 3729-3738.

Mariotte, P., Vandenberghe, C., Meugnier, C., Rossi, P., Bardgett, R.D., Buttler, A. (2013) Subordinate plant species impact on soil microbial communities and ecosystem functioning in grassland: findings from a removal experiment. *Perspectives in Plant Ecology, Evolution and Systematics*, 15, 77-85.

Meier, C. L. & Bowman, W. D. (2008) Links between plant litter chemistry, species diversity, and below-ground ecosystem function. *Proceedings of the National Academy of Sciences of the United States of America,* 105, 19780-19785.

Michalet, R., Gandoy, C., Joud, D., Pagès, J.-P. & Choler, P. (2002) Plant community composition and biomass on calcareous and siliceous substrates in the northern French Alps: Comparative effects of soil chemistry and water status. *Arctic, Antarctic, and Alpine Research,* 34, 102-113.

Milbau, A., Shevtsova, A., Osler, N., Mooshammer, M. & Graae, B. J. (2013) Plant community type and small-scale disturbances, but not altitude, influence the invasibility in subarctic ecosystems. *New Phytologist,* 197, 1002-1011.

Molenda, O., Reid, A. & Lortie, C. J. (2012) The Alpine Cushion Plant Silene acaulis as Foundation Species: A Bug's-Eye View to Facilitation and Microclimate. *Plos One,* 7.

Monson, R. K., Rosenstiel, T. N., Forbis, T. A., Lipson, D. A. & Jaeger, C. H. (2006) Nitrogen and carbon storage in alpine plants. *Integrative and Comparative Biology,* 46, 35-48.

Monteiro, J. A. F., Hiltbrunner, E. & Korner, C. (2011) Functional morphology and microclimate of Festuca orthophylla, the dominant tall tussock grass in the Andean Altiplano. *Flora,* 206, 387-396.

Moore, J. C., Ruiter, P. C. de & Hunt, H. W. (1993) Influence of productivity on the stability of real and model ecosystems. *Science*, 261, 906–908.

Naeem, S. (1998) Species Redundancy and Ecosystem Reliability. *Conservation Biology*, 12, 39–45.

Novak, J., Petr, L. & Treml, V. (2010) Late-Holocene human-induced changes to the extent of alpine areas in the East Sudetes, Central Europe. *Holocene,* 20, 895-905.

Oeggl, K., Schmidl, A. & Kofler, W. (2009) Origin and seasonality of subfossil caprine dung from the discovery site of the Iceman (Eastern Alps). *Vegetation History and Archaeobotany,* 18, 37-46.

Oksanen, L. & Ranta, E. (1992) Plant strategies along mountain vegetation gradients: a test of two theories. *Journal of Vegetation Science,* 3, 175-186.

Olofsson, J. (2004) Positive and negative plant-plant interactions in two contrasting arctic-alpine plant communities. *Arctic Antarctic and Alpine Research,* 36, 464-467.

Onipchenko, V. G., Semenova, G. V. & van der Maarel, E. (1998) Population strategies in severe environments: Alpine plants in the northwestern Caucasus. *Journal of Vegetation Science,* 9, 27-40.

Onipchenko, V. G., Makarov, M. I., Akhmetzhanova, A. A., Soudzilovskaia, N. A., Aibazova, F. U., Elkanova, M. K., et al., (2012) Alpine plant functional group responses to fertiliser addition depend on abiotic regime and community composition. *Plant and Soil,* 357, 103-115.

Ozenda, P. (1985) *La végétation de la chaîne alpine dans l'espace montagnard européen.* Masson, Paris.

Parolo, G. & Rossi, G. (2008) Upward migration of vascular plants following a climate warming trend in the Alps. *Basic and Applied Ecology,* 9, 100-107.

Peratoner, G. (2006) Extreme sites and their restoration requirements in Arctic-Alpine climatic regions. *Soil-engineering: ecological restoration with native plant and seed material* (eds B. Krautzer & E. Hacker), pp. 37-42. Research and education centre for agriculture Raumberg-Gumpenstein

Perrussel, O. (1986) *La cicatrisation végétale des surfaces terrassées dans le domaine supraforestier: cas de la station de La Plagne (Savoie).*

Peterson, G., Allen, C. R. & Holling, C. S. (1998) Ecological Resilience, Biodiversity, and Scale. *Ecosystems*, 1, 6–18.

Pierce, S., Luzzaro, A., Caccianiga, M., Ceriani, R. M. & Cerabolini, B. (2007) Disturbance is the principal α-scale filter determining niche differentiation, coexistence and biodiversity in an alpine community. *Journal of Ecology,* 95, 698-706.

Pohl, M., Graf, F., Buttler, A. & Rixen, C. (2012) The relationship between plant species richness and soil aggregate stability can depend on disturbance. *Plant and Soil,* 355, 87-102.

Pohl, M., Stroude, R., Buttler, A. & Rixen, C. (2011) Functional traits and root morphology of alpine plants. *Annals of Botany,* 108, 537-545.

Pohl, M., Alig, D., Korner, C. & Rixen, C. (2009) Higher plant diversity enhances soil stability in disturbed alpine ecosystems. *Plant and Soil,* 324, 91-102.

Prach, K. & Pysek, P. (1999) How do species dominating in succession differ from others? *Journal of Vegetation Science,* 10, 383-392.

Price, J. S., Rochefort, L. & Quinty, F. (1998) Energy and moisture considerations on cutover peatlands: surface microtopography, mulch cover and Sphagnum regeneration. *Ecological Engineering,* 10, 293-312.

Quiroz, C. L., Cavieres, L. A. & Pauchard, A. (2011) Assessing the importance of disturbance, site conditions, and the biotic barrier for dandelion invasion in an Alpine habitat. *Biological Invasions,* 13, 2889-2899.

Reid, A. M. & Lortie, C. J. (2012) Cushion plants are foundation species with positive effects extending to higher trophic levels. *Ecosphere,* 3.

Rixen, C., Huovinen, C., Huovinen, K., Stockli, V. & Schmid, B. (2008) A plant diversity x water chemistry experiment in subalpine grassland. *Perspectives in Plant Ecology Evolution and Systematics,* 10, 51-61.

Rixen, C., Stoeckli, V., Ammann, W. & Rixen, C. (2003) Does artificial snow production affect soil and vegetation of ski pistes? A review. *Perspectives in Plant Ecology, Evolution and Systematics,* 5, 219-230.

Rose, J. & Malanson, G. (2012) Microtopographic heterogeneity constrains alpine plant diversity, Glacier National Park, MT. *Plant Ecology,* 213, 955-965.

Roux-Fouillet, P., Wipf, S. & Rixen, C. (2011) Long-term impacts of ski piste management on alpine vegetation and soils. *Journal of Applied Ecology,* 48, 906-915.

Rydgren, K., et al., (2011) Restoration of alpine spoil heaps: Successional rates predict vegetation recovery in 50 years. *Ecological Engineering,* 37, 294-301.

Rydgren, K., Halvorsen, R., Odland, A. & Skjerdal, G. (2011) Restoration of alpine spoil heaps: Successional rates predict vegetation recovery in 50 years. *Ecological Engineering,* 37, 294-301.

Scheffer, M., Carpenter, S., Foley, J. A., Folke, C. & Walker, B. (2001) Catastrophic shifts in ecosystems. *Nature*, 413, 591–596.

Schob, C., Armas, C., Guler, M., Prieto, I. & Pugnaire, F. I. (2013) Variability in functional traits mediates plant interactions along stress gradients. *Journal of Ecology,* 101, 753-762.

Schroder, R. & Prasse, R. (2013) From nursery into nature: A study on performance of cultivated varieties of native plants used in re-vegetation, their wild relatives and evolving wild x cultivar hybrids. *Ecological Engineering,* 60, 428-437.

Semelová, V., Hejcman, M., Pavl, V., Vacek, S. & Podrázský, V. (2008) The Grass Garden in the Giant Mts. (Czech Republic): Residual effect of long-term fertilization after 62 years. *Agriculture, Ecosystems and Environment*, 123, 337–342.

Simmons, I. G. (2005) The mountains of Northern Europe: towards an environmental history for the last ten thousand years. *Mountains of Northern Europe: conservation, management, people and nature* (eds D. B. A. Thompson, M. Price & C. A. Galbraith), pp. 420. Stationery Office Books, Edinburgh.

Snyder, R. E. & Chesson, P. (2004) How the spatial scales of dispersal, competition, and environmental heterogeneity interact to affect coexistence. *American Naturalist,* 164, 633-650.

Sohlberg, E. H. & Bliss, L. C. (1984) Microscale pattern of vascular plant distribution in two high arctic plant communities. *Canadian Journal of Botany,* 62, 14-20.

Song, M.-H., Yu, F.-H., Ouyang, H., Cao, G.-M., Xu, X.-L. & Cornelissen, J. H. C. (2012) Different inter-annual responses to availability and form of nitrogen explain species coexistence in an alpine meadow community after release from grazing. *Global Change Biology,* 18, 3100-3111.

Special Issue: Proceedings of the Symposium of the International Geographical Union Commission on High Altitude Geoecology, 5, A19.

Spiegelberger, T., Deléglise, C., DeDanieli, S. & Bernard-Brunet, C. (2010) Resilience of acid subalpine grassland to short-term liming and fertilisation. *Agriculture, Ecosystems & Environment,* 137, 158-162.

Spiegelberger, T., Hegg, O., Matthies, D., Hedlund, K. & Schaffner, U. (2006) Long-term effects of short-term perturbation in a subalpine grassland. *Ecology*, 87, 1939–1944.

Stoate, C., Baldi, A., Beja, P., Boatman, N. D., Herzon, I., van Doorn, A., et al., (2009) Ecological impacts of early 21st century agricultural change in Europe-a review. *Journal of environmental management,* 91, 22-46.

Stöcklin, J. & Bäumler, E. (1996) Seed rain, seedling establishment and clonal growth strategies on a glacier foreland. *Journal of Vegetation Science,* 7, 45-56.

Tasser, E. & Tappeiner, U. (2002) Impact of land use changes on mountain vegetation. *Applied Vegetation Science,* 5, 173-184.

Tasser, E., Mader, M. & Tappeiner, U. (2003) Effects of land use in alpine grasslands on the probability of landslides. *Basic and Applied Ecology,* 4, 271-280.

Tilman, D. & Downing, J. A. (1994) Biodiversity and stability in grasslands. *Nature*, 367, 363–365.

Troll, C. (1973) High Mountain Belts between the Polar Caps and the Equator: Their Definition and Lower Limit. *Arctic and Alpine Research.*

Urbanska, K. M. & Fattorini, M. (2000) Seed rain in high-altitude restoration plots in Switzerland. *Restoration Ecology*, 74-79.

Urbanska, K. M., Erdt, S. & Fattorini, M. (1998) Seed Rain in Natural Grassland and Adjacent Ski Run in the Swiss Alps: A Preliminary Report. *Restoration Ecology,* 6, 159-165.

Urbanska, K. M. (1997a) Restoration ecology research above the timberline: colonization of safety islands on a machine-graded alpine ski run. *Biodiversity and Conservation,* 6, 1655-1670.

Urbanska, K. M. (1997b) Safe-sites-interface of plant population ecology and restoration ecology. *Restoration ecology and sustainable development.* (eds K. M. Urbanska, N. R. Webb & E. P.J.), pp. 81-110. Cambridge University Press, Cambridge.

Urbanska, K. M. & Schütz, M. (1986) Reproduction by seed in alpine plants and revegetation research above timberline. *Botanica Helvetica,* 96, 41-60.

van Andel, J. & Aronson, J. (2005) Chapter 1. *Restoration ecology the new frontier* (eds J. van Andel & J. Aronson), pp. 340. Wiley-Blackwell, Chichester.

Wenk, E. H. & Dawson, T. E. (2007) Interspecific differences in seed germination, establishment, and early growth in relation to preferred soil type in an Alpine community. *Arctic, Antarctic and Alpine Research,* 39, 165-176.

Wipf, S., Rixen, C., Fischer, M., Schmid, B. & Stoeckli, V. (2005) Effects of ski piste preparation on alpine vegetation. *Journal of Applied Ecology,* 42, 306-316.

In: Grassland Biodiversity and Conservation ...
Editors: Pierre Mariotte and Paul Kardol

ISBN: 978-1-61761-619-8

Chapter 10

THE ROLE OF BRYOPHYTES IN CENTRAL EUROPEAN GRASSLANDS

Valentin H. Klaus[1,*] and Jörg Müller[2*]
[1]Institute of Landscape Ecology, University of Münster, Germany
[2]Institute of Biochemistry and Biology, University of Postdam, Germany

ABSTRACT

Bryophytes are well known from boreal regions, extreme habitats and as typical components of forest vegetation, but they are likewise important elements of temperate grassland ecosystems. As most studies ignore these special plants, knowledge of the ecology of bryophytes in grasslands is scarce and scattered. Nevertheless, bryophytes significantly contribute to grassland diversity and facilitate fundamental ecosystem functions. Due to their small growth form, bryophytes constitute a separate vegetation layer below the canopy of the vascular vegetation. As bryophytes are inferior to vascular plants in terms of productivity and competition for light, grasslands with exceptional high cover and richness of bryophytes are generally restricted to nutrient poor and particularly dry or wet habitats. However, even in mesic grasslands, several bryophyte species can be found. Similar to vascular plants, bryophytes are significantly influenced by land-use intensity, partly positive but mostly negative. While soil disturbance by grazing and mowing can increase bryophyte diversity, mechanical impacts, drainage, toxicity of high nitrogen applications and increased vascular plant productivity lead to a drastic reduction of bryophyte diversity under intensive land use.

Bryophytes play an important role for many ecosystem processes such as facilitating productivity and carbon storage in nutrient-poor environments. Similarly, bryophyte layers affect soil humidity, increase water retention capacity and interact with vascular plants by promoting or hampering germination and seedling establishment. Due to their special physiological characteristics, bryophytes are sensitive to environmental changes such as nutrient enrichment, pollutant deposition and climatic shifts and, thus, serve as ecological indicators for specific environmental conditions. To advance our understanding of bryophytes and their importance for the ecosystem functioning in grasslands, experimental approaches testing effects of the pre- or absence of bryophytes

[*] Corresponding authors. E-mail: v.klaus@uni-muenster.de.; muellerj.potsdam@googlemail.com.

as well as their diversity on e.g. nutrient cycling and microclimatic conditions are urgently needed. Furthermore, to assess the role of bryophytes in grassland community assembly, interactions with vascular vegetation such as the control of seedling recruitment, establishment and facilitation by bryophytes should be studied in more detail.

Keywords: Bryophyta, cryptogams, mosses, liverworts, hornworts, morphology and physiology of bryophytes, temperate grasslands, land-use intensity, ecological restoration, interactions, ecosystem functioning

INTRODUCTION

Why Considering Bryophytes in Grasslands?

Bryophytes occur in all terrestrial biomes and are well-known inhabitants of boreal regions, forests and extreme habitats such as rocks. However, bryophytes are also a typical component of temperate grasslands and other grassland-like habitats around the world (e.g. Virtanen et al., 2000; Zechmeister et al., 2002; Müller et al., 2012). Up to now, including bryophyte records in grassland studies was seen as not significantly increasing explanatory power of ecological analyses and therefore, surveys widely ignored their occurrence (Herben 1987; Cornelissen et al., 2007). So, why should we change this old habit? If we take a closer look at these rather primitive but unique organisms and their role in grassland ecosystems, we will see that bryophytes make a significant contribution to grassland diversity and influence important ecosystem functions. In many nutrient-poor grassland types in Central Europe, bryophytes occur in high densities and build up a significant part of the vegetation biomass. For example, about 200 bryophyte and lichen species are known from European calcareous dry grasslands (Willems 1983).

Bryophytes often directly interact with vascular plants and serve as a habitat for microbes and other small animal species (Lindo & Gonzales 2010). Due to their physiological sensitivity, bryophytes are regarded as indicators for certain environmental conditions such as moisture, eutrophication or pollution. However, at the same time, environmental changes and land-use intensification severely threaten bryophytes in grasslands (Zechmeister et al., 2003; Müller et al., 2012), making detailed knowledge of their ecology as well as their role in ecosystem functioning even more important. Vascular plants and bryophytes respond differently as well as to a different set of environmental factors. Thus, an insufficient assessment of relevant factors and drivers is the reason why including bryophytes in grassland surveys often gave only minor additional insight (Herben 1987). This leads us to the conclusion that, when seeking for an in-depth assessment of the diversity and ecology of grassland, bryophytes should definitively be included.

Ecology of Grassland Bryophytes

Bryophytes: Unique and Special

Together with ferns, fern allies, fungi, algae and lichens, bryophytes compose the broad and diverse group of so-called “cryptogams”, contrasting the group of all seed plants. The designation as cryptogams already illustrates the distanced view of several botanists on these small, non-flowering plants and it reveals that we know comparatively little about them. Cryptogams are "plants, that marry in secret" [Greek: *kryptós:* concealed, hidden, in secret; *gamós*: marriage]. Bryophytes represent a very species-rich group among the cryptogams with worldwide 12,500 - 15,000 bryophyte species (Frahm 2001). Bryophytes can be further divided into mosses, liverworts and hornworts, all of which can potentially occur in grassland ecosystems. Morphologically, bryophytes are small plants that lack many of the complex structures found in vascular plants, such as meristematic tissue, lignin, sclerenchyma, tracheids and sieve cells (Vanderpoorten & Goffinet 2009). Thus, bryophytes are strongly limited in vertical growth. Bryophytes, as close descendants of early land plants, have a dominant vegetative gametophyte supporting the generative sporophyte for sexual reproduction (Shaw et al., 2011). The gametophyte is a thallus and in case of mosses and leafy liverworts it consists of stems, leafs and rhizoids. A rhizoid is a simple root-like structure, which is used for adherence to soil or other substrates but without an active nutrient and water supply function. Gametophytes carry out photosynthesis and, due to the lack of roots with active transport mechanisms, take up water and nutrients by simple diffusion through the cell walls over their entire surface (Vanderpoorten & Goffinet 2009; Tuba et al., 2011). Bryophytes usually reproduce with spores distinguishing them from so-called seed plants. When the gametophytes produce sexual organs (sperm and eggs) for generative reproduction, these melt into so-called “sporophytes”, which finally produce spores in a capsule. The sporophyte remains attached to the longer-living gametophyte, which supplies the sporophyte with nutrients.

Although bryophytes have no internal mechanism for transporting water or nutrients, they contain chlorophyll and carry out photosynthesis (Smith 1982). Therefore, bryophytes developed various structural features assisting water conduction such as specific cell surfaces, or tomentums along the stem. However, both metabolic activity and reproduction still crucially depend on liquid water in the surrounding (Frahm 2001). This mechanism is called “poikilohydry”, a typical feature of bryophyte species in grasslands. It means that bryophytes cannot actively control their cell water content and depend on simple diffusion. In phases of drought, the obligatory equilibration of the cell water content to that of the environment causes far-reaching desiccation and the cessation of the metabolic activity, while after rehydration, recovery is generally very rapid with most bryophytes reaching their full photosynthetic capacity within a few hours. This mechanism has changed only little from the manifestation of desiccation tolerance in the earliest land plants and it allows bryophytes to survive long-lasting de-hydration, sometimes for months. Although this mechanism restricts productivity, it is essential to survive harsh environmental conditions (Oliver et al., 2005). Additionally, most bryophyte species are also highly tolerant against low temperatures enabling them to expand the active vegetation period. Consequently, bryophytes in grasslands often show a different phenological pattern than vascular plants, with increasing abundance (and activity) within the moist and cool conditions of spring and autumn (Al-Mufti et al., 1977).

Dispersal in Grasslands

Like vascular plants, bryophytes need dispersal vectors for two main reasons: to maintain gene flow between populations and to reach and colonize suitable new sites (Pharo & Zartman 2007). Generally, bryophytes have various dispersal strategies: generative (sexual) via spores, but also vegetative (asexual) via specialized propagules (gemmae), unspecialized gametophyte fragments and ramets (clonal reproduction; Frey & Kürschner 2011). Potentially, every part of the gametophyte is able to produce protonema as the preliminary stage of a completely new bryophyte plant. Thus, accidental fragmentation of a gametophyte is the simplest vegetative dispersal mode. In contrast, the main advantage of generative reproduction is the high number of spores produced. Single capsules can contain a vast number of spores, ranging from 220 (*Riccia glauca*), 5,500 (*Phascum cuspidatum*) and 243,000 (*Sphagnum squarrosum*) up to 280,000 – 700,000 (*Eurhynchium*) (Glime 2007). Spores are very small and can be easily transported by wind, enabling bryophytes to disperse over very long distances. In addition, spores have a high longevity resulting in a long dormancy in the diaspore bank (During 1979; Vanderpoorten & Goffinet 2009).

Certain grassland bryophyte species produce spores only rarely and some others, such as *Thuidium abietinum* and *Rhytidium rugosum,* are known to not produce spores at all (Pfeiffer et al., 2006; Lieske 2010). Therefore, these species strongly depend on vegetative reproduction and dispersal strategies. A group of ruderal species, which are also common in grasslands such as *Leptobryum pyriforme, Dicranella staphylina* and species of the *Bryum erythrocarpum* complex can also produce vegetative belowground rhizoid gemmae. Similar to spores, rhizoid gemmae also contribute to the soil diaspore bank. This mechanism is particularly important for more ruderal bryophyte species, quickly spreading after a disturbance event (During 2001). Due to their larger size, the dispersal distance of vegetative diaspores is much shorter than for wind-dispersed spores. Nevertheless, especially in grasslands, zoochory is also an important dispersal vector when bryophyte fragments are attached to (migratory) grazing animals. This mechanism strongly increases the probability to reach and colonize a suitable new habitat. Studies revealed that grassland bryophyte fragments are easily dispersed attaching to the fur and the hooves of sheep (Pauliuk et al., 2011). Additionally, successful endozoochorous dispersal by large ungulates has also been shown (Jaroszewicz et al., 2011). Nevertheless, bryophyte communities in grasslands are often diaspore and/or dispersal limited, entailing problems when restoring typical grassland communities that have been lost (Kooijman & Beltman 1994; Löbel et al., 2006; see section below).

Figure 1. Bryophyte species of different growth forms: dense turfs of a) *Calliergonella cuspidata* and b) *Thuidium philibertii*; c) irregular branched creeping stems of *Amblystegium humile*; d) turf of small *Physcomitrium pyriforme* with capsules; e) thallose rosettes of liverworts *Riccia sorocarpa* (left) and *Riccia cavernosa* (right); and f) the hornwort *Anthoceros agrestis* with long capsules. Photos: J. Müller.

Bryophyte Growth Forms

Bryophytes occuring in grasslands can be categorized into four different growth forms(During 1979; Smith 1982; Hill et al., 2007). The first category integrates dominant tufts of branching moss species (mainly pleurocarpous species) and peat mosses, which are several centimeters high, occurring mostly in unimproved semi-natural grasslands (Figure 1a, b). Where moisture and nutrient conditions are suitable, bryophytes of this category can build up an extensive and dense sward. Examples for species of this category are *Rhytidiadelphus squarrosus*, *Climacium dendroides*, *Rhytidium rugosum* and, in near-natural peatlands and fen grasslands, some *Sphagnum* species (peat mosses).

The second category comprises a horizontally creeping growth form with long branches and stems (Figure 1c). This growth form is characteristic for many nitrophilous species of pleurocarpous mosses (e.g. *Brachythecium* spp., *Eurhynchium* spp., *Amblystegium* spp.) and liverworts (*Lophocolea bidentata*), which are able to tolerate high nutrient levels, shading and overgrowing by vascular plants. These species are typically perennials, which often remain in their vegetative stage and seldom reproduce sexually in grassland habitats.

The third category is the short erect stature of mosses (mainly acrocarpous species) forming small, more or less dense turfs (Figure 1d). These simple forms show a colonizing strategy accompanied with a short life cycle including rapid development of generative organs (capsules with spores) or vegetative propagules (especially rhizoid gemmae). Prevalently, such acrocarpous species are small or even very small and constitute micro-populations of few individuals of only few millimeters height (e.g. *Acaulon* spp., *Bryum* spp., *Phascum* spp., *Pleuridium* spp.).

The fourth category consists of patchy thallose statures such as rosettes or soil attached mats. This is typical for some liverworts and hornworts (e.g. *Riccia* and *Anthoceros*) (Figure 1e, f). Species of this growth form mainly occur on frequently disturbed soils e.g. on agricultural fields, but they can also be found in some grassland habitats. Similar to acrocarpous mosses, these species have a short life cycle with ephemeral appearance with rapid capsule and spore production to endure intensive disturbance regimes.

Competitive Interactions

Similar to vascular plants, competition is a generally important mechanism in bryophyte communities (Rydin 1997; Zamfir & Goldberg 2000). In grasslands with an extensive and dense bryophyte layer and few open soil patches, competition among bryophyte species is likely to be strong. However, During & Lloret (2001) suggested that competitive exclusion rarely occurs in bryophyte communities, not due to the general lack of competition, but to a balance between competition and facilitation. In productive grasslands, effects of superior vascular plants will widely overrule competitive interactions among bryophyte species. Unfortunately, only few studies inquired competition among bryophytes in grasslands but focused on interactions with vascular plants.

Bryophytes compete with vascular plants for nutrients, light and space. While dense bryophyte layers on the soil surface can immobilize nutrients that would otherwise have been available to vascular plants (Rydin 1997), vascular plants use their complex root system to acquire nutrients from the deeper soil. Thus, as both groups have different strategies for nutrient uptake due to barely overlapping uptake zones, direct competition for nutrients between bryophytes and vascular plants is weak. In contrast, competition is much stronger for space and light (Slack 1997; Ayres et al., 2006; Rydin 2009). The higher growth potential and biomass production of vascular plants clearly favors them over bryophytes at optimal environmental conditions (but see Box 1). Thus, both abundance and diversity of bryophytes suffer strongly from high productivity of vascular plants (Bergamini et al., 2001a), although Ingerpuu et al. (2005) showed in a pot experiment that (sparse) vascular plant cover can also facilitate bryophyte growth by improving microclimatic conditions. In contrast, if nutrient availability is not optimal, Ingerpuu & Kupper (2007) found that on dry calcareous grasslands, the cover of vascular plants could be significantly reduced by an extensive moss cover, which was favored in years with high precipitation (see also Box 1).

Box 1. A moss on the march

Generally, the proportion of invasive bryophyte species is low compared to invasive vascular plants (Nebel & Philippi 2000). However, in many regions of Western and Central Europe, the neophytic moss *Campylopus introflexus* (Figure 2) invades disturbed peatlands as well as dunes, heathlands and dry sand grasslands. This pioneer species is native to the Southern Hemisphere (Frahm 1984) and was recorded for the first time in 1941 in Great Britain (Equihua & Usher 1993). In The Netherlands, *Campylopus introflexus* is called "tank moss" as it is supposed to have spread with the tanks during the Second World War. Today, this species is rolling over the indigenous vegetation, especially in sandy dry grasslands and heathlands. *Campylopus introflexus* can become particularly dominant by building thick dense mats over hundreds of square meters, in which virtually no other bryophyte or lichen species can survive (Biermann & Daniëls 1997). Furthermore, *Campylopus introflexus* strongly alters environmental conditions such as water availability and nutrient status of these rare habitats (Biermann & Daniëls 1997). In Britain and the Czech Republic, *Campylopus introflexus* is also known to spread in disturbed peatlands, although its invasive pressure is less strong compared to sand ecosystems (Mikulášková et al., 2012).

Figure 2. The invasive moss species *Campylopus introflexus* forms dense cushions in dry sand grasslands and heathlands. The long grayish hair points, reflexed to a right angle, give the species a cactus-like appearance.

Additionally, due to special morphological features, bryophytes occupy ecological niches that are unfavorable or even intolerable for vascular vegetation. For example, at nutrient poor and acidic sites such as bogs, bryophytes gain advantage over vascular plants by holding as much as twenty-times their own weight in water and using their complete surface to catch nutrients from atmospheric depositions, the major pathway of nutrient influxes in this ecosystem (Ellenberg & Leuschner 2010). Meanwhile, vascular plants are restricted to nutrient releases from litter and peat mineralization. Consequently, bryophytes have the ability to dominate this harsh and nutrient-limited environment, e.g. *Sphagnum* mosses in bogs. As outlined above, bryophytes can also shift their growth period in time by tolerating cold conditions to avoid (or reduce) competition with vascular vegetation (Al-Mufti et al., 1977; Ingerpuu & Kupper 2007). Tolerating temporal dryness and cold is thus a very important characteristic of bryophytes enabling coexistence with strongly competitive vascular vegetation.

There is also a strong direct effect of bryophytes on vascular vegetation during the germination and establishment phase of vascular plants (see also Box 1). Bryophyte layers trap seeds and influence microsite hydrology, temperature and nutrient cycles (Glime 2007; Bates 2009). Studies on the effects of bryophyte layers on the generative regeneration of vascular plants showed both inhibitory but also facilitative effects (e.g. Keizer et al., 1985; van Tooren 1988). On the one hand, bryophytes can inhibit germination of seeds by growing in thick mats without gaps available to vascular plants. For example, during restoration of a dry sandy grassland, a German study showed that the carpet building (pleurocarpous) moss *Hypnum cupressiforme* var. *lacunosum* severely inhibited vascular plant species regeneration by building dense mats (Eichberg et al., 2010). These mats had to be broken up by grazing and trampling of livestock to ensure vascular plant recovery. Similar inhibitory effects of bryophyte layers on seedling recruitment were shown in permanently wet meadows (Špačková et al., 1998; Kotorová & Lepš 1999; Eckstein et al., 2004), semi-arid steppe grasslands (Serpe et al., 2006) as well as in xerophytic grassland on calcareous gravel (Jeschke & Kiehl 2008). The latter study was carried out over 26 months in order to show effects of moss layer removal on germination and establishment of vascular plants. Results suggested that shading and the barrier of the bryophyte layer inhibited germination and were assumed the most important negative effects on vascular vegetation. Meanwhile, during a harsh winter, seedling mortality in plots with an intact bryophyte layer was significantly lower compared to plots with bryophyte layer removal (Jeschke & Kiehl 2008). This indicated an additional protective role of the bryophytes against heavy frosts (Oostermeijer et al., 1994). Forming a dense matrix below vascular plants, bryophytes can thus provide safe conditions for seeds as the bryophyte matrix minimizes seed predation risk and prevents desiccation of seedlings (Špačková et al., 1998). Several authors (van Tooren 1988; During & van Tooren 1990; Ryser 1993; Bissels et al., 2004) suggested that the protective role of a moist moss layer is especially important in grasslands subjected to a severe summer drought or distinct seasonal wet-dry cycles. Bryophytes absorb nutrients during their growth period and can thus subsequently act as an important nutrient source to vascular plants due to N, C and P rich leachates from their decaying biomass (Turetsky 2003; Startsev & Lieffers 2006). In contrast, bryophytes may also leak cell solutes, including terpenoids and phenolics (Inderjit 1996; Asakawa 2007), which primarily suppress herbivory on bryophyte plants, but also have allelopathic effects on vascular plants by inhibiting shoot and root growth (Rice 1984; Bates 2009; Oliver 2009; Kato-Noguchi et al., 2010). Liverworts might have even stronger inhibitory effects on seed germination and root development than mosses due to the large oil bodies in their leaf cells containing terpenes but also other species-specific phytotoxic secondary metabolites (Asakawa 2007; Xie & Lou 2009). However, further evidence for such effects in natural grassland systems is scarce, but see Tsubota et al. (2006) and Michel et al. (2011).

BRYOPHYTE VEGETATION IN GRASSLANDS

In Central Europe, except for areas where forest growth is inhibited by harsh environmental conditions, e.g. in the high mountains, all semi-natural grasslands are created and shaped by grazing and/or mowing and thus depend on continuous management. After abandonment, succession begins and, on the long run, forest re-growth replaces grassland vegetation (Ellenberg & Leuschner 2010). Originally, these semi-natural grasslands were extremely low-input systems and their productivity depended strongly on the natural resource supply. Thus, except for sites with natural nutrient influxes such as floodplains, semi-natural grasslands have mostly been depleted of nutrients due to the constant removal of biomass (Dieterich & Beinlich 2009). Over the centuries and under such low-productive conditions, Central European grasslands have become an extremely species-rich habitat for both flora and fauna. Beside high numbers of vascular plant species, grasslands can contain a large diversity and proportion of bryophytes, for example in dry grasslands or fen grasslands.

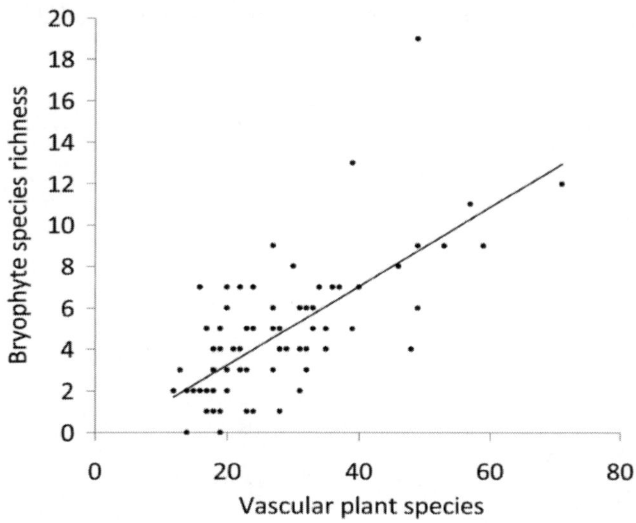

Figure 3. Positive relationship between species richness of bryophytes and vascular plants per 16 m^2 in 85 agricultural used grasslands in Germany ($R^2 = 0.50$; $p < 0.001$; modified after Müller et al., 2012).

Moreover, high plant species richness promotes the diversity of herbivorous species and subsequent species of higher trophic levels via bottom-up effects (e.g. Balvanera et al., 2006). As coexistence and persistence mechanisms are generally similar for vascular plants and bryophytes, a generally positive relationship between species richness` of these plant taxa exists, with highest species numbers in low-productive and low-intensively used grasslands (Figure 3). Consequently, unimproved semi-natural grasslands of the temperate zone are among the most species-rich habitats in the world and of outstanding conservation value (Wilson et al., 2012). In the last decades, intensification of land use and landscape scale eutrophication caused severe losses in grassland bryophytes and overall grassland diversity (Tscharntke et al., 2005; Müller et al., 2012; see section below).

In temperate grasslands, bryophytes form a distinct vegetation layer located under the canopy of tall-growing herbaceous vascular plants (Dierschke & Briemle 2002). As this bryophyte layer is situated on the soil surface, all bryophytes are directly affected by soil pH and site hydrologic conditions. Virtanen et al. (2000) observed a strong negative relationship of bryophyte occurrence with decreasing soil pH, resulting in virtually no grassland bryophyte at values below pH 4. Due to the physiological peculiarities of bryophytes, such as poikilohydry and the wide variability in desiccation tolerance, relationships between bryophyte vegetation and moisture conditions are rather complex, with habitat specialists at both ends of the moisture gradient (Oliver et al., 2005; Ellenberg & Leuschner 2010). Thus, although bryophytes are found in grasslands along the whole moisture gradient, they successfully compete with vascular plants only at rather extreme abiotic conditions, especially in very moist and very dry grasslands (Figure 4). Otherwise, vascular vegetation will widely suppress bryophytes and decrease their diversity tremendously. Similar to moisture conditions, the nutrient status of the respective grassland is of great importance for the dominance of vascular plants and the persistence of bryophyte vegetation (Figures 4 and 5). Consequently, diversity, species richness and density of bryophytes considerably differ among different grassland types (Table 1 and references therein). In the following, we relate "grassland-like" bryophyte habitats to water levels and refer to differences in nutrient availability within these categories.

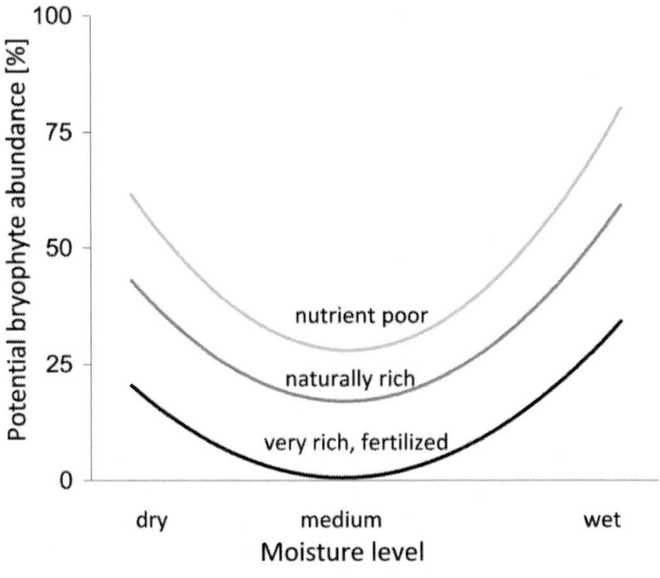

Figure 4. Potential abundance of bryophytes in grassland ecosystems along a gradient of soil moisture. Different curves represent conditions of different nutrient levels. Upper grey curve: nutrient-poor conditions, lower grey curve: naturally more nutrient-rich conditions, black curve: disturbed and very nutrient-rich conditions e.g. under rather intensive agricultural use. Please note that intensive land-use under very wet or very dry conditions is unlikely.

Table 1. Total species numbers and species densities per plot in bryophytes surveys in different grassland types throughout Europe. Numbers given can be either the reported span or a mean value of the respective study. Please note that both values depend on grassland type but also on study size as well as plot size

Study	Location	Grassland type	Area	Total species number	Cover (%)	Plot size	Bryophyte density per plot
Bergamini et al. (2001b)	Switzerland	Grazed & mown fen grasslands	36 sites, five 2 m × 1m plots in each	126	-	2 m^2	4 to 22
Hydbom et al. (2012)	Sweden	Grazed calcareous sandy grasslands	4 control sites with transects of 24 suplots of 1 m^2 each	16	-	1 m^2	-
Ingerpuu et al. (1998)	Estonia	Mown dry calcareous grasslands	2 control plots of fertilization experiment, three 1 m × 1m subplots in each	21	60	1 m^2	4.7 (3-6)
Keizer et al. (1985)	Netherlands	Mown dry calcareous grasslands	200 plots of 0.1 m × 0.1 m at 6 ha site	47	30 (9-57)	0.01 m^2	-
Löbel et al. (2006)	Sweden	Dry calcareous grasslands, mostly grazed	452 plots of 2 m × 2 m (southern Öland)	153	31 (0-85)	4 m^2	11.4 (0-40)
Müller et al. (2012)	Germany	Mown & grazed grasslands from low to high land-use intensity	85 sites, one 4 m × 4 m plots in each	45	11 (0-60)	16 m^2	4.6 (0-19)
Peintinger & Bergamini (2006)	Switzerland	Calcareous fen grasslands, mostly abandoned	21 sites, four 0.2 m × 0.2 m plots in each	49	-	0.04 m^2	7 (0-12)
Takala et al. (2012)	Finland	Grazed mesic grasslands	21 sites with twenty 0.6 m × 0.6 m plots in each	42	0.1-64	0.36 m^2	4 (0.3-5.6)
Vanderpoorten (2004)	Belgium	Mown calcareous grassland	Six 100 m^2 plots with ten 1 m^2 subplots in each	-	-	1 m^2	6.8
Zechmeister et al. (2003)	Austria	Mesic grasslands	31 sites with ten 1 m^2 plots in each	28	-	2 m^2	-

Grasslands with High to Very High Soil Moisture

Bryophytes are superior to vascular plant vegetation at the extreme end of the moisture gradient (Figure 4). Thus, they are a crucial part of peatland ecosystems dominating the vegetation of many fens and bogs. As bryophytes are highly efficient in nutrient acquisition at very nutrient poor (ombrotrophic) conditions, their productivity and biomass can easily exceed those of vascular vegetation. Therefore, bryophytes are the most important primary producers, which significantly contribute to carbon fixation in undisturbed peatlands (Vitt 2000). Moist heathlands are quite similar to nutrient-rich peatlands with a higher proportion of vascular plants, causing a multiplicity of transitions from heathlands to mires, degraded bogs and fen grasslands. Many of these habitats have a potentially rich and dense layer of bryophytes, sometimes more than 60% of the total plant species richness (Hobbs & Grimingham 1984). In moist heathlands, many large (pleurocarpous) perennial species such as *Hylocomium splendens, Pleurozium schreberi and Scleropodium purum* occur, forming thick mats and tufts, which cannot endure disturbance very well (Vandvik et al., 2005).

Because of habitat loss, widespread nutrient enrichment, anthropogenic acidification and drainage of wet habitats (see below), many mosses and liverworts typical for peat- and heathlands are threatened in Central Europe (Kooijman 1992). Today, the vast majority of these ecosystems is under agricultural use, often as intensively used grasslands or arable fields (Ellenberg & Leuschner 2010). While near natural wet grasslands such as calcareous fen meadows are suitable habitats for bryophyte species (Bergamini et al., 2001b), drained wet grasslands are charaterized by strong fluctuations of water table leading to alternate phases of drought in summer and flooding in winter. Together with direct impacts of increased land-use intensity and indirect impacts of nutrient releases from peat mineralization, such extremely fluctuating hydrologic conditions widely inhibit the establishment of rich bryophyte layers and cause generally species-poor grass-dominated vegetation (Hellberg & Kundel 1995). Müller et al. (2012) found in northeastern Germany a mean bryophyte cover of less than 1% and a mean number of 1.7 species on 16 m^2 plots in drained fen grasslands, where mostly ubiquitous species such as *Brachythecium rutabulum, Ditrichum cylindricum* and *Physcomitrium pyriforme* occurred. In contrast, from near-natural fen grasslands in Switzerland, Bergamini et al. (2001b) reported more than 120 different bryophytes species with up to 22 species at 2 m^2 plots (see also Table 1).

In regularly inundated floodplain grasslands along large rivers, the vegetation comprises only occasional bryophyte vegetation, because naturally fluctuating water levels due to changing river discharge but also to regular depositions of sediments hamper the establishment of bryophyte vegetation (Ellenberg & Leuschner 2010; Klaus et al., 2011). Similar to peatlands, most floodplain grasslands were strongly improved for agricultural purposes or converted into arable fields (Nebel & Philippi 2000). This situation might be somewhat better in smaller grasslands along brooks and in mountain areas, which on the contrary often suffer from abandonment. Generally, while permanently wet conditions with balanced soil moisture and low nutrient status favor bryophyte vegetation, strongly fluctuating water tables of drained wet grasslands going along with increased nutrient availability are severely detrimental to bryophyte species.

Mesic Grasslands

Under suitable moisture conditions and with sufficient nutrient supply, vascular plants widely dominate the vegetation of mesic grassland, with bryophytes at the most constituting a sparse and comparable species poor layer under the canopy of the tall-growing vascular plants (Figure 4). The degree to which bryophytes are part of mesic grassland communities is significantly affected by the vascular plant pressure, as modified by nutrient levels. In nutrient rich grasslands, nearly no bryophytes can be found, while in comparably nutrient-poor and less productive mesic grasslands, a reasonably dense moss layer can develop (Bergamini et al., 2001a; Dierschke & Briemle 2002; Ellenberg & Leuschner 2010). Müller et al. (2012) found up to 20 species on 16 m^2 plots in extensively used mesic grasslands in Germany (Table 1). On very intensively used grasslands, bryophyte communities adapted to a high disturbance regime occur, as they are otherwise typical of arable land. These specialist species can quickly regenerate and colonize open soil patches due to the diaspore bank with vegetative tubers or bulbils and spores, giving at least some bryophyte species the ability to endure in this extreme environment. Thus, compared to vascular plant communities, bryophyte communities show a higher fluctuation. Consequently, bryophytes in mesic grasslands show a primarily opportunistic behavior with rather poor contribution to ecosystem functioning (Herben 1987).

As all gardeners know, bryophytes can be a significant part of urban grasslands, in both private and public lawns. Relatively high humidity (e.g. by tree shading) together with insufficient aeration of the soil, frequent short mowing above the heads of the mosses and rather acidic conditions can strongly foster certain bryophyte species. This often results in a dense and thick moss layer and a simultaneous drastic decline in the cover of vascular vegetation. However, richness of bryophytes is often quite low in urban grasslands with few dominant species yielding most of the biomass. In Central Europe, *Rhytidiadelphus squarrosus* is often the prevailing species forming dense mats and excluding other plant species. In areas that are too shady and moist to support dominant grass growth, using moss as an alternative ground cover is a slowly growing gardening trend, which is already used in Japan for centuries (Nebel & Philippi 2000). However, such moss layers often consist of just one or two species. A further factor favoring bryophytes in some urban grasslands is a high density of free-living rabbits, which cause exceptionally short vegetation and create open gaps for regeneration, both strongly favoring bryophytes (Porley 1999; Vanderpoorten et al., 2004).

Dry Grasslands

All grasslands characterized by periodical drought and rather nutrient-poor conditions are potentially important bryophyte habitats. Together with lichens, bryophytes can dominate the vegetation of these low productive dry grasslands. For example, Ingerpuu et al. (1998) found up to 63 species in Estonian calcareous dry grasslands, while Löbel et al. (2006) found even up to 153 species in similar grasslands on Öland/ Sweden (Table 1). A key for the bryophyte dominance is the low nutrient availability of the system, but also the high desiccation tolerance of some bryophyte species withstanding long phases of drought but recovering within a few hours after rewetting (see section above). This mechanism is crucial to enable bryophytes to compete successfully with vascular plants in particularly dry environments (Oliver et al., 2005). In temperate zones, different factors can cause drought conditions: (micro) climate, extreme soil conditions like permeable substrates or thin soil layers, deep groundwater tables and topographic exposure on steep slopes with quick run-off and high irradiation. Mostly, combinations of these factors provoke exceptionally low stand productivity, giving bryophytes the possibility to establish in the comparably sparse vegetation layer and to compete with vascular plants "at eye level". Consequently, several different types of dry grasslands can be found, differing in soil substrate, altitude, exposition, species pool, site history and management (Ellenberg & Leuschner 2010). Two major groups of dry grasslands can be distinguished: siliceous versus base-rich dry grasslands, with the latter usually having a higher mean plant species richness (Ellenberg & Leuschner 2010). For example, in calcareous dry grasslands, about 200 bryophyte and lichen species are known, including a significant number of rare species restricted to this habitat type (Willems 1983; Vanderpoorten et al., 2004). A further subdivision is possible into dry grasslands on rock substrate, mostly going along with distinct inclination, grasslands on shallow skeletal soils and grasslands on sandy substrate such as xeric grasslands on dunes. Many of these semi-natural grazed grasslands reveal an outstanding diversity and cover of bryophytes, especially if sites were low intensively managed for decades or centuries. In the east of Central Europe, with overall lower annual precipitation, dry grasslands gradually turn into steppe grasslands, which can also have significant occurrences of bryophyte vegetation (Kämpf, unpublished data). However, dry continental steppes generally contain only few bryophyte species (Hallingbäck & Hodgetts 2000).

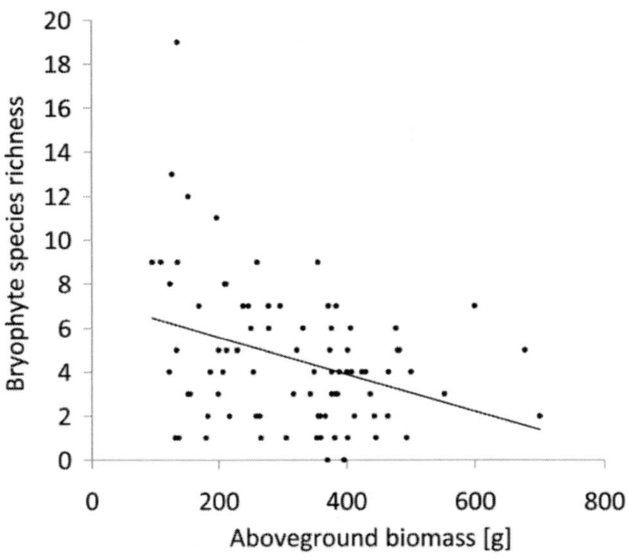

Figure 5. Relationship between bryophyte species richness and aboveground biomass of vascular plants per square meter in 85 agricultural used grasslands on 16 m^2 plots in Germany ($R^2 = 0.12$; $p = 0.0012$; modified after Müller et al., 2012).

Grasslands in abandoned stone quarries are a further suitable habitat for bryophytes, benefiting from the thin soil layer and the low nutrient status while coping with seasonally dry conditions (Sieg & Gilhaus, unpublished data). Furthermore, dry heathlands are a typical habitat with potentially high bryophyte diversity, harboring many rather fugitive small colonizing species (Vandvik et al., 2005). As heathlands in central Europe have to be managed by fire, grazing or sod cutting to inhibit succession towards forest, they contain several highly specialized species (Ellenberg & Leuschner 2010). For example, in heathlands managed by burning, fire-dependent species such as *Ceratodon purpureus* and *Leptodontium flexifolium* can be found (Vandvik et al., 2005). After burning a heath, certain bryophyte species act as important early stage colonizers paving the way for subsequent species by moderating extreme site conditions (Piessens et al., 2008). However, today's remaining dry heathlands are often quite depleted in bryophyte richness due to acidification as well as landscape-scale eutrophication and subsequent vegetation changes (Ellenberg & Leuschner 2010).

TRENDS, DRIVERS AND DYNAMICS

Temporal Dynamics of Bryophyte Communities

Community assembly of bryophyte vegetation in grasslands strongly depends on light, moisture and temperature conditions, which often show considerable shifts with time. Therefore, bryophyte communities naturally fluctuate among seasons and years (Nebel & Philippi 2000). In dry grasslands, the growth of mosses is clearly enhanced by occasionally high annual precipitation (Ingerpuu & Kupper 2007). As bryophytes show a rather opportunistic behavior, bryophyte communities exhibit higher fluctuations and a rather immediate response to environmental changes and seasonal shifts in disturbed grasslands, compared to grasslands with low disturbance regimes (Herben 1987). For many bryophyte species, early spring days following snow melt constitute the most important vegetation period of the year (Glime 2007). Later in the year, increased growth of vascular plants accompanied with reduced humidity due to generally higher temperatures leads to decreased bryophyte growth and eventually also to the point of bryophyte die off during summer. Thus, bryophyte vegetation is exposed to a high seasonality with great differences in abundance and appearance of certain small bryophytes species. By using the time niche when vascular plants cannot tap their full growth potential, bryophytes are able to remain even in productive grasslands (Zechmeister et al., 2003; Müller et al., 2012). These pronounced seasonal differences in community assembly have also to be taken into account when recording bryophytes in grasslands. Optimally, this is done in early spring whereas later in the season bryophytes species richness and abundance can be easily underestimated (Figure 6).

Due to their physiological sensitivity, bryophyte communities will be significantly affected by long-term climatic change. A study by Bates et al. (2005) on effects of climate change on bryophytes in a calcareous grassland included different treatments combining drought, warming and supplementary rain. Outcomes revealed that responses to the climate change scenarios varied considerably among bryophyte species. While some species increased, e.g. as a result of supplementary water supply, others decreased due to summer drought and/or winter warming. In general, experimental summer drought resulted in a drop of bryophyte cover, in line with Ingerpuu & Kupper (2007), while winter warming decreased bryophyte species richness. However, the study also showed that natural environmental factors were more important to explain the major variation in the bryophyte community, overruling effects of the simulated climate change treatments (Bates et al., 2005).

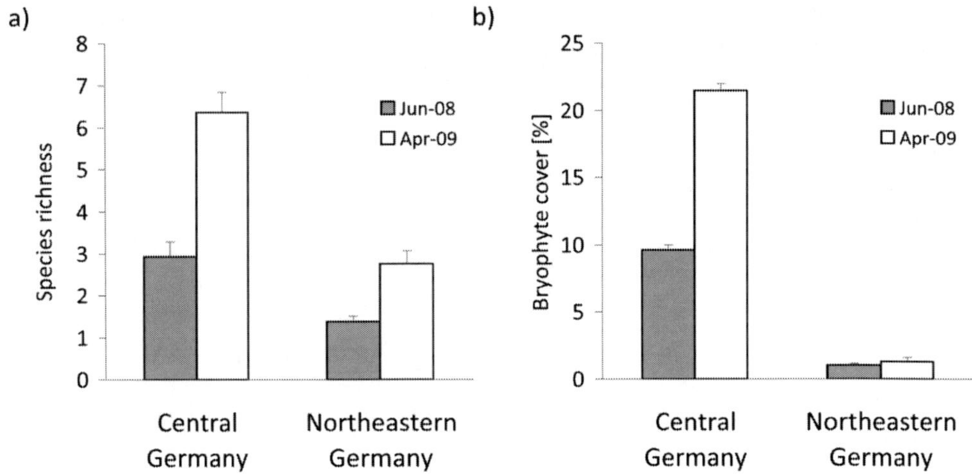

Figure 6. Seasonal differences in a) observed bryophyte species richness and b) cover on 16 m^2 plots in agricultural grasslands in Central (n = 43) and Northeastern Germany (n = 42). Data was derived by repeating bryophyte recordings in June 2008 and April 2009 by the same person (Müller et al., unpublished data).

Effects of Land-Use on Grassland Bryophytes

As grasslands with rich bryophyte vegetation are primarily restricted to low-productive, nutrient-poor habitats, they mostly contrast demands of modern agricultural land-use. If such grasslands are not under strict nature conservation, three main developments threaten these species-rich habitats: i) conversion into arable land, ii) abandonment or afforestation and iii) land-use intensification of remaining grasslands. Thus, in Central Europe, the area covered by permanent grasslands has been in constant decline and within the last 20 years, about four million hectares of grasslands have been abandoned or converted to arable land in the EU. This has led to a great loss of habitats for the majority of grassland species (Carlier et al., 2005; Smit et al., 2008). Abandonment of grasslands generally results in accumulation of litter and dead plant material as well as establishment of tall-growing vegetation. Both factors reduce light availability for bryophytes on the ground. However, the progress of abandonment differs significantly between different types of grassland. In wet and moist grasslands such as fen meadows, tall-growing vegetation and litter fall quickly start shadowing out bryophytes (Nebel & Philippi 2000; Peintinger & Bergamini 2006), whereas in nutrient-poor and dry habitats, vegetation communities are more stable and succession of tall growing vegetation is considerably slower. Vanderpoorten et al. (2004) found that bryophyte species richness of calcareous grasslands in southern Belgium did not differ between mowed and unmanaged plots even after 15 years. However, progressive enrichment of grass litter (e.g. of *Brachypodium pinnatum)* and shrub encroachment are assuredly detrimental for bryophytes on the very long run (Bobbink 1991).

As bryophyte communities are generally quite sensitive to environmental conditions, they directly respond to land-use intensification by shifts in species richness, compositions and cover. Studies recorded multiple negative impacts of land use on bryophytes provoking a shift towards few nitrogen tolerant generalist species (Zechmeister et al., 2002). For example, Nebel & Philippi (2000) found that occurrences of *Weissia squarrosa*, *W. rostellata*, *Ephemerum cohaerens*, *Bryum weigelii* and *Plagiomnium elatum* in grasslands strongly decreased due to intensification of land use in Germany. Fertilizer applications and high mowing frequencies are the most adverse factors for bryophyte diversity (Zechmeister et al., 2003; Müller et al., 2012). Fertilization has both direct and indirect negative effects on bryophytes. First, direct toxicity of nitrogen leads to the dieback of sensitive species, while secondly reduced light availability arises from enhanced aboveground growth of vascular plants (van der Wal et al., 2005). This was also shown by experiments, which found vascular plant dry biomass above 400 $g*m^{-2}$ to be particularly detrimental for bryophyte vegetation (Aude & Ejrnæs 2005; Hejcman et al., 2010). A fertilization study by Bergamini & Pauli (2001) in semi-natural fen grasslands revealed that NPK-fertilization was particularly harmful for bryophytes resulting in a decrease of bryophyte biomass by more than 50% and a decrease in species richness by more than 15% within two years. A more indirect effect of today's land-use intensification on bryophytes in regions with enhanced stockbreeding are aerial depositions of nitrogen at the landscape scale, which can easily exceed the critical load of nutrient-poor habitats and alter plant community composition in wet and dry grasslands (e.g. Bobbink 1991; Bergamini et al., 2009). Drainage of wet grasslands is also severely harmful for grassland bryophytes due to subsequent fluctuations of the water table and nutrient releases from decomposing organic soils (Lamers et al., 2002). While high water table in winter causes O_2 stress, drainage leads to severe droughts in summer, discriminating against both flooding tolerant and mesic species. Such extreme hydrologic conditions of drained wet grasslands generally cause species-poor grass-dominated vegetation with only occasional bryophyte occurrences (Hellberg & Kundel 1995; Müller et al., 2012; Klaus et al., 2013).

Mowing has both direct and indirect effects on bryophytes, but in opposed directions to each other. The direct effect of mowing can be positive due to higher light penetration in nutrient-rich environments arising from the removal of aboveground biomass of vascular plants (Aude & Ejrnæs 2005). However, mowing intensity in agricultural systems is also associated with higher fertilizer applications and soil compaction due to the use of heavy machines (Müller et al., 2012). Thus, the latter two factors may overrule positive direct effects of mowing resulting in an overall negative relationship. While bryophyte diversity generally benefits from grazing by cattle or sheep, comparisons of grazing and mowing treatments in fen grasslands did not always reveal significant differences in bryophyte species richness (Bergamini et al., 2001b; Stammel et al., 2003). Positive effects of grazing were found in calcareous dry grasslands (van Tooren 1990), montane fen grasslands (Bergamini et al., 2001a; Peintinger and Bergamini 2006); heathlands (Piessens et al., 2008) and mesic grasslands (Takala et al., 2012). This beneficial effect can be mainly attributed to the creation of gaps in the dense vegetation layer by i.e. trampling and grubbing but also to the heterogeneous removal of biomass (Aude and Ejrnæs 2005; Mayer et al., 2009; Preston et al., 2009). In mesic grasslands in Finland, the degree of grazing pressure within the studied range was positively related to the average cover of bryophytes (Takala et al., 2012), while this will not be the case when stocking rates reach very high densities (Müller et al., 2012). In general, patchy disturbance of the soil surface supports bryophytes in vascular plant dominated stands.

In addition to natural disturbances by digging and burrowing animals (bioturbation), disturbance is mostly driven by agricultural land use. While intermediate soil disturbance (10-15% bare ground) increases bryophyte diversity in grasslands, high level of soil disturbance (>15%) negatively affect bryophyte diversity (Müller et al., 2014). Thus, the relationship between open soil and bryophyte species richness in grasslands is approximately hump-shaped (Figure 7).

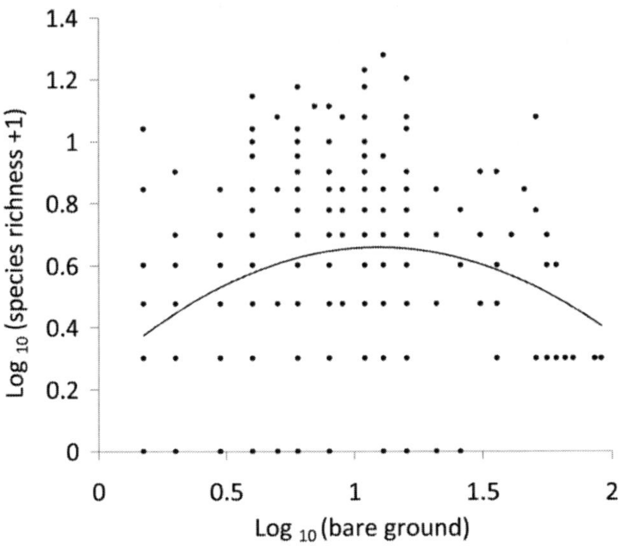

Figure 7. Relationship between bare ground and bryophyte species richness on 3 m^2 sized plots in a soil-disturbance experiment in temperate grasslands ($R^2 = 0.07$; $p < 0.001$; redrawn from Müller et al., 2014).

Effects of Bryophytes on Ecosystem Functioning

Bryophytes definitively play a relevant role for ecosystem functioning of grasslands. Besides interactions with other plants, which are discussed above, the retention of nutrients and water as well as the regulation of the microclimate and contributions to overall productivity are among the most important impacts of bryophyte vegetation in grasslands. Obviously, the more abundant bryophytes are, the more they are important for ecosystem functioning in the particular grassland. Thus, especially in nutrient poor environments, bryophytes are very important functional components.

The special tissue structure of bryophytes provokes a very high water holding capacity of up to 1400% of their own dry mass (Glime 2007; Proctor 2008). In contrast to vascular plants, bryophytes can obtain water directly by absorbing moisture from the atmosphere as dew, fog or rainfall. Additionally, they can also absorb condensed water vapor from the ground (Anderson & Bourdeau 1955; Glime 2007). Thus, bryophytes can act as storage for intercepted water, slowly re-delivering it into the surrounding (Schulten 1985; Belnap 2006; Michel et al., 2013). For this reason, area-based water storage capacity is increasingly considered as an important feature of bryophyte vegetation, especially in sparsely vegetated environments with high seasonal precipitation (Bowker et al., 2010). However, the relative contribution of bryophytes to the water holding capacity of ecosystems is poorly known for grassland communities.

Bryophytes also contribute to nutrient cycling by taking up and storing nutrients in seasons when vascular plants are widely inactive and nutrients in the soil are prone to washout. Later in the season, bryophytes can thus increase N availability for vascular plants, when N is slowly released from the decomposition of bryophyte litter (Startsev & Lieffers 2006). Although this mechanism is irrelevant in intensively used grasslands, which are usually strongly fertilized (Blüthgen et al., 2012), bryophytes play an important role in keeping nutrients in systems of low and medium fertility (Turetsky 2003). Moreover, not only the presence but also the diversity of bryophytes may have positive effects on ecosystem functioning. However, while experimental studies highlighted diverse benefits of vascular plant diversity for several ecosystem processes (e.g. productivity and nitrogen cycling) in artificially created grassland communities (e.g. Tilman et al., 2001; Hooper et al., 2005; Weigelt et al., 2009) and in semi-natural grasslands (Stein et al., 2008; Kleinebecker et al., 2014), detailed studies on benefits of bryophyte diversity on ecosystem functioning are lacking. Although many plant diversity experiments in grasslands have been conducted with enormous financial efforts, neither bryophyte presence nor bryophyte diversity has been studied.

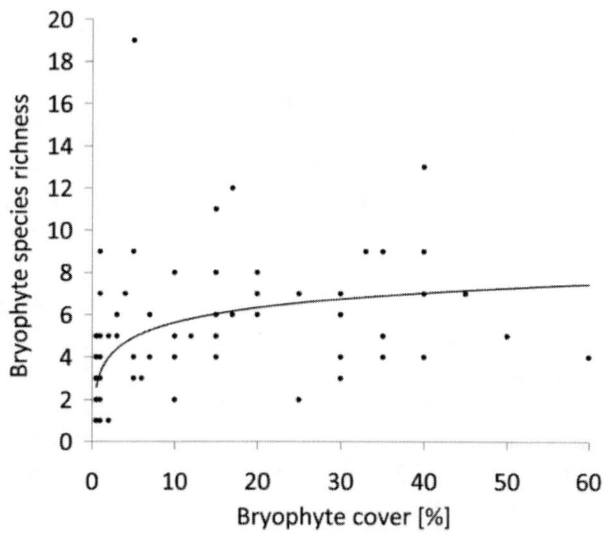

Figure 8. Relationship between bryophyte cover and species richness in 83 grassland plots (4 m x 4 m) in Germany with species numbers > 0 (Müller et al., unpublished). Bryophyte cover can be used as proxy for bryophyte biomass (MacDonald et al., 2012).

In wet and dry grasslands, where vascular plants have to cope with harsh environmental conditions, bryophytes can significantly contribute to stand productivity. However, on the other hand, extensive bryophyte mats can also reduce overall productivity of a grassland site by suppressing vascular vegetation, for example in urban grasslands. Use of bryophyte cover as a surrogate for biomass production, which is otherwise difficult to measure (MacDonald et al., 2012), revealed the sufficiency of 6 to 8 species to sustain maximum bryophyte biomass (i.e. related to maximum cover) in grasslands (Figure 8). Together with productivity, further ecosystem processes are likely to depend not only on bryophyte presence but also on bryophyte diversity in a similar way, for example contributions to carbon fixation (Bortoluzzi et al., 2006), nutrient cycling or decomposition dynamics (Turetsky 2003; Garcia-Pausas 2004). High bryophyte diversity (accompanied with high abundance) has also been shown to improve water retention capacity and microclimate regulation (Beringer et al., 2001).

RE-CREATING BRYOPHYTE HABITATS: ECOLOGICAL RESTORATION

Although bryophytes significantly contribute to plant diversity and ecosystem functioning, neither grassland conservation nor restoration has seriously taken bryophyte diversity into account (Vanderpoorten et al., 2004). Though grassland restoration is a broad and well-developed research field, very few projects explicitly considered bryophytes as a typical and important element of grasslands (Jeschke & Kiehl 2006). For peatlands, the re-establishment of bryophyte vegetation, especially of *Sphagnum*, but also other species, is well applied by using gametophyte fragments as propagules to stimulate carbon fixation or biomass production after perturbation (Mälson & Rydin 2007). However, so far, bryophyte reintroductions are widely neglected in grassland restoration projects. This might be partly linked to differences in ecological demands of bryophytes versus vascular plants. Indeed, while for example topsoil removal is often necessary to lower nutrient availability to enhance bryophyte diversity, this cost-intensive measure is rather rarely applied because it is not necessary to conserve most vascular plant species. Jeschke & Kiehl (2006) used topsoil removal during experimental grassland restoration and increased bryophyte and lichen species richness from 12 to 19 including five red-list species. Hydbom et al. (2012) found that increased pH due to soil perturbation resulted in increased bryophyte species richness in restored calcareous sand grasslands.

Where bryophyte species richness decreases due to abandonment, the re-introduction of grazing animals can quickly raise bryophyte diversity (Takala et al., 2012). However, Takala et al. (2012) also showed that re-grazed sites did not reach the full bryophyte diversity compared to continuously grazed grasslands, pointing out the constraints of dispersal limitation when restoring bryophyte vegetation. Unfortunately, restoration techniques to re-introduce plant species to new habitats usually employ plant material poor in bryophyte and lichen propagules such as commercial seed mixtures or common hay material. By contrast, using sods or topsoil as well as raked or harrowed material have been proven to foster the successful re-establishment of typical bryophyte-rich dry grassland vegetation (Eichberg et al., 2010; Kiehl et al., 2010). However, greater efforts have to be made in developing bryophyte-friendly ecological restoration techniques and in monitoring their outcomes, because bryophytes are typical and important components of grassland ecosystems.

FUTURE DIRECTIONS

As mentioned above, many open questions should be tackled in the future to increase our limited knowledge on bryophytes and their role in the functioning of grassland ecosystems (Turetsky 2003; Cornelissen et al., 2007).

- Several fundamental research questions relate to diversity effects of bryophytes on ecosystem functions, mechanisms of nutrient re-delivery from bryophyte mats to surrounding vegetation as well as trade-offs between benefits of bryophyte biomass for productivity versus oppression of vascular vegetation. Including bryophytes in plant diversity experiments in grasslands would be a first step, to give bryophytes the relevance they deserve.
- Research should also focus on comparably easy approaches such as removal experiments testing effects of bryophyte presence and bryophyte diversity on several ecosystem processes. Such experiments must be carried out for a longer period of time to provide insights in inter- and intra-annual differences of such effects and relationships.
- There are also important remaining questions on the effect size of bryophytes on the recruitment of vascular vegetation. Here, long-term climate data should be used to account for potentially opposed effects of moist and dry years. As pointed out above, competitive interactions among bryophyte species have also hardly been studied.
- Furthermore, bryophytes in grasslands could be used as bio-indicators ("phytometer") to track nutrient fluxes, pollutants or climate change in grasslands.
- More research should also be done on bryophyte effects on microclimate such as buffering low and/or high temperatures. Connected to this, severe concerns of climate change effects on bryophytes in grasslands, but also possible buffer functions of bryophyte vegetation such as water storage and water re-delivery should be taken into account in ecological research.
- Finally, as many bryophyte species are currently declining, restoration approaches have to be developed (or refined) and their outcomes should be thoroughly monitored.

Even if we do not know all effects of bryophytes in grasslands, we have to regard these small plants as essential members of most grassland ecosystems in the temperate zone, with a unique diversity and various contributions to ecosystem functioning.

REFERENCES

Al-Mufti, M.M., Sydes, C.L., Furness, S.B., Grime, J.P., Band, S.R., (1977). A quantitative analysis of shoot phenology and dominance in herbaceous vegetation. *J. Ecol.* 65, 759-791.

Anderson, L.E., Bourdeau, P.F., (1955). Water relations in two species of terrestrial mosses. *Ecology* 36, 206-212.

Asakawa, Y.A., (2007). Biologically active compounds from bryophytes. *Pure Appl. Chem.* 79, 557-580.

Aude, E., Ejrnæs, R., (2005). Bryophyte colonisation in experimental microcosms, The role of nutrients, defoliation and vascular vegetation. *Oikos* 109, 323-330.

Ayres, E., van der Wal, R., Sommerkorn, M., Bardgett R.D., (2006). Direct uptake of soil nitrogen by mosses. *Biol. Lett.* 2, 286-288.

Balvanera, P., Pfisterer, A.B., Buchmann, N., He, J.-S., Nakashizuka, T., Raffaelli, D., Schmid, B., (2006). Quantifying the evidence for biodiversity effects on ecosystem functioning and services. *Ecol. Lett.* 9, 1146-1156.

Bates, J.W., Thompson, K., Grime, J.P., (2005). Effects of simulated long-term climatic change on the bryophytes of a limestone grassland community. *Glob. Change Biol.* 11, 757-769.

Bates, J.W., (2009). Mineral nutrition and substratum ecology. In: Goffinet, B., Shaw, A.J., [eds]. *Bryophyte biology*. Cambridge Univ. Press, Cambridge, 300-356.

Belnap, J., (2006). The potential roles of biological soil crusts in dryland hydrology cycles. *Hydrol. Process.* 20, 3159-3178.

Bergamini, A., Pauli, D., (2001). Effects of increased nutrient supply on bryophytes in montane calcareous fens. *J. Bryol.* 23, 331-339.

Bergamini, A., Pauli, D., Peintinger, M., Schmid, B., (2001a). Relationships between productivity, number of shoots and number of species in bryophytes and vascular plants. *J. Ecol.* 89, 920-929.

Bergamini, A., Peintinger, M., Schmid, B., Urmi, E., (2001b). Effects of management and altitude on bryophyte species diversity and composition in montane calcareous fens. *Flora* 196, 180-193.

Bergamini, A., Ungricht, S., Hofmann, H., (2009). An elevational shift of cryophilous bryophytes in the last century – an effect of climate warming? *Diversity Distrib.* 15, 871-879.

Beringer, J., Lynch, A.H., Chapin, F.S., Mack, M., Bonan, G.B., (2001). The representation of arctic soils in the land surface model: the importance of mosses. *J. Climate* 14, 3324-3335.

Biermann, R., Daniëls, F.J.A., (1997). Changes in a lichen-rich dry sand grassland vegetation with special reference to lichen synusia and *Campylopus introflexus*. *Phytocoenologia* 27, 257-273.

Bissels, S., Hölzel, N., & Otte, A., (2004). Population structure of the threatened long-lived perennial *Serratula tinctoria* in relation to vegetation and management. *Appl. Veg. Sci.* 7, 267-274.

Blüthgen, N., Dormann, C., Prati, D., Klaus, V.H., Kleinebecker, T., Hölzel, N., et al., (2012). A quantitative index of land-use intensity in grasslands: integrating mowing, grazing and fertilization. *Basic Appl. Ecol.* 13, 207-220.

Bobbink, R., (1991). Effects of nutrient enrichment in Dutch chalk grassland. *J. Appl. Ecol.* 28, 28-41.

Bortoluzzi, E., Epron, D., Siegenthaler, A., Gilbert, D., Buttler, A., (2006). Carbon balance of a European mountain bog at contrasting stages of regeneration. *New Phytologist* 172, 708-718.

Bowker, M.A., Maestre, F.T., Escolar, C., (2010). Biological crusts as a model system for examining the biodiversity–ecosystem function relationship in soils. *Soil Biol. Biochem.* 42, 405-417.

Carlier, L., De Vliegher, A., Van Cleemput, O., Boeckx, P., (2005). Importance and functions of European grasslands. *Commun. Agric. Appl. Biol Sci.* 70, 5-15.

Cornelissen, J.H.C., Lang, S.I., Soudzilovskaia, N.A., During, H.J., (2007). Comparitive cryptogam ecology: a review of bryophyte and lichen traits that drive biogeochemistry. *Ann. Bot.-London* 99, 987-1001.

Dierschke, H., Briemle, G., (2002). *Kulturgrasland.* Ulmer, Stuttgart.

Dieterich, M., Beinlich, B., (2009). The calcareous grasslands of the Schwäbische Alb, Geramany: cultural and natural heritage. In: Van Veen, P., Van Der Straaten, J., [eds.]. *Grasslands in Europe of high nature value.* KNNV Publishing, Zeist, 104-111.

During, H.J., Lloret, F., (2001). The species-pool hypothesis from bryological perspective. *Folia Geobot.* 36, 63-70.

During, H.J., van Tooren, B.F., (1990). Bryophyte interactions with other plants. *Bot. J. Linn. Soc.* 104, 79-98.

During, H.J., (1979). Life strategies of bryophytes: a preliminary review. *Lindbergia* 5, 2-17.

During, H.J., (2001). Diaspore banks. *The Bryologist*, 104, 92-97.

Eckstein, R.L., Danihelka, J., Hölzel, N. & Otte, A., (2004). The effects of management and environmental variation on population stage structure in three river-corridor violets. *Acta Oecol.* 25, 83-91.

Eichberg, C., Storm, C., Stroh, M., Schwabe, A., (2010). Is the combination of topsoil replacement and inoculation with plant material an effective tool for the restoration of threatened sandy grassland? *Appl. Veg. Sci.* 13, 425-438.

Ellenberg, H., Leuschner, C. (2010). *Vegetation Mitteleuropas mit den Alpen.* Ulmer-Verlag, Stuttgart.

Equihua, M., Usher, M.B., (1993). Impact of carpets of the invasive moss *Campylopus introflexus* on *Calluna vulgaris* regeneration. *J. Ecol.* 81, 359-365.

Frahm, J.P., (1984). Phytogeography of European *Campylopus* species. In: Vána J. [ed.], Proceedings of the Third Meeting of Bryologists of Central and East Europe, Karolinum, Prague, 191-212.

Frahm, J.-P. (2001). *Biologie der Moose.* Springer, Heidelberg, Berlin.

Frey, W., Kürschner, H., (2011). Asexual reproduction, habitat colonization and habitat maintenance in bryophytes. *Flora* 206, 173-184.

Garcia-Pausas, J., Casals, P., Romanya, J. (2004). Litter decomposition and faunal activity in mediterranean forest soils: Effects of N content and the moss layer. *Soil Biol. Biochem.* 36, 989–997.

Glime, J.M., (2007). *Bryophyte Ecology.* Volume 1. Physiological Ecology. Ebook sponsored by Michigan Technological University and the International Association of Bryologists. Accessed on 14.10.2013 at <http://www.bryoecol.mtu.edu/>.

Hejcman, M., Száková, J., Schellberg, J., Šrek, P., Tlustoš, P., Balík, J. (2010). The Rengen Grassland Experiment: bryophytes biomass and element concentrations after 65 years of fertilizer application. *Environ. Monit. Assess.* 166, 653-662.

Hellberg, F., Kundel, W., (1995). Entwicklung winterlich überfluteter Grünlandvegetation. *NNA-Berichte* 8, 22-34.

Herben, T., (1987). Bryophytes in grassland vegetation sample plots: what is their correlation with vascular plants? *Folia Geobot. Phytotax.* 22, 35-41.

Hill, M.O., Preston, C.D., Bosanquet, S.D. S., Roy, D.B. (2007). BRYOATT: attributes of British and Irish mosses, liverworts and hornworts. Centre for Ecology and Hydrology.

Hobbs, R.J., Gimingham, C.H., (1984). Studies on fire in Scottish heathland communities, II. Post-Fire Vegetation. *J. Ecol.* 72, 585-610.

Hooper, D.U., Chapin, F.S., Ewel, J.J., Hector, A., Inchausti, P., Lavorel, S., et al., (2005). Effects of biodiversity on ecosystem functioning: a consensus of current knowledge. *Ecol. Monogr.* 75, 3-35.

Hydbom S., Ödman A.M., Olsson P.A., Cronberg, N. (2012). The effects of pH and disturbance on the bryophyte flora in calcareous sandy grasslands. Nord J Bot 30: 1-7.

Inderjit (1996). Plant phenolics in allelopathy. *Bot. Rev.* 62, 186-202.

Ingerpuu, N., Kull, K., Vellak, K. (1998). Bryophyte vegetation in a wooded meadow: relationships with phanerogam diversity and responses to fertilization. *Plant Ecol* 134:163-171.

Ingerpuu, N., Kupper, T., (2007). Response of calcareous grassland vegetation to mowing and fluctuating weather conditions. *J. Veg. Sci.* 18, 141-146.

Ingerpuu, N., Liira, J., Pärtel, M., (2005). Vascular plants facilitated bryophytes in a grassland experiment. *Plant Ecol.* 180, 69-75.

Hallingbäck, T., Hodgetts, N., (2000). *Mosses, Liverworts, and Hornworts. Status Survey and Conservation Action Plan for Bryophytes.* IUCN/SSC Bryophyte Specialist Group. IUCN, Gland, Switzerland and Cambridge, UK.

Jaroszewicz, B., Sondej, I., Stebel, A., Pirożnikow, E., Kwiatkowska-Falińska, A.J., (2011). Biodiversity of bryophytes growing on the faeces of ungulates. *Cryptogam. Bryol.* 32, 221-231.

Jeschke, M., Kiehl, K., (2008). Effects of a dense moss layer on germination and establishment of vascular plants in newly created calcareous grasslands. *Flora* 203, 557-566.

Jeschke, M., Kiehl, K., (2008). Restoration and management of calcareous grasslands: Is there a trade-off between species diversity of cryptogams and vascular plants? *Extended abstract of the 6th European Conference on Ecological Restoration Ghent*, Belgium, 8-12/09/2008.

Kato-Noguchi, H., Seki, T., Shigemori, H., (2010). Allelopathy and allelopathic substance in the moss *Rhynchostegium pallidifolium*. *J. Plant Physiol.* 167, 468-471.

Keizer, P.J., van Tooren, B.F., During, H.J., (1985). Effects of bryophytes on seedling emergence and establishment of short-lived forbs in chalk grassland. *J. Ecol.* 73, 493-504.

Kiehl, K., Kirmer, A., Donath, T., Rasran, L. & Hölzel, N., (2010). Species introduction in restoration projects - evaluation of different techniques for the establishment of semi-natural grasslands in Central and Northwestern Europe. *Basic Appl. Ecol.* 11, 285-299.

Klaus, V.H., Hölzel, N., Boch, S., Müller, J., Socher, S.A., Prati, D., et al., (2013). Direct and indirect associations between plant species richness and productivity in grasslands: regional differences preclude simple generalization of productivity-biodiversity relationships. *Preslia* 85, 97-112.

Klaus, V.H., Sintermann, J., Kleinebecker, T., Hölzel, N., (2011). Sedimentation-induced eutrophication in large river floodplains - an obstacle to restoration? *Biol. Conserv.* 144, 451-458.

Kleinebecker, T., Hölzel, N., Prati, D., Schmitt, B., Fischer, M., Klaus, V.H., (2014). Evidence from the real world: ^{15}N natural abundances reveal enhanced nitrogen use at high plant diversity in Central European grasslands. *J. Ecol.* [doi:10.1111/1365-2745.12202]

Kooijman, A.M., (1992). The decrease of rich fen bryophytes in The Netherlands. *Biol. Conserv.* 35, 139-143.

Kooijman, A.M., Beltman, B., (1994). Extinction and reintroduction of the bryophyte *Scorpidium scorpioides* in a rich-fen spring site in the Netherlands. *Biol. Conserv.* 69, 87-96.

Kotorová, I., Lepš, J. (1999). Comparative ecology of seedling recruitment in an oligotrophic wet meadow. *J. Veg. Sci.* 10, 175-186.

Lamers, L.P.M., Smolders, A.J.P., Roelofs J.G.M., (2002). The restoration of fens in the Netherlands. *Hydrobiologia* 478, 107-130.

Lieske, K., (2010). Vegetative reproduction and clonal diversity in pleurocarpous mosses (Bryopsida) of xeric habitats. A combined molecular and morpho-anatomical study in the three mosses *Abietinella abietina* (Hedw.) Fleisch. (Thuidiaceae), *Homalothecium lutescens* (Hedw.) Robins. (Brachytheciaceae) and *Homalothecium sericeum* (Hedw.) Schimp. (Brachytheciaceae). Diss. Freie Univ. Berlin, Berlin.

Lindo, Z., Gonzalez, A., (2010). The bryosphere: an integral and influential component of the Earth's biosphere. *Ecosystems* 13, 612-627.

Löbel, S., Dengler, J., Hobohm, C., (2006). Species richness of vascular plants, bryophytes and lichens in dry grassland: The effects of environment, landscape structure and competition. *Folia Geobot.* 41, 377-393.

Mälson, K., Rydin, H., (2007). The regeneration capabilities of bryophytes for rich fen restoration. *Biol. Conserv.* 135, 435-442.

MacDonald, RL, Burke, JM., Chen, HYH, Prepas, EE. (2012) Relationship between aboveground biomass and percent cover of ground vegetation in Canadian boreal plain riparian Forests. *For. Sci.* 58: 47-53.

Mayer, R., Kaufmann, R., Vorhauser, K., Erschbamer, B., (2009). Effects of grazing exclusion on species composition in high-altitude grasslands of the Central Alps. *Basic Appl. Ecol.* 10, 447-455.

Michel, P., Burritt, D.J., Lee, W.G., (2011). Bryophytes display allelopathic interactions with tree species in native forest ecosystems. *Oikos* 120, 1272-1280.

Michel, P., Payton, I.J., Lee, W.G., During, H.J., (2013). Impact of disturbance on above-ground water storage capacity of bryophytes in New Zealand indigenous tussock grassland ecosystems. *New Zeal. J. Ecol.* 37, 114-126.

Mikulášková, E., Fajmonová, Z., Hájek, M., (2010). Invasion of central-European habitats by the moss *Campylopus introflexus*. *Preslia* 84, 863-886.

Müller, J., Heinze, J., Joshi, J., Klaus, V.H., Boch, S., Fischer, M., Prati, D., (2014). Influence of experimental soil disturbances on the diversity of plants in agricultural grasslands. *J. Plant Ecol.-UK* [doi: 10.1093/jpe/rtt062]

Müller, J., Klaus, V.H., Kleinebecker, T., Prati, D., Hölzel, N., Fischer, M., (2012). Impact of land-use intensity and productivity on bryophyte diversity in agricultural grasslands. *PLoS ONE* 7(12), e51520.

Nebel, M., Philippi, G., (2000). Die Moose Baden-Württembergs. Eugen Ulmer, Stuttgart.

Oliver, M.J., Velten, J., Mishler, B.D., (2005). Desiccation tolerance in bryophytes: A reflection of the primitive strategy for plant survival in dehydrating habitats? *Integr. Comp. Biol.* 45, 788-799.

Oliver, M.J., (2009). Biochemical and molecular mechanisms of desiccation tolerance in bryophytes. In: Goffinet, B., Shaw, A.J., [eds.]. *Bryophyte biology*. Cambridge Univ. Press, Cambridge, 269-297.

Oostermeijer, J.G.B., van't Veer, R., den Nijs, J.C.M., (1994). Population-structure of the rare, long-lived perennial *Gentiana pneumonanthe* in relation to vegetation and management in the Netherlands. *J. Appl. Ecol.* 31, 428-438.

Pauliuk, F., Müller, J., Heinken, T., (2011). Bryophyte dispersal by sheep on dry grassland. *Nova Hedwigia* 92, 327-341.

Peintinger M, Bergamini A (2006) Community structure and diversity of bryophytes and vascular plants in abandoned fen meadows. *Plant Ecol.* 185, 1-17.

Pfeiffer, T., Fritz, S., Stech, M., Frey, W., (2006). Vegetative reproduction and clonal diversity in *Rhytidium rugosum* (Rhytidiaceae, Bryopsida) inferred by morpho-anatomical and molecular analyses. *J. Plant. Res.* 119, 125-135.

Pharo, E.J., Zartman, C.E. (2007). Bryophytes in a changing landscape: The hierarchical effects of habitat fragmentation on ecological and evolutionary processes. *Biological Conservation*, 135, 315-325.

Piessens, K., Stieperaere, H., Honnay, O., Hermy, M., (2008). Effects of management and adjacent forest on the heathland bryophyte layer. *Basic Appl. Ecol.* 9, 253-262.

Porley, R.D., (1999). Bryophytes of chalk grassland in the Chiltern Hills, England. *J. Bryol.* 21, 55-66.

Preston, C.D., Hill, M.O., Pilkington, S., Pywell, R.J. (2009). The effect of disturbance on the bryophyte flora of Salisbury Plain, western Europe's largest chalk grassland. *J. Bryol.* 31, 255-266.

Proctor, M.C.F., (2008). Physiological ecology. In: Goffinet, B., Shaw, A.J., [eds.]. *Bryophyte biology*. Cambridge Univ. Press, Cambridge, 237-268.

Rice, E.L., (1984). *Allelopathy*. Academic Press, Waltham.

Rydin, H., (1997). Competition among bryophytes. *Adv. Bryol.* 6, 135-168.

Rydin, H., (2009). Population and community ecology of bryophytes. In: Goffinet, B., Shaw, A.J., [eds.]. *Bryophyte biology*. Cambridge Univ. Press, Cambridge, 393-444.

Ryser, P., (1993). Influences of neighbouring plants on seedling establishment in limestone grassland. *J. Veg. Sci.* 4, 195-202.

Schulten, J.A., (1985). Soil aggregation by cryptogams of a sand prairie. *Am. J. Bot.* 72, 1657-1661.

Serpe, M.D., Orm, J.M., Barkes, T., Rosentreter, R., (2006). Germination and seed water status of four grasses on moss-dominated biological soil crusts from arid lands. *Plant Ecol.* 185, 163-178.

Shaw, A.J., Szövényi, P., Shaw, B. (2011). Bryophyte diversity and evolution: Windows into the early evolution of land plants. *Am. J. Bot.* 98, 352-369.

Slack, N.G., (1997). Niche theory and practice: bryophyte studies. *Adv. Bryol.* 6, 169-204.

Smit, H.J., Metzger, M.J., Ewert, F., (2008). Spatial distribution of grassland productivity and land use in Europe. *Agr. Syst.* 98, 208-219.

Smith, A.J.E. [eds] (1982). *Bryophyte ecology.* London: Chapman & Hall.

Špačková, I., Kotorová, I., Lepš, J., (1998). Sensitivity of seedling recruitment to moss, litter and dominant removal in an oligotrophic wet meadow. *Folia Geobot. Phytotyxon* 33, 17-30.

Stammel, B., Kiehl, K., Pfadenhauer, J., (2003). Alternative management on fens: Response of vegetation to grazing and mowing. *Appl. Veg. Sci.* 6, 245-254.

Startsev, N.A., Lieffers, V.J., (2006). Dynamics of mineral nitrogen released from feathermoss after dehydration or handling stress. *Bryologist* 109, 551-559.

Stein, C., Auge, H., Fischer, M., Weisser, W.W., Prati, D. (2008) Dispersal and seed limitation affect diversity and productivity of montane grasslands. *Oikos* 117, 1469-1478.

Takala, T., Tahvanainen, T., Kouki, J., (2012). Can re-establishment of cattle grazing restore bryophyte diversity in abandoned mesic semi-natural grasslands? *Biodivers. Conserv.* 21, 981-992.

Tilman, D., Reich, P., Knops, J., Wedin, D., Mielke, T., Lehman, C., (2001). Diversity and productivity in a long-term grassland experiment. *Science* 294, 843-845.

Tscharntke, T., Klein, A.M., Kruess, A., Steffan-Dewenter, I. & Thies, C., (2005). Landscape perspectives on agricultural intensification and biodiversity – ecosystem service management. *Ecol. Lett.* 8, 857-874.

Tsubota, H., Kuroda, A., Masuzaki, H., Nakahara, M., Deguchi, H., (2006). Preliminary study on allelopathic activity of bryophytes under laboratory conditions using the sandwich method. *J. Hattori Bot. Lab.* 100, 517-525.

Tuba, Z., Slack, N.G., Stark L.R., [eds] (2011). *Bryophyte ecology and climate change.* Cambridge Univ. Press, Cambridge.

Turetsky, M.R., (2003). The role of bryophytes in carbon and nitrogen cycling. *Bryologist* 106, 395-409.

van der Wal, R., Pearce, I.S.K., Brooker, R.W., (2005). Mosses and the struggle for light in a nitrogen-polluted world. *Oecologia* 142, 159-168.

van Tooren, B.F., (1988). The fate of seeds after dispersal in chalk grassland: the role of the bryophyte layer. *Oikos* 53, 41-48.

van Tooren, B.F., (1990). Effects of a bryophyte layer on the emergence of seedlings of chalk grassland species. *Acta Oecol.* 11, 155-163.

Vanderpoorten, A., Delescaille, L.-M., Jacquemart, A.-L., (2004). The bryophyte layer in a calcareous grassland after a decade of contrasting mowing regimes. *Biol. Conserv.* 117, 11-18.

Vanderpoorten, A., Goffinet, B., (2009). *Introduction to bryophytes*. Cambridge Univ. Press, Cambridge. UK.

Vandvik, V., Heegaard, E., Måren, I.E., Aarrestad, P.A., (2005). Managing heterogeneity: the importance of grazing and environmental variation on post-fire succession in heathlands. *J. Appl. Ecol.* 42, 139-149.

Virtanen, R., Johnston, A.E., Crawley, M.J., Edwards, G.R., (2000). Bryophyte biomass and species richness on the Park Grass Experiment, Rothamsted, UK. *Plant Ecol.* 151, 129-141.

Vitt, D.H., (2000). Peatlands: ecosystems dominated by bryophytes. In: Shaw, A.J., Goffinet, B., [eds.]. *Bryophyte biology*. Cambridge Univ. Press, Cambridge, 312-343.

Weigelt, A., Weisser, W.W., Buchmann, N., Scherer-Lorenzen, M. (2009). Biodiversity for multifunctional grasslands: equal productivity in high-diversity low-input and low-diversity high-input systems. *Biogeosciences* 6, 1695-1706.

Willems, J.H., (1983). Species composition and above ground phytomass in chalk grassland with different management. *Vegetatio* 52, 171-180.

Wilson, J.B., Peet, R.K., Dengler, J., Pärtel, M. (2012). Plant species richness: the world records. *J. Veg. Sci.* 23, 796-802.

Xie, C.-F., Lou, H.-X., (2009). Secondary metabolites in bryophytes: An ecological aspect. *Chem. Biodiv.* 6, 303-311.

Zamfir, M, Goldberg, D. E. (2000) The effect of initial density on interactions between bryophytes at individual and community levels. *Journal of Ecology* 88: 243-255.

Zechmeister H.G., Schmitzberger, I., Steurer, B., Peterseil, J., Wrbka, T., (2003). The influence of land-use practices and economics on plant species richness in meadows. *Biol. Conserv.* 114, 165-177.

Zechmeister, H.G., Tribsch, A., Moser, D., Wrbka, T., (2002). Distribution of endangered bryophytes in Austrian cultural landscapes. *Biol. Conserv.*103, 173-182.

INDEX

D

E

F

N

O

S

T

U

V

W

X

Y